THE ... N AND
... Y

The Social Study of Information and Communication Technology

Innovation, Actors, and Contexts

Edited by

CHRISANTHI AVGEROU, CLAUDIO CIBORRA, AND FRANK LAND

This book has been printed digitally and produced in a standard specification in order to ensure its continuing availability

OXFORD
UNIVERSITY PRESS

Great Clarendon Street, Oxford OX2 6DP

Oxford University Press is a department of the University of Oxford.
It furthers the University's objective of excellence in research, scholarship,
and education by publishing worldwide in

Oxford New York

Auckland Cape Town Dar es Salaam Hong Kong Karachi
Kuala Lumpur Madrid Melbourne Mexico City Nairobi
New Delhi Shanghai Taipei Toronto
With offices in
Argentina Austria Brazil Chile Czech Republic France Greece
Guatemala Hungary Italy Japan South Korea Poland Portugal
Singapore Switzerland Thailand Turkey Ukraine Vietnam

Published in the United States
by Oxford University Press Inc., New York

Reprinted 2007

ISBN 978-0-19-925352-4

Acknowledgements

This book is the result of a collective effort of the Department of Information Systems of the London School of Economics and Political Science (LSE). Our understanding of the research area we set out to present in this volume also owes much to our exposure to the specialized research insights of many of our colleagues whose work did not make it to this volume. And we are indebted to the lively community of our Ph.D. students, who are a constant source of challenge to existing knowledge. The presentations of their research efforts and the following debates at the weekly Doctoral seminars of the Department have offered us valuable opportunities for tracing emerging ideas and theories.

We would like to thank Lynne Rienner Publishers, the Information Systems Research in Scandinavia (IRIS) Association, and Sage Publishers and for their permission to include, respectively, Chapter 4 (Saskia Sassen), Chapter 7 (Eric Monteiro), and Chapter 10 (Robert Hunter Wade).

We are especially grateful to Malcolm Peltu for his excellent editorial work and we thank David Musson and Matthew Derbyshire of the OUP who trusted us and provided helpful support for the completion of this book.

Chrisanthi Avgerou,
Claudio Ciborra,
Frank Land
Information Systems Department, LSE

Contents

List of Figures, Tables, and Boxes

Figures

Tables

Boxes

Notes on Contributors

Ian O. Angell has been the Professor of Information Systems at the London School of Economics (LSE) since 1986. His original degrees were in Pure Mathematics, and until coming to the LSE he taught Computer Science at the University of London. Now a researcher in Information Systems, he concentrates on strategic information systems, computers and risk, and organizational and national IT policies. Within this broad field, he has two particular and related specialities: 'systemic risk' and 'the global consequences of information technology'. He has published 14 books and is working on *The Black Art of Management*, a philosophical questioning of the use of information and measurement in management.

Chrisanthi Avgerou is Professor of Information Systems at the LSE. Her main research interests concern the relationship of information technology to organizational change, and the role of IT in socio-economic development. She is vice-chair of the IFIP Technical Committee 9 on social implications of IT and past chair of IFIP WG 9.4 on computers in developing countries. Among her latest publications is *Information Systems and Global Diversity* (Oxford: Oxford University Press, 2002). She earned her Ph.D. at the LSE.

Richard L. Baskerville is Professor of Information Systems and Chairman in the Department of Computer Information Systems, College of Business Administration, Georgia State University. His research specializes in the security of information systems, methods of information systems design and development, and the interaction of information systems and organizations. His interests in methods extend to qualitative research methods. He is the author of *Designing Information Systems Security* (New York: John Wiley, 1988) and more than a hundred articles in scholarly journals, professional magazines, and edited books. He serves on the editorial boards of *The Information Systems Journal, MIS Quarterly*, and *The European Journal of Information Systems*. He holds a Ph.D. from the LSE.

Richard J. Boland, Jr. is Professor of Information Systems at the Weatherhead School of Management, Case Western Reserve University, and also serves as a Senior Research Associate at the Judge Institute of Management at the University of Cambridge. His research emphasizes interpretive studies of how individuals experience the design, implementation, and use of information systems in organizations. He has published extensively and his more recent publications include *Managing as Designing* (Palo Alto, CA: Stanford University Press, forthcoming).

Claudio Ciborra is Professor of Information Systems, Convenor of the Department of Information Systems, and PwC Risk Management Professor at the LSE. His research has focused on issues of technology, organization, strategy, and innovation. He also teaches at IULM University, Milan and is Visiting Professor at Oslo University. He is on the editorial board of a dozen academic journals, has consulted widely, and is a member of a number of IFIP working groups, including WG 8.2. He has published many articles and books on information systems. His recent books include *The Labyrinths of Information—Challenging the Wisdom of Systems* (Oxford: Oxford University Press, 2002).

ROBERT D. GALLIERS, FBCS, was appointed as Provost of Bentley College, Massachusetts in 2002. He had previously been Professor of Information Systems at the LSE and Lucas Professor of Business Management Systems and Dean of Warwick Business School. His main research focus is on IS strategy and the management of change associated with the adoption and appropriation of ICT-based systems within and between organizations. He is Editor-in-Chief of the *Journal of Strategic Information Systems* and a past president of the Association for Information Systems. He has published widely in the leading international IS journals and co-authored a number of books, such as *Strategic Information Management* (Oxford: Butterworth-Heinemann, 2003). He holds a Ph.D. in Information Systems from the LSE.

OLE HANSETH is Professor in the Department of Informatics at the University of Oslo and Visiting Professor in the Department of Information Systems at the LSE. He worked within applied research and business until he moved to academia in 1997. His main research interests are related to large scale, complex, global information systems and infrastructures. He has published in numerous journals, including: *Accounting, Management and Information Technology; Computer Supported Cooperative Work; Data Base for MIS; Information and Organization; Scandinavian Journal of Information Systems; Science, Technology & Human Values*; and *Knowledge, Technology and Policy*.

FERNANDO M. ILHARCO is Assistant Professor at the School of Human Sciences of the Catholic University of Portugal, Lisbon. His central research interests are the philosophical and social dimensions of information and communication technologies. He has written a regular column in the Portuguese daily newspaper *Público* since 1996. He gained his Ph.D. in 2002 at LSE's Department of Information Systems and has an MBA from the Catholic University of Portugal, specializing in Information Management.

JANNIS KALLINIKOS is a Reader in the Department of Information Systems at LSE. His major research interests involve the institutional construction of 'predictable worlds' (the practices, technologies, and formal languages by which organizations are rendered predictable and manageable) and the investigation of the modes by which current institutional and technological developments challenge the organizational forms that dominated modernity. His publications include *In the Age of Flexibility: Managing Organizations and Technology* (Lund, Sweden: Academia-Adacta, 2001) and numerous articles in journals and chapters in books of collected works. He has a Ph.D. from Uppsala University, Sweden.

FRANK LAND, FBCS, started his career in computing in 1953, working on the pioneering LEO Computer for J. Lyons. He joined the LSE in 1967 to establish teaching and research in systems analysis and was appointed Professor of Systems Analysis in 1982. In 1986, he joined the London Business School as Professor of Information Management. He retired in 1991 and received the title Emeritus Professor at the LSE's Department of Information Systems in 2002. He is also Visiting Professor at Leeds Metropolitan University and his many other visiting professorships have included the Wharton School and the University of Sydney. He is past chairman of IFIP WG 8.2 and is on the editorial board of a number of academic journals. In 2003, he received from the Association for Information Systems the LEO Award for lifetime contribution to the IS discipline.

BRUNO LATOUR is Professor at the Centre de Sociologie de l'Innovation at the Ecole Nationale Supérieure des Mines in Paris and Visiting Professor at the LSE's Department of Information Systems. He was trained first as a philosopher and then an anthropologist. After field studies in Africa and California, he specialized in the analysis of scientists and engineers at work. In addition to work in philosophy, history, sociology, and the anthropology of science, he has collaborated in many studies in science policy and research management. His latest book is *The Politics of Nature: How to Bring the Sciences into Nature* (Boston: Harvard University Press, forthcoming). In a series of new books in French, he is exploring the consequences of science studies on different traditional topics of the social sciences.

SHIRIN MADON is a Lecturer in Information Systems at the LSE. Her main research interest is studying the impact of information systems on planning and administration in developing countries. This has involved extensive fieldwork in India, including work sponsored by the UK Department for International Development, that has studied the impact of different structures of local government on the implementation of information systems. From 1998 to 2000, she worked on a project funded by the British Academy that studied information flows for local governance in the city of Bangalore, focusing on reform of the property tax system. She gained a Ph.D. from Imperial College, London.

ERIC MONTEIRO is Professor at the Department of Computer and Information Systems at the Norwegian University of Science and Technology. He is broadly interested in the socio-technical aspects of the design, use, and diffusion of IS in private and public organizations. As part of this, he has for some time been interested in, and inspired by, debates around Actor Network Theory. He has published his work in: *Accounting, Management and Information Technology; Computer Supported Cooperative Work; Information and Organization; IT for Development; Scandinavian Journal of Information Systems*; and *The Information Society*.

SASKIA SASSEN is the Ralph Lewis Professor of Sociology at the University of Chicago, and Centennial Visiting Professor at the LSE. She is currently completing the book *Denationalization: Territory, Authority and Rights in a Global Digital Age* (Princeton, NJ: Princeton University Press, forthcoming), which is based on her five-year project on governance and accountability in a global economy. Her most recently published book is the edited *Global Networks, Linked Cities* (London: Routledge, 2002). She serves on several editorial boards, is an advisor to several international bodies, and is Chair of the Information Technology and International Cooperation Committee of the Social Science Research Council in the USA.

STEVE SMITHSON is a Senior Lecturer in Information Systems at the LSE. He is Vice President of the UK Academy of Information Systems and was Editor of the *European Journal of Information Systems* from its inception in 1991 until the end of 1999. His research interests lie in IS management, the evaluation of information systems, and developments in e-business. He has published numerous journal articles and conference papers, as well as four books. He was recently involved with a national UK e-business project, focusing on the organizational and management issues of e-business. He has a Ph.D. in Information Systems from the LSE.

PRODROMOS TSIAVOS is a Ph.D. candidate in the LSE's Department of Information Systems and is currently funded by the Greek Scholarship Foundation for conducting research on Information Technology Law. He is a member of the Athens Bar Association and has worked for the Directorate-General Information Society in the European Commission. He has conducted research on Information Systems evaluation and e-government issues and his most recent work looks at the interaction between IS design and regulation, with a particular focus on Intellectual Property Rights.

ROBERT HUNTER WADE is Professor of Political Economy at the LSE. A New Zealand citizen, he worked as an economist at the World Bank and at the Office of Technology Assessment (US Congress). He has taught at Princeton University, Massachusetts Institute of Technology, and Brown University. He has done field work in the Pitcairn Islands, Italy, India, South Korea, Taiwan, and inside the World Bank. He is the author of *Governing the Market: Economic Theory and the Role of Government in East Asia's Industrialization* (Princeton, NJ: Princeton University Press, 2003) and has published recent essays on world income distribution, financial crises, World Trade Organization agreements, US hegemony, and how to create capitalism.

Abbreviations

AI	Artificial Intelligence
ARPA	Advanced Research Projects Agency
B2B	Business-to-Business
BPR	Business Process Redesign (or Re-engineering)
BSP	Business Systems Planning
CBD	Central Business District
CNC	Computer Numerical Control
CSCW	Computer Supported Cooperative Working
CSF	Critical Success Factor
DOT Force	Digital Opportunity Task Force, sponsored by the G-7
DP	Data Processing
EDI	Electronic Data Interchange
EIS	Executive Information System
ERP	Enterprise Resource Planning
G-7	Group of Seven leading industrialized nations
GDSS	Group Decision Support System
GSM	Global System for Mobile communications
ICANN	Internet Corporation for Assigned Names and Numbers
ICT	Information and Communication Technology
IP	Internet Protocol
IRD	Integrated Rural Development
IS	Information System
ISP	Internet Service Provider
IT	Information Technology
ITU	International Telecommunication Union
KMS	Knowledge Management System
LDC	Less-Developed Countries
MBA	Masters of Business Administration
MCT	Multipurpose Community Telecenters
MIS	Management Information System
MIT	Massachusetts Institute of Technology
NGO	Non-Government Organization
OECD	Organization for Economic Cooperation and Development
PARC	Palo Alto Research Center (Xerox)
PC	Personal Computer
PDA	Personal Digital Assistant
PQM	Programme Quality Management

SBU	Strategic Business Unit
STS	Science and Technology Studies
TCP/IP	Transmission Control Protocol/Internet Protocol
UNDP	United Nations Development Programme
USAID	US Agency for International Development
WTO	World Trade Organization

Introduction

Chrisanthi Avgerou, Claudio Ciborra, and Frank Land

The social study of information and communication technology

Information systems (IS) is an area of research positioned between management studies and applied computing, where it is influenced by numerous kindred and reference disciplines. In this book, we bring together a collection of papers that exemplify the current state of one of the strands of this hybrid field, the social study of information and communication technology (ICT).[1] The book broadly reflects the continuing engagement of the Information Systems Department at the London School of Economics and Political Science (LSE) in this area of research.

The direction our research has taken has been grounded, on the one hand, in our understanding of the world of practice gained through empirical work, action research, and work experience; on the other hand, crucially, it is grounded in the social science environment of the LSE. Our colleagues in the departments of economics, sociology, anthropology, social psychology, politics, philosophy, and industrial relations have always been a constant source of influence. More recently, as innovations in ICT continue to proliferate, increasing both its reach and scope, the technology's significance has been the source of wide ranging interdisciplinary research. The LSE has provided a challenging setting for that debate, presenting the IS group with opportunities to expose, defend, and develop our perspectives. Thus, while taking stock of the social studies of ICT in the IS field—both at the LSE and the wider IS research community—we have also included chapters in this book by LSE colleagues whose concerns and intellectual positions we have found particularly influential: the sociologist Saskia Sassen, the political economist Robert Wade, and the anthropologist/sociologist Bruno Latour.

No university department or school, however big and prestigious, is an island; rather, each is a node of a wider scientific community. Hence, we felt that in collecting contributions for this book we should not be limited by the physical boundaries of our London location. We have therefore not hesitated to invite some of our closer colleagues in the IS field to send us their most recent contributions. This has allowed us to present to the reader a rich and contemporary collection of international contributions in the social study perspective.

The chapters in this volume are diverse in terms of the issues addressed and theories used. Social studies of ICT expand the frame of the mainstream corpus of IS research, which has been centred typically on the assumption of technical/rational behaviour in the setting of professionally-managed business organizations.

Such a widening of perspective introduces new dimensions of enquiry, where research is confronted with an arena of various interested actors. It reveals that, in addition to the ICT specialist, the 'user', and the business executive conventionally studied in IS research, there are many other legitimate actors in the ICT innovation scene: the operator, the customer, the citizen, the gendered individual, the government, the entrepreneur, the 'poor', the developing country, etc. Each of these actors voices different interests, and defends the logic of different strategies of action. Moreover, action that is political, unintended, and inexplicable by the techno-economic logic stands out as being at least as significant as action that is designed, planned, and driven by the normative models of professional control.

Consequently, a research perspective looking beyond the calculative behaviour of decision makers cannot assume that the benefits and risks of the spreading of ICT are self-evident, and therefore cannot aspire to produce knowledge to govern them. It also cannot reduce the processes of ICT innovation, and the economic, organizational, and social changes in which it is implicated, to trajectories obeying laws inherent in the materiality of the technology, or in human nature or social structure. Instead, we are obliged to search for conceptual vocabularies and epistemological fundamentals that allow us to capture and interpret socio-technical phenomena.

There is no uncontested theoretical ground for such questions in the social sciences. An understanding of society does not cluster in paradigms that supersede each other by accumulating knowledge and by producing more accurate and truthful accounts of a reality waiting to be deciphered. Rather, competing theories offer accounts of relations and processes of change, while at the same time contributing to their realization. Unlike the physical sciences, the social sciences do not claim to establish irrefutable 'natural laws'. Different theoretical frameworks attempt to interpret phenomena from different perspectives. This does not imply that we are absolved from the obligation to apply rigorous criticism to each new notion as it is presented and tested (Popper 2002), but it does assert that viewing the phenomenon through different theoretical perspectives may provide the enquirer with a richer picture.

In each epoch of social history, the dominant perspectives that shape orthodox common sense and fuel movements, policies, or trends co-exist with (and to a large extent define themselves in relation to) alternative, marginal, or apparently discredited views and theoretical constructs. It is part of the mission of social sciences to scrutinize established orthodoxy, to sense the criticisms of the (often silent) objectors, and to juxtapose alternatives. Such juxtaposition is not arbitrary or random; it is patterned by the unfolding of the phenomena that social theory simultaneously reflects upon and contributes towards. Thus, the essays in this volume can be best understood as part of the ongoing research efforts to address ICT in relation to society that emerged when computers became visible components in government and business organizations. A comprehensive account of the unfolding of such studies is beyond the scope of this Introduction. However, we believe that tracing some of the most influential preceding research will help the reader to put into perspective the contributions contained in this book.

The unfolding of the social study of ICT

A seminal publication that captured the early discourse on 'computers and society' is the proceedings edited by Mumford and Sackman (1974) of the 'Human Choice and Computers' conference, which took place in Vienna in 1974. The overall tone of the authors in that volume suggests uneasiness about 'the machine' and the way it was perceived to affect key social institutions: the social order of industrial democracy at the workplace, education, democracy in broader society, and the privacy rights of the individual. Nevertheless, the book manifests, optimistically, an underlying conviction that the computer and its social effects are a matter of human choice. The editors explain that the conference and the selection of papers were:

> motivated by the belief that all levels of society should take stock of the growing use of computers in all walks of life and decide what it wants to do with computers for the benefits of everyone. Society should deliberately lead and direct the application of computers in the image of its most cherished values and ideals rather than be the unwitting victim of the vagaries of technology and the fluctuations of the market-place (Mumford and Sackman 1974: v).

The authors in these proceedings called for democratic society at large to be vigilant while computer technology is developed and deployed in its institutions, and to be prepared to exercise choice and enact processes of social accountability. Nevertheless, their discussion involved or referred to specific social categories: trade unionists, management, computer technologists, and social scientists. Many chapters were dedicated to the confrontation of interests between management and trade unions. It is interesting to note how computer technology was seen to be implicated in this conflict and what roles were perceived for the computer technologists and social scientists. In a nutshell, 'choice' was seen to refer to the task of designing the features of the computer and its applications at the workplace—the core of a democratic industrial society. The editors, Mumford and Sackman, pointed out that a consensus view emerged at the conference that the design of computer systems ought to include more formal and explicit values and ideas from both workers and management, while broader community and national interests should not be neglected. The workplace was viewed as the battleground that would decide whether computers would have humanizing or de-humanizing effects, and on that battleground neither computer technologists nor social scientists could afford to be neutral specialists: they were called on to be engaged combatants.

The editors' reflections in the concluding chapter reveal the way the conference discussions problematized the roles of social scientists:

> The social scientists, as might be expected, took relatively academic and analytical views of problems of human choice. They tended to speak from research rather than practice in relation to the problems of computers, and attempted to argue on the basis of what they believed to be a logical and objective assessment of the computer environment and its impact on human beings. Those scientists, such as the Swedes and the British, who have assumed an action research role and attempted to get human goals—such as an improved quality of working life—incorporated into the design of computer systems were surprised and somewhat shocked at being accused by the trade unions of acting as the tools of management... This

led to a great deal of useful discussion in which the social scientists attempted to convince the unions of their neutrality and good intentions (Mumford and Sackman 1974: 325–26).

There are two aspects in this quotation concerning the conflict about the perceived role of the social scientist that are remarkable in relation to the different course that the social study of ICT subsequently followed. Firstly, there is the critical theoretical orientation that called on 'social scientists' to take an active role in support of the trade unions; and then there is the justification of this research/action mission in terms of the perception that social scientists pursue the development of 'logical objective knowledge' on ICT and society. As the chapters in this book demonstrate, neither of these views remains prevalent. Rather than seeing their task as producing logical, objective knowledge, researchers on ICT and society have increasingly tended to adopt interpretive and phenomenological epistemologies.

Change in the perceived mission of the social study of ICT

The view that ICT has a potential impact that social science can assist to harness by informing correct courses of action has come to be seen generally as being naive. There has been a change from a perception of the researcher's role as an expert activist to the emergence of a greater interest in theoretical analyses, as exemplified in this volume. This trend towards unravelling the relationship between technological innovation and the social condition—or the technical artefact and the human condition—took place gradually, but was clearly articulated over the 1990s.

This change emerged from a background in which there was a diminishing legitimacy of the critical researcher as an activist in the socio-technical design approach of the 1970s and 1980s, and where a research tradition prevailed in the broader information systems field that was aligned single-mindedly with the mission of business success in a competitive market regime. More specifically, the 1970s saw a spread of socio-technical projects across the industrial world (Mumford 2003), peaking with the redesign of automobile manufacturing at the Volvo plant in Sweden. By the 1990s, however, many of these projects had not survived the changing economic conditions and realignment of industry by mergers and acquisitions. The humanistic ideals of the socio-technical pioneers had paid too little attention to the problems of power imbalance between different classes of stakeholders, and the feeling amongst management that the kind of change required to meet socio-technical ideals were risky and not to be attempted in times of economic decline.

Instead of aspiring to guide action for a humanistic use of ICT, research effort was gradually redirected towards elaborating satisfactory conceptual vocabularies and theoretical underpinnings to help comprehend the nature of the complex phenomena of ICT innovation in organizations and society at large (e.g. Walsham 1993). The seminal work in this respect has been Shoshana Zuboff's (1988) book, *In the Age of the Smart Machine*, which—backed in an unprecedented way by social theories of power, organization, and human psychology—is still a relevant and illuminating study of the social and personal consequences of many managerial choices about ICT.

This change in direction of the social research on ICT does not mean that the researcher has lost interest in 'putting the world to rights'. But the emphasis has shifted from engaging in action research as a kind of quasi-consultancy towards empirical research that aims to test and illustrate theoretical concepts and help in the refining and framing of new concepts. Action research may still be used, but in the context of a theoretical conception that acts both as a way of studying the dynamics of the world of practice and of feeding back to test the usefulness of the conception in that particular context.

The critical social theoretical orientation has not been lost, but the social study of ICT abandoned simplifying dichotomies such as 'empowerment versus managerial control' and 'liberation or domination' by ICT. Literature with a critical theoretical stance has begun to identify a much more complex set of conflicting interests and power dynamics in a variety of settings of ICT development and use, including: the changing workplace in Western society (Bloomfield *et al.* 1997; Howcroft and Truex 2001); the transformation of modern state institutions; and the world's diversity of cultures and historically-developed rationalities (Walsham 2001; Avgerou 2002). The fundamental preoccupation of critical theoretical research with the threat that ICT may be deployed to serve as an instrument of blatant authoritarianism has also changed, with critical social studies addressing more subtle issues regarding a person's experience with ICT in the modern social context (Ciborra 2002).

Change of epistemological orientation

The seeds of doubt about the validity of the mission of objective knowledge are detectable even in Mumford and Sackman's (1974) *Human Choice and Computers* proceedings. For instance, in the first chapter on 'the human being and the automation', Heinz Zemanek (1974) juxtaposed the rational and emotionless logic built into the computer with the 'real' human being, thus highlighting some of the limitations of the former. Zemanek briefly attempted to draw from a critique of Wittgenstein's early formalistic and later revisionist work on language, but he was unable to go further than voicing concerns on potential dangers and to reach conclusions on the kind of knowledge that research should aspire to develop. In his concluding paragraphs, Zemanek (ibid.: 29) summarizes the contradiction between human values and scientific method that he was unable to overcome: 'While human properties and values are in danger of being destroyed by the physical and technical mentality, the protection of the individual and of human society can be achieved only by the methods and tools provided by science and technology.'

Zemanek was not alone at that time in reflecting on issues of knowledge in what were then new computer-mediated social settings and in being concerned about the way the human was portrayed in technology research. The efforts to build 'intelligent' machines in the field of artificial intelligence and the hype about the prospects of artefacts reaching and surpassing human cognitive behaviour prompted the development of phenomenological positions on cognition in general, and the cognitive potential of computers in particular. Such positions did not invest the methods and tools provided by science and technology with the capacity to protect the properties of the individual and the values of human society. On the contrary, they saw a very limited formal cognitive capacity in such methods and tools. Instead, they acknowledged that the most important aspects of human cognition are its capacity to cope haphazardly with messy, ill-defined situations—a capacity derived from the flesh-and-blood experiences of humans in their social world (Winograd and Flores 1986; Suchman 1987).

The critiques of the grand artificial intelligence project were directly associated with, and inevitably influenced, the epistemology of social studies of ICT. The empirical observations and theoretical analyses that discredited the ambition of artificial intelligence and computer science to capture human behaviour unambiguously and objectively in a formal system of knowledge left little room for the search for objective knowledge that would direct the correct human choice towards technologies that would enhance good society values. For example, Lucy Suchman's (1987) theoretical critique of the narrowness of the formal logic of artificial intelligence was continued in the anthropological studies of work and technologies carried out since the 1980s at Xerox's PARC, its Palo Alto Research Center (Brown and Duguid 2000).

The need for epistemological perspectives of technology and society that are more capable of accounting for human social experience found support from a wave of new social theory that has gained increasing momentum since the early 1980s. Long-established rigid theoretical conceptions of the world in terms of the structuralist traditions in sociology, anthropology, and political theory were challenged. They gave way to perspectives of society that acknowledged the plurality of concerns involved in human interaction and emphasized human agency in the shaping of social conditions. The perspectives that have been particularly influential for the social studies of ICT include Giddens' structuration theory (Giddens 1984) and the interdisciplinary field of the study of the history and sociology of science and technology (e.g. Hughes 1983; Law 1991; Bijker and Law 1992).

In the 1990s, concerted efforts were made to import such theoretical perspectives into the information systems field, for example in Wanda Orlikowski's work (Orlikowski 1992, 2000, 2002; Orlikowski *et al.* 1996) and in the series of analyses that advocated the merits of using Actor Network Theory (ANT) from the sociology of technology area (Monteiro and Hanseth 1996; Walsham 1997; Hanseth and Monteiro 1998). Moreover, IS research took a closer look at the philosophical underpinnings of existing IS knowledge. Phenomenology provided new ground for critiques of the 'best practice' knowledge of the field (Introna 1997). It soon became obvious that such theoretical perspectives marked a new understanding of the nature of socio-technical change (Bowker and Leigh Star 2000; Ciborra 2000).

In short, the social study of ICT has devoted much effort since the early 1990s to investigating the conceptual relationship of ICT and society and to the kind of knowledge and practice that is valid to pursue in this area of human activity.

This effort is necessary in order to tackle the multiple substantive issues that are associated with ICT (Dunlop and Kling 1991), particularly with the new type of technologies such as the Internet and mobile phones.

Structure of this book

Chapters in this book are clustered into three parts:

(a) *Foundations (Chapters 1 to 4).* This part addresses broad questions about the kind of basic line of enquiry that the social study of ICT should be pursuing. Its chapters can be seen as efforts to establish the research scene in terms of the basic concepts used to express the phenomena under study, the legitimate mission of the research, appropriate methods, and research questions worth studying.
(b) *Theories at work (Chapters 5 to 9).* These chapters cover discussions on currently debated research themes. The authors examine conceptual relationships in some of the most enduring areas of IS research, such as on the nature of organizational knowledge and the way it is affected by ICT innovation, or on the nature of ICT innovation and the way it is associated with structures and processes of social context.
(c) *Substantive issues and applications (Chapters 10 to 13).* The aim here is to demonstrate how the field we call the social study of ICT deals with specific questions regarding ICT in organizational and social transformations.

Part I: Foundations

This part contains four chapters. Each one employs a distinctive style in arguing for a particular combination of philosophical or socio-theoretical principles, but collectively they suggest quite similar orientations for the study of ICT that depart from the technical or managerial perspectives that have been in general currency.

In the first chapter, Ciborra sets the scene for the studies contained in this volume by taking issue with the established tradition of information systems studies that aims at equipping practitioners with capabilities to control their object of knowledge—IS and organizational change—by design and method, that is, by mimicking the natural sciences. His more specific task is to present a phenomenological perspective that brings forward a world of IS practice and management based on an understanding of lived human experience rather than abstractions of ideal rational behaviour. He proposes metaphors such as bricolage, improvisation, tinkering, hospitality, and care in order to understand the effort made to accommodate the non-human technology as an element in an organization's life. As we explained above, a phenomenological epistemology is not new in information systems research, but here Ciborra points out that earlier phenomenological accounts have been limited to cognitive aspects only, ignoring the emotional aspects of human existence and action. He opens new scope for IS research to tackle what had been previously dismissed as 'irrational' by adding 'moods', such as panic, boredom, anxiety, and frustration, to the cognitive notions through which phenomenological accounts of information systems have been construed so far.

A similar perspective is presented in Chapter 2 by Angell and Ilharco, in a polemic that echoes Nietzsche, one of the sources of their philosophical underpinnings. In the first part of the chapter, they provide a critique of the positivism that forms the basis of much of IS and management knowledge. They mock several aspects they find naive and misguided in this kind of knowledge, such as: the neat classification schemes that squeeze the variation experienced in organizational life into two-by-two matrices; the faith placed on numbers and statistical methods for discussing the true state of 'reality'; the reductionist efforts involved in trying to understand a complex whole by detailing and measuring the parts; the engineering rationality that promises control of organizations as if they were machines; and the belief that technology can control risks and uncertainty implicit in the world. In the second part of their chapter, Angell and Ilharco compose an alternative philosophical perspective that draws on various phenomenologist and critical thinkers. The view they propose privileges action that takes advantage of the emergent and the transitory to lead to useful states of affair, rather that 'right solutions'.

In Chapter 3, Latour responds to a fictitious student's bewilderment about the use of ANT in IS research. Since the 1990s, ANT has been more influential than any other theory on the sociology of technology—yet it remains in many respects elusive and controversial.[2] Is it a theory or a method that should be accompanied by other substantive theories? Are its relativism and apparently a-political and a-moral stance serious faults or do these particular features make it suitable for tracing events and actions that matter, without the prejudices of categories rooted in social circumstances that attracted attention in the social sciences in the past? Latour addresses these questions in Chapter 3 by discussing a series of epistemological features and relationships relevant to ANT, such as descriptions and explanations, interpretivism and objectivism, relativism and structuralism, and the role of social theory and research. The Socratic-like dialogue he employs shakes deeply-rooted distinctions and meanings of concepts that have become cliché labels, rather than offering authoritative advice to researchers. Latour's playful confrontation between the 'professor' and the 'student' wouldn't surprise readers familiar with his critique of Plato's manipulative staging of the Gorgias dialogue in the way it makes the wisdom of the 'Master' shine over the 'mistaken' views of his naive opponents (Latour 1999).

Chapter 4 by Sassen elaborates a conceptual perspective for understanding ICT (or 'digital technologies', in the term she uses) rooted in the tradition of sociology. Although without directly addressing philosophical questions concerning the knowledge and practice of ICT, this chapter explores fundamental concepts for the social study of ICT. Sassen addresses the question of what makes a sociological reading of ICT, that is, how should ICT be understood if not simply in terms of its technical properties but by also recognizing its embeddedness in social order? The way she examines ICT as potentially constitutive of new social dynamics is particularly interesting and revealing for information systems research. She does this by pursuing three lines of analysis: the interaction between the 'digital' and the material world, the mediating cultures, and the destabilizing of existing social hierarchies.

Part II: Theories at work

The five chapters in this part provide conceptual perspectives which, in general, are congruent with the socio-theoretical orientation suggested in Part I. They highlight the significance of the study of action and historically-shaped logics and attitudes, rather than the analysis of static and abstract structural conditions for understanding IS phenomena. The authors in this second part also consider technology and knowledge issues as being inseparable from the human lived experience. The chapters elaborate in more detail on some of the general concepts hinted at in Part I, such as 'knowledge' and 'context', and focus more closely on the relationship between ICT and society. They show how debate and contestation among partly overlapping and partly conflicting perspectives bring forward alternative interpretations of empirical evidence and reveal neglected facets of complex socio-technical phenomena.[3]

In Chapter 5, Hanseth elaborates on the nature of knowledge as ICT-mediated organizational practice. He considers knowledge as a network of interdependent parts or pieces, drawing insights from network economics, such as the occurrence of externalities, increasing returns for the value of knowledge achieved by the growth of the network, path dependencies, and lock-in effects resulting from the standards necessary to support interdependencies. However, unlike the prevailing view that a network structure is light and flexible, Hanseth argues that a knowledge network has the systemic nature of an infrastructure comprising not only 'pure' knowledge but also practices that include knowledge manipulation, technologies, and organizational structures. Such a network is similar to the systemic interdependencies of theories, methods, and institutionalized practices that form the difficult-to-change 'paradigms' in scientific traditions. This perspective suggests that radical innovation of an organization's information systems is difficult, and indeed rare, because it is confronted by the inertia of a system of interdependent elements that constitute an organization's knowledge infrastructure: knowledge instruments, work practices, and organizational and physical structures.

Boland presents a different perspective on knowledge in Chapter 6, in which he looks at the experience of human beings in an ecology of ICT-supported distributed practice that consists of knowledge workers, mediations, knowledge objects, documents, and data repositories. His core argument is that distributed practice in such an ecology is confronted by a tension between the local meaning-making of face-to-face interaction and the abstract relations of concepts, calculations, and institutional rules, which he calls 'global knowledge'. On the basis of this schematic analysis, he suggests an approach to study ICT-mediated knowledge work that takes into account the individual's subjectivity and the tension it is put under due to the oppositions of local versus global knowledge.

In Chapter 7, Monteiro criticizes the way IS research privileges calculated action, neglecting the everyday practices of technology use that are not goal directed. He points out that this limiting bias (already identified by Ciborra in Chapter 1) is prevalent in research using ANT. Monteiro argues that although the rational analyses for the design of goal-targeting information systems and the analyses of power dynamics of ANT translation processes may have been useful platforms for understanding earlier purpose-built technologies, they are not as good in explaining the use and significance of new technologies, such as mobile phones or the Internet. He suggests that an understanding of such technologies can be achieved by taking into account their ascribed symbolic meaning in the everyday life of the community that 'domesticates' them in particular patterns of use. He recommends that new approaches and concepts are needed in such research, such as cultural analyses that explore processes of identity formation.

In Chapter 8, Kallinikos also puts forward a critique of ANT. Here, he questions more generally the validity of such social constructivist approaches, which tend to emphasize the local shaping of technology through the interpretive flexibility of human agency. Instead, he points out the recalcitrant nature of technology, which he perceives as complex systems of technical, organizational, and social arrangements that are formed cumulatively over time. He argues that these kinds of systems impose their own ways of operating and allow little scope for local adaptation. Situated analyses and practices may manipulate and tinker with particular features and functions—usually of the interface—but are unable to challenge the complex interdependencies and institutional relations that sustain a technology. In particular, Kallinikos expresses his scepticism about the adequacy of ANT analyses to capture the historical and cultural processes through which technologies are constituted as socially-embedded systems.

Avgerou and Madon share Kallinikos' perspective of technology as historically and culturally constituted, and thus intertwined with deeply-rooted social institutions. In Chapter 9, they argue for a research perspective that makes the social context explicit. Motivated by the need to address the endemic difficulties of accommodating ICT effectively in the organizations of developing countries, they are interested in understanding cases where both built-in inscriptions of ICT prove incapable of imposing the technology's institutional and behavioural characteristics and where local actors are not able to shape ICT to their interests and to appropriate its functionality. Contrary to Latour's aphorism in Chapter 3 that 'as a rule, context stinks', and his suspicion that contextual explanation allows the creeping in of structuralist ideas that are incompatible with ANT's focus on actors' agency, Avgerou and Madon suggest that actor network analysis can be a useful starting point—but should be followed by the tracing of the historical/cultural processes and circumstances that justify actors' attitudes and behaviour. It is in such descriptions going backward in time, and wider in tracing networks of relations and actors' allegiances, that they believe the logic of apparently irrational action and paradoxical results of rendering ICT an incompetent actor rather than a powerful ally can be revealed.

Part III: Substantive issues and applications

The chapters in the final part engage with specific issues of concern regarding the spread of ICT in modern society. A total of four chapters here can hardly give an idea of the vast range of such issues, but they are indicative of the kind of research being undertaken to address the issues. Specifically, these authors suggest ways in which theory and epistemology may be blended implicitly (as in Wade's Chapter 10) or explicitly (e.g. in Smithson and Tsiavos' Chapter 11) into social studies of ICT. Part III generally shows how, without aiming at discovering truths and correct solutions, such studies can identify situations of concern, analyse their nature, and seek their probable causes. The findings and conclusions are consistent, in varying degrees, with the epistemological orientation that acknowledges the significance of lived and historically-shaped experience. They do not aim to equip professionals with instrumental knowledge for getting ICT 'right'; rather, their value lies in fuelling the critical judgement of diverse actors by providing a better understanding of underlying processes.

Chapter 10 by Wade is critical of the prevailing view on the role ICT can play in promoting development, which portrays ICT as a solution to problems plaguing developing countries and a mechanism to leapfrog over those difficulties. He strongly criticizes the way the emphasis on e-governance by international development organizations in the early twenty-first century not only considers ICT as a means of overcoming endemic inefficiencies in developing countries' public sector bureaucracies, but also links it with a political agenda for reducing the role of the state and promoting a global free-market regime. Wade points out the contradiction between the developmental potential attributed to ICT and the disadvantages confronting developing countries in the international ICT market, as well as in the international regulatory regime. He voices concern that, in the current circumstances, ICT in effect brings a new form of dependency to developing countries, while its developmental effects are over-estimated and remain questionable. It is interesting to note that no particular theory on ICT and society or on socio-economic change is explicitly employed in Wade's discussion of these issues. Nevertheless, the intellectual roots of the chapter in the political economy of development are easily discernible. Wade's challenge to the current wisdom on ICT and development reflects ongoing debates on development, market versus government, and global regulatory regimes.

Smithson and Tsiavos focus in Chapter 11 on the thorny issue of information systems evaluation, which they consider as a process of representation that selectively inscribes and translates the interests of diverse actors into decisions for future action. They use ANT and a plethora of concepts from contemporary sociology to trace the significance and consequences of making such a selection based on consideration of just a small number of issues, out of a potentially much wider range of aspects with which an IS innovation process may be associated. Their analysis highlights how the methods and tools employed in an evaluation constitute the simplified representation of a complex situation through a manageable set of criteria. Yet, the world that is compressed in an evaluation instance will have to be de-compressed subsequently through follow-up action—as the manager will have to deal with the full complexity of the world. This, they suggest, may explain the reluctance of managers to go through such representation exercises. Smithson and Tsiavos clarify their conceptual analysis with a case study of the development and use of an evaluation method in a business environment. They also demonstrate its validity for addressing a set of common-sense questions in evaluation practice, such as why, when, and how evaluation should be conducted.

Chapter 12 by Galliers traces the way the literature on IS strategy has changed over the years. This overview of a key topic of the IS field shows how business-driven practice has been problematized, revised, and extended, mainly in response to the unfolding of technologies and managerial concerns, but also through the influence of related theoretical discourse, as for example on organizations, organizational behaviour, decision making, and the nature of information and knowledge. It is interesting to note how the concepts, relations, and processes of this area are captured in diagrammatic models and frameworks, in the typical fashion of business and management fields. Nevertheless, as Galliers explains, their role is a sensitizing device, rather than instruments leading to 'correct' decisions.

Finally, Baskerville and Land identify and probe into a type of IS risk that is rarely acknowledged in IS theory or practice, namely that ICT systems may subtly trigger processes by which the organizational structures they are meant to support gradually and unintentionally begin to fall apart. In Chapter 13, they study this problem through a range of concepts. For instance, they see ICT-based systems as structural artefacts that convey abstract representations of the world, and they examine their relationship with the changing structure of organizations. Baskerville and Land distinguish between inwardly-destructive systems that come apart themselves—the kind of problem typically studied in IS failure research—and outwardly-destructive systems that take apart their environment.

The journey continues

This book is our contribution to the growth of the IS field outside the domains of management and computer science. Its rich and varied collection of chapters has been the result of a rather long journey in gathering, sifting, rejecting, and including further contributions. The final result celebrates the diversity and liveliness of the social study of ICT. But the journey has obviously not come to an end, and it is hoped that the book will support the further exploration of the theoretical issues and further study of the empirical evidence we have highlighted and discussed.

References

Avgerou, C. (2002). *Information Systems and Global Diversity*. Oxford: Oxford University Press.

Bijker, W. E., and Law, J. (1992) (eds.). *Shaping Technology/Building Society*. Cambridge, MA: MIT Press.

Bloomfield, B. P., Coombs, R., Knights, D., and Littler, D. (1997) (eds.). *Information Technology and Organizations: Strategies, Networks, and Integration.* Oxford: Oxford University Press.

Bowker, G. C., and Leigh Star, S. (2000). *Sorting Things Out: Classification and its Consequences.* Cambridge, MA: MIT Press.

Brown, J. S., and Duguid, P. (2000). *The Social Life of Information.* Boston: Harvard Business School Press.

Ciborra, C. U. (2000) (ed.). *From Control to Drift: The Dynamics of Corporate Information Infrastructures.* Oxford: Oxford University Press.

—— (2002). *The Labyrinth of Information.* Oxford: Oxford University Press.

Dunlop, C., and Kling, R. (1991) (eds.). *Computerization and Controversy: Value Conflicts and Social Choices.* San Diego, CA: Academic Press.

Giddens, A. (1984). *The Constitution of Society. Outline of the Theory of Structuration.* Cambridge: Polity Press.

Hanseth, O., and Monteiro, E. (1998). *Changing Irreversible Networks.* Paper presented at the 6th European Conference on Information Systems, Aix en Provence, France.

Howcroft, D., and Truex, D. (2001). 'Special Issue on Critical Analyses of ERP Systems: The Macro Level'. *The Data Base for Advances in Information Systems,* 32/4: 14–18.

Hughes, T. P. (1983). *Networks of Power: Electrification in Western Society,* 1800–1930. Baltimore, MD: Johns Hopkins University Press.

Introna, L. D. (1997). *Management, Information and Power: A Narrative of the Involved Manager.* Basingstoke, UK: Macmillan.

Latour, B. (1999). *Pandora's Hope. Essays on the Reality of Science Studies.* Cambridge, MA: Harvard University Press.

Law, J. (1991) (ed.). *A Sociology of Monsters: Essays on Power, Technology and Domination.* London: Routledge.

Monteiro, E., and Hanseth, O. (1996). 'Social Shaping of Information Infrastructure: On Being Specific about the Technology', in W. Orlikowski, G. Walsham, M. R. Jones, and J. I. DeGross (eds.), *Information Technology and Changes in Organizational Work.* London: Chapman & Hall, 325–43.

Mumford, E. (2003). *Redesigning Human Systems.* Hershey, PA: IRM Press.

—— and Sackman H. (1974) (eds.). *Human Choice and Computers.* Amsterdam: North-Holland.

Orlikowski, W. J. (1992). 'The Duality of Technology: Rethinking the Concept of Technology in Organizations'. *Organization Science,* 3/3: 398–427.

—— (2000). 'Using Technology and Constituting Structures: A Practice Lens for Studying Technology in Organizations'. *Organization Science,* 11/4: 404–28.

—— (2002). 'Knowing in Practice: Enacting a Collective Capability in Distributed Organizing'. *Organization Science*: 13/3: 249–73.

—— Walsham, G., Jones, M. R., and DeGross, J. I. (1996) (eds.). *Information Technology and Changes in Organizational Work.* London: Chapman & Hall.

Popper, K. (2002). *Conjectures and Refutations: The Growth of Scientific Knowledge.* London: Routledge.

Suchman, L. (1987). *Plans and Situated Action.* Cambridge: Cambridge University Press.

Walsham, G. (1993). *Interpreting Information Systems in Organizations.* Chichester, UK: John Wiley.

—— (1997). 'Actor-network Theory and IS Research: Current Status and Future Prospects', in A. S. Lee, J. Liebenau, and J. DeGross (eds.), *Information Systems and Qualitative Research.* London: Chapman & Hall, 466–80.

Walsham, G. (2001). *Making a World of Difference: IT in a Global Context.* Chichester, UK: John Wiley.

Winograd, T., and F. Flores (1986). *Understanding Computers and Cognition: A New Foundation for Computer System Design.* Chichester, UK: John Wiley.

Zemanek, H. (1974). 'The Human Being and the Automation: Selected Aspects of a Philosophy of Information Processing', in E. Mumford and H. Sackman (eds.), *Human Choice and Computers.* Amsterdam: North-Holland, 3–30.

Zuboff, S. (1988). *In the Age of the Smart Machine: The Future of Work and Power.* New York: Basic Books.

Notes

1. In this book, authors use varying terminology, for example some authors refer to IT, some to ICT, and others to IS. In general, this variety and inconsistency does not harm the essence of the work presented here and we believe that editorial interventions for a uniform language across the chapters would be inappropriate. When the choice of one of the three terms does matter, the authors themselves explain their terminology.
2. In this volume, see also Monteiro (Chapter 7), Kallinikos (Chapter 8), Avgerou and Madon (Chapter 9), and Smithson and Tsiavos (Chapter 11).
3. Although the chapters in Part II have little scope to present empirical research, they draw on, and refer to, background to their own empirical work and that of other researchers.

PART I
Foundations

1

Encountering information systems as a phenomenon

CLAUDIO CIBORRA

Introduction

When dealing with the deployment and management of complex information and communication infrastructures, managers at the turn of the twenty-first century seem to lack the words to describe, let alone capture, the multiplicity of unexpected consequences, serendipitous occurrences, and emergent, disappointing features of the new technological systems they are busy installing. In this, they sound very much like the early twentieth century poet, dramatist, and essayist Hugo von Hofmannstahl (1995) in a letter to Lord Chandos: 'Briefly, my case is this: I've completely lost the ability to think or speak coherently about anything ... The abstract words which the tongue has to employ in order to express any kind of daily opinion decompose in my mouth like rotten mushrooms.'

A key reason for managers' bafflement and uncertainty lies in the ungrounded expectations created by widely-used managerial and consulting models. Leveraging on the belief that ICT is a powerful means to control processes, people, and resources, these business models and systems methodologies promise a variety of ways in which top managers can 'align' ICT with strategy by re-engineering processes and creating entirely new, competitive e-businesses. And that's not all: even knowledge can now be formalized and managed; workflows centralized; transparency enhanced; and data mined wherever they hide within the enterprise's procedures and the departmental files.

The vacuity and boastfulness of these promises should not fool anyone. They are not the result of just another managerial fad. Business schools are filled with industrious Ph.D. students busily trying to measure 'ICT strategic alignment', or the variables correlated with the successful implementation of ICT infrastructures, and the like. Then, with neo-positivist ardour, they are obliged by the leading faculty to test hypotheses and identify success factors. This results in an incredibly consistent outcome that cuts across the rigorous quantitative analyses of the Ph.D. candidates and the sleek PowerPoint presentations of the consultants. The one panacea that invariably emerges is the need for 'managerial foresight' and 'top management involvement', including bringing the Chief Information Officer into the top management team. In short, the recommendation is: 'more command and control'.

Sadly, our latter-day von Hofmannstahls know in their bones that, although these sorts of factors might be necessary (justified at the limit by sheer common sense), they are definitely not sufficient. Model systems often do not roll out as planned, leaving managers with an array of puzzles and a myriad of related issues, which make it hard to make any sense of what went wrong during the implementation. When practitioners next attend an executive course or hear about the latest consulting methodology, initially they are likely to feel their own usually problematic case is an exception, as others have succeeded—or will soon triumph—thanks to the most recent, ultimate methodology. But this game has been going on for so long now in the ICT field that managers typically fall into an even deeper frustration in these sessions. This is a sign of crisis (Ciborra 2002), and this chapter's starting point in addressing it goes back to a series of lectures and conferences more than 70 years ago by the founder of phenomenology, Edmund Husserl.

The importance of the everyday

Husserl (1970) discussed the 'crisis of European sciences' in a way that is highly relevant for the dilemmas many of the management disciplines have been facing recently. He talked about science and method in a way that offered a number of insights into how to address the operational dilemmas which ICT practitioners are currently stumbling into more and more frequently. Husserl's relevant contribution lies in his analysis of the origins and motives of the crisis; his way of dealing with the trade-offs between objectivity and subjectivity in the sciences; his valuable reflection upon the emergence of the modern scientific method; and, critically, his highlighting of the central and foundational role played by everyday life practices.

The very definition of information systems as a set of technical/scientific and human resources devoted to the management of information in organizations spells out the composite nature of the field. The same could apply to a frequently-used definition of information as 'data': physical or logical signs (natural-scientific) endowed with interpretation by a user (human or social-scientific). What should concern us here is that a common, unified paradigm has been adopted in the IS discipline to deal with its human, social, and natural dimensions, despite the hybrid and—at the limit ambiguous—nature of the problem. The common solution is the paradigm of the natural sciences and the relevant methodologies of measurement, formalization, and calculation. However, some aberrations are encountered by most of the consultants and academics who try to adopt, extend, and sometimes mimic the natural-science way of proceeding. This arises because business organizations get analysed in the same fashion, and even using the same means of representation, as the physical computer systems or quasi-mathematical abstractions of their data flows, entity-relationship graphs, or object-oriented languages.

The consequence is that disciplines, such as IS, that are inspired by the paradigm of the Galilean method tend to disregard the fundamental role of the everyday 'life-world' of the relevant agents, users, designers, and managers—and of the messiness and situatedness of their acting. At the same time, such disciplines privilege the geometric worlds created by systems methodologies and business process models. It is in this way that a key element is neglected: human existence, which is an essential ingredient of what information is, how the lifeworld gets encountered, defined, and reshuffled, and—last but not least—how technology gets used every day. Ironically, the main lesson that could be extracted from the major developments sustaining the current success of the Internet is that there is strategic importance precisely in the ordinary modes of operation, such as bricolage, heuristics, serendipity, and make-do, rather than in the idealizations of the scientific method.

As an antidote to the prevailing emphasis on formalization and reification of business and systems, Husserl's phenomenology urges us to go back to the basics and encounter the world as it presents itself in our everyday experience. We rely on evidence, intuition, and empathy. We listen to practitioners and we participate in their dealings with puzzles and riddles; on the other hand, we do not confer any particular relevance to words like 'strategy', 'processes', 'data', or 'system'. In so doing, and in putting aside the models and methods of management science, we come closer to the everyday life of the manager, which is made up of frustrations, accomplishments, gossip, confusion, tinkering, joy, and desperation.

If we are able to accept the messiness of the everyday world's routines and surprises without panicking, we may encounter business phenomena that deeply enrich the current 'objective' and reified models of organization and technology. We can then start to build a new vocabulary around notions closer to human existence and experience. As an invitation to this phenomenological style of reflection, this chapter discusses some key metaphors, such as bricolage, drift, hospitality and care, improvisation, and moods as a fresh way of revisiting the well-known management issues of strategy, technology, project management, knowledge management, and business processes.[1]

Strategy: the virtues of bricolage and improvisation

The Internet, e-business, and e-commerce have made many organizations aware of the strategic role that ICTs can play in restructuring their internal processes, and in establishing better and new links with suppliers, customers, and the public at large. Since the early efforts in the 1970s in building strategic information systems, methods put forward to design strategic ICT applications have generally contained a standard set of prescriptions and guidelines on:

- supporting the business *vis-à-vis* the competition, together with an approach to planning and implementation;
- obtaining top management awareness and commitment;
- aligning ICT applications with the business strategy; and
- implementing applications that may generate competitive advantage.

Unfortunately, strict adherence to the prescriptions, guidelines, and strategy models included in the methods leads to a series of conceptual paradoxes. To begin with, since imitation has always been the driving force behind the diffusion of any technological innovation, any competitive advantage evaporates if every major player in the industry adopts the same or similar applications, such as the spread since the late 1990s of enterprise resource planning (ERP) systems. Only normal economic returns can be generated by systems that can be copied and built by a large number of firms in an environment where no firm enjoys any distinctive or sustainable advantage in implementation. Small firms, in particular, are likely to lose out in applying standard planning methods and solutions because, as they do not set prices in their markets, they will find it very difficult to manipulate the industry structure to their advantage through the strategic use of ICT.

Some inter-organizational systems, for example electronic data interchange (EDI), require the interconnection of all the major firms in an industry, which undermines the competitive advantage to an individual firm claimed by advocates of such systems. More generally, market analyses and the identification of strategic applications are research and consultancy services that can be purchased: they are carried out according to common frameworks, use standard data sources, and—if performed professionally—will reach similar results and recommend similar applications to similar firms. In sum, competition tends to force the standardization of solutions and equalization of production and coordination costs among participants. These dynamics will certainly unfold, unless a firm's strategy cannot be imitated perfectly, as the more difficult it is for other firms to imitate a successful system or application, the longer can a firm obtain increased returns.[2]

Hence, two alternative ICT strategies can be compared, which contain different sets of ideas about business strategy and competition relevant to designing a strategic information system. According to the first, strategy gets formulated before the fact, is based on careful industry analysis, and consists of a series of moves that can be planned and implemented to gain advantage through manipulation of the industry structure. According to the second, strategy formulation is difficult to plan before the event, and competitive advantage stems from the exploitation of unique, intangible characteristics of the firm (including its networks of relations), and the unleashing of its innovative capabilities. We have seen that the first package is bound to encounter a deadlock. We need, then, to explore the second one further.

To avoid easy imitation, the quest for a strategic application must be based on intangible, even opaque, areas such as organizational culture. Thus, it is the enactment of unique sources of practice, knowhow, and culture at firm and industry level that can be the source of sustained advantage, rather than the structured analysis of internal assets and market structures. This departs from the world of strategy to enter the workaday world, in particular what people actually do every day.

As soon we leave the realm of method, procedure, and systematic ways of organizing and executing work according to rational study, planning, and control, we enter the murky world of informal, worldly, and everyday modes of operation and practice. It is the realm of hacking; practical intelligence; the artistic embroidery of the prescribed procedure; the shortcut; and the transgression of the established organizational order as embedded in systems and formalized routines. 'Bricolage', 'improvisation', and 'hacking' are terms that are often used to describe such modes of operation in various domains of activity. The terms differ slightly in meaning, but they have many aspects in common, of which these modes are among the most important. Since these approaches diverge from the formalized, preplanned ways of operating, their outcomes can well lead to serendipity and to the possibility of finding something valuable that was not sought for at the outset, as exemplified by many of the key technical choices made during the development of the Internet (Abbate 1999).

Another characteristic of these modes of operating is that they have an uncertain status, always at the boundary between highly competent behaviour and incompetence. As this is at the margins of the formalized procedures, improvisers are granted an added element of liberty, and sometimes of play, in the choices made about which resources to harness and how to do it. These approaches might look rough compared to neat and tidy formal procedures, but they are highly situated. This means they tend to include an added element of ingenuity, experience, and skill, which belongs to individuals and their communities of practice (Brown and Duguid 1991) rather than to the organizational systems. They also all seem to share the same way of operating, by unfolding in a dance that always includes the key aspects of: localness and time (the 'here and now'); small forces and modest interventions; and add-ons made on-the-fly, but deeply rooted in personal and collective skill and experience. When performed skilfully and with close attention to the local context, the dance leads to momentous consequences, unrelated to the speed and scope of the initial intervention. Last, but not least, these activities tend to be invisible, because they are both marginalized and unfold in a way that is small in scope. The results are modes of operating that are latent and not easy to imitate. Thus, the smart bricolage, imaginative improvisation (Weick 1993), or the good hack cannot be easily replicated outside the cultural bed from which it has emerged.

If one considers many strategic applications—ranging from the Internet, to airline reservation systems, to new operating systems such as Linux—the phenomenological perspective helps us recognize and value the importance of bricolage versus the over-inflated role of method and planning in strategic applications. Practices that are considered by the positivist school of thinking as marginal play a strategic function, and *vice versa*. What is put forward as structured, top-down planning approaches appear to lead to non-sustainable strategic solutions which, in general, end up playing a marginal role in organizational innovation.

Between procedure and drift

The phenomenological gaze at what happens in organizations, outlined in the previous section, has unveiled the strategic importance of the ubiquitous processes of tinkering, hacking, and improvisation around the implementation and use of new technology. Given the naturally-occurring, or triggered, learning processes surrounding any innovation, and the unavoidable breakdowns that punctuate the trajectory of any complex system when deployed, one main outcome of all these practices and events is that the technology does not seem to work completely according to plan.

For example, consider the results of a qualitative empirical study in a sample of large multinationals using groupware, software providing support for teams of people working collaboratively (Ciborra 1996). The researchers found that the development and use of groupware in large, complex organizations is variable and context-specific. In particular, in almost all the cases considered, groupware presented itself as a technology that tended to 'drift' when put to use. Drifting here describes a slight—or sometimes significant—shift of the role and function of usage in concrete situations. This contrasts with the planned, predefined, and assigned objectives and requirements that the technology is called on to play in more structured perspectives. Drifting should not be considered as a negative phenomenon per se, as it can occur both for applications that are seen as successes and for those that are not.

For example, an extensive application of a Group Decision Support System (GDSS) within the World Bank led to a surprising result in use. The initial goal of the application was to improve collective decision making during important policy meetings within the complex environment of the Bank, where different cultures and political interests have to converge on highly sensitive investment decisions. The GDSS system offered a range of possible uses, besides its voting functionality. Note, however, that in the daily dynamics at the Bank, highly sensitive decision making is usually being prepared outside the meetings where the decision has to be formally deliberated. Users of the GDSS tacitly agreed to keep the present practice, while using the facility only to brainstorm and focus on issues. So, the system was successful in being heavily used, but not as a decision support system.

Drifting can be looked at as the outcome of two intertwined processes. One is given by the openness of the technology: its plasticity in response to the reinventions carried out by users and specialists, who gradually learn to discover and exploit features, affordances, and potentialities of systems. On the other hand, there is the sheer unfolding of the actors' 'being-in-the-workflow' and the continuous stream of interventions, tinkering, and improvisations that colour the entire system life cycle. The outcome of these two processes led the researchers on the groupware research project to the surprising conclusion that 'what groupware is' can only be ascertained *in situ*, where the matching takes place between plasticity of the artefact and the multiform practices of the actors involved. Such a matching is open, situated, and continuously unfolding.

The previous section on bricolage showed that the use of applications is always shaped by hacks, short cuts, and twists, or is punctuated through unpredictable processes of re-invention. Drifting is the result of these processes, which range from sabotage, to passive resistance, to learning by doing, to astonishing micro discoveries and radical shifts—or just plain serendipity. Usage, maintenance, redevelopment, and continuous—or sometimes fortuitous—improvements take place simultaneously in these processes. Schön (1983) suggests that those who promote or adhere to methodologies have a view from the high ground, and tend to ignore what goes on in the daily 'swamp': the fluid territory of actual projects and everyday life where such an unfolding takes place in organizations. Drifting characterizes life in the swamp. It denotes the dynamics of an encounter, of pasting up a hybrid composed of technology, organizations, people, and artefacts. Drifting is a way of capturing the unfolding of the intrinsic openness of such an encounter.

We are therefore confronted with the methodological challenge of interpreting drifting in terms of the swamp, not of the high ground. The human dimension of drifting belongs to the world of everyday practices: ubiquitous but anonymous; made of ruses and short cuts; improvised; marginal; relying on age-old, timeless skills. As such, it needs to be captured. Plans and methods belong to the high ground dominated by the notion of space, where time is clock time. It is in this space that the flow diagrams, and the activity and sequence maps, are currently expressed. In contrast, drifting is made of bricolage and improvisation. These can hardly be 'hosted' by, or find a meaningful context in, the large, chilly spaces of the high ground. They are local, short, and sudden, and do not exist outside the specific situation where they appear. They belong to the opaque, shapeless 'boxless' world of the swamp, where time is fluid or out of joint (such 'extemporaneity' is also discussed later in the chapter).

Procedures, on the other hand, unfold according to clock time and their execution relies upon prepackaged knowledge, which is placed in front of the actor as deadlines, goals, and planned actions. Models and methods of military more than scientific origin belong to those strategies that focus on the space dimension in multiple ways, since they want to pursue:

- neatness: by establishing proper space for activities to be performed in an orderly fashion;
- structure: by elaborating theoretical places through systematic discourses; and
- articulation: by identifying appropriate physical spaces, from overhead transparencies to the estate of screens throughout the office buildings and laboratories.

In this way, the models and methods 'attempt to reduce temporal relations to spatial ones, through the analytical attribution of a proper place to each particular element and through the combinatory organization of the movements specific to units or groups of units' (De Certeau 1988: 38). However, the tactics, ruses, and improvisations that lead to drifting as the outcome are contingent procedures indexed by the here and now, and meaningless outside a specific time-tagged situation. Given a linear, preplanned procedure made of a sequence of actions, then tactics are precisely those scrambling interventions, multiple variations, and fleeting creative acts which transform the expected neutral situation into one perceived as favourable or pleasant. In a nutshell: from command and control to the aesthetics of systems and organizations.

In general, the plans and stable procedures that comprise models and methods want to defy time by their robustness and stability, that is they want to announce organizations as 'Pyramids'. Tactics, instead, are rapidly moving, with their mobile interventions being dictated by, and forcing, the seizing of the moment. The models and methods bet on space and order, tactics on the appropriate time and the contingent situation; models adhere to a fixed structure, tactics are condemned to be ephemeral and transformative.

The practices found in drifting also tell us something about knowledge in organizations. Drifting stems from those mundane, invisible practices that, compared to the crisp world of procedure and method, could be seen to represent the dark side of organizational work. They are intelligent practices, the expression of a practical intelligence (Scribner 1984). Far from what has become the conventional distinction between tacit and explicit knowledge, practical intelligence is the *metìs* of the Greek—the intelligence of the octopus: flexible, polymorphic, ambiguous, oblique, twisted, circular. This is the opposite of the straight, direct, rigid, and univocal character of the knowledge embedded in method. To orient oneself in the complex and changing world, dealing with forces that are too strong to be fully controlled, one needs to leverage the situation at hand by zig-zagging behaviour, which never faces such forces up-front but accomplishes the project at hand with a sudden move of improvisation.

Of course, these two modes of operating can be, and have been, coexisting and complementary, at the same time excluding each other (e.g. business process re-engineering wanting to eliminate improvisation) and supporting each other (e.g. when tactics grow like lichen over procedures and models when put in use). Large systems and models seem to be able to take off only when surrounded by a regimen of tactics that fills the gaps between complex formalized procedures and the floating world of complex organizations and turbulent economies.

Systems development: hospitality

The notion of 'hospitality' offers the opportunity to explore further what the buzz is all about in the design, development, and implementation of ICT platforms, where technology and local practices intertwine in organizations. This new perspective is consistent with recent findings from the social studies of technology, in particular the symmetric role played by humans and non-humans (Latour 1999).

Rethinking what is known about IS development and use

Before jumping in to find better ways of improving, streamlining, and re-engineering what we think is known about systems development and implementation processes, it is necessary to 'put in brackets' what is already known about them. This is a first step toward grasping systems development (and more generally our relationship to technology and design) from a phenomenological perspective. Bracketing what we tend to take for granted allows us to dispose of those apparently self-evident appearances—such as goals, plans, control procedures, measurement techniques, and the vaguely pervasive and seductive notion of technology as a familiar, domesticated tool—which often serve to obscure our understanding of the phenomenon under investigation and discourage any alternative approaches. Most probably, this phenomenon will not actually be like its appearances.

On the other hand, we need to reflect upon the puzzling evidence provided by the continuous apparitions that punctuate any IS development effort and the system in use. These include: unexpected consequences and the drifting of the technology resulting from frequent tinkering and improvisations; and partial acceptance or even resistance to continuous improvement methodologies, if not open critique of their scientific foundations—even by specialists and practitioners. Such apparitions are symptoms of a malaise in the current ways of understanding and approaching systems development and use.

The phenomenon to which we want to come closer generates both the (false) appearances and the apparitions (the actual symptoms), but itself usually remains hidden. A way to unveil it is to start from what both managers and practitioners have carefully left out of current approaches to systems design and management: human existence. This relates to the designers' and users' practical dealings in the lifeworld of systems in development and in use, particularly to their concerns as humans being thrown into situations facing uncertainty; the intertwining between their personal trajectories and the project execution; their identities as subjects; and the unavoidable openness of any project or innovation, which rarely fails to resonate with the existential openness of the participants' own life projects. Indeed, the successful completion of any initiative may well depend upon the marrying up of such existential traits with the objectivized characteristics of the project.

Human life as a whole cannot be separated from what one can achieve during an innovation, the launch of a project, or a new development. Although such initiatives are all future-oriented, and the accompanying methodologies put exclusive emphasis on the management and execution of the 'in-order-to's of the project, they inevitably share a lot with the participants' experience and personal history. The injunctions of the in-order-tos are supposed to mobilize the attention and resources of the project members towards a future state of affair, but it is the members' biographic, historic, and ethical 'because-of' motives that can endow the innovation or the project with meaning and momentum.

If the project goals and plans do not make sense to those called on to implement them, only perfunctory or distracted compliance will follow. To disregard the complex chemistry and balance between the because-of and in-order-to motives of action may lead to much of the unexpected consequences for both successful or failed innovations. But such an existential balance is extremely precarious, if not contradictory. On the one hand, any development is supposed to lead along a carefully planned trajectory to a better future state. It is expected that any deviation can be controlled and the course restored or improved by feedback and learning. Thus, it is an endeavour full of rational promises laid out in front of the actor by the structured methodologies. On the other hand, the actor's personal past experience and life trajectory remain blurred, and the actor has to cope with the fact of being 'thrown' into the project or use situation almost by chance, or by a series of circumstances largely outside the actor's control. Furthermore, the actor's past is made up of cognitive frames and scripts that can hamper the person's ability to learn, in ways the individual is hardly aware of. As a result, any development will result in an inextricable mix of failure and success, with many minor or major unintended consequences that can trigger new learning and innovations, or just lead to frustrating vicious circles. The sense of achievement and discovery will always be intertwined with the anxiety of failure, falling, and drifting.

Alas, none of these preoccupations that are closer to our being human gets featured in the countless methodologies developed by software engineering organizations all over the industrialized world. This opens the door to the notion of hospitality (Derrida 1997), which was first deployed as a promising notion in this context as a way of interpreting the case of a surprising evolution of a groupware system in a multinational consumer goods company (Ciborra 1996). The case dealt with a fairly large application of Lotus Notes used for new product development within a worldwide team-based organization. Dedicated Notes applications were built to allow multidisciplinary and multinational teams to work jointly on common projects, regardless of distance between locations. Implementation of the new system was carried out according to a participative methodology, an incremental introduction of the system, and comprehensive training. Usage was immediate, ubiquitous, and successful. One day, however, a cheering message broadcasted over the network by a top marketing manager in London made every user realize that the new transparent platform could be deployed by headquarters as a powerful controlling eye, able to access any working document and local bulletin board of all the distant teams. Such a possibility made usage fall immediately and significantly. Attempts to revamp the system succeeded only when the applications were redesigned to replicate the pre-existing routines and organizational structures, thus losing much of the original innovative design, transparency, and collaboration opportunities.

This case provides evidence, among other things, about the ambiguity of new technology. Despite the careful planning and design, and the extensive training, the new groupware technology appeared to the user suddenly as an ambivalent, threatening *stranger*. The latent tensions between the professional dimension and the existential one exploded as a consequence of a small incident; underlying anxieties about the new ways of working and the new powerful tool could not be tamed even by an advanced design concept and a careful project management plan.

Reaching out to technology as a guest

Hospitality describes the phenomenon of dealing with new technology as an ambiguous stranger. Hospitality is a human institution. It is about being receptive and adopting, and about managing boundaries between what or whom is known, and what or whom is unknown. Hospitality is a first step in accepting the Other.

This existential knot is carefully avoided by methodologies too abstract and high-flying to deal with such a human and worldly moment, lest the methodologies risk being caught by surprises when such events and forces creep in and burst out unexpectedly—and when sudden apparitions, such as apparently irrational users' reactions, appear from nowhere to disrupt irreversibly the carefully crafted appearances of rationality, planning, and control. This raises the possibility that hospitality is the hidden phenomenon generating both the false appearances of systems development methodologies and the array of symptoms pointing to the limited role such methodologies ultimately play in actual systems development, despite their claims to the contrary.

Hospitality is about crossing a boundary, reaching out to the Other, the Stranger—but without abolishing such a boundary. The host must deal with the ambiguity of the stranger, who can be a friend or an enemy. If hosting is about weakening one's own identity to enrich it, reaching out to the Other means establishing the new symmetry: recognizing and accepting the identity of the Other, at least on a temporary basis. Hospitality is the human process of making the Other a human like oneself. Hosting the new technology is then seen to mean accepting a paramount symmetry between humans and non-humans (Latour 1999). This is exemplified by considering systems development as a process by which the organization hosts the technology. Such a perspective of hospitality reveals a number of surprising features of this process, which is usually looked at as a structured technical activity or as a political struggle:

- Being able to host the technology will redefine our identities.
- Unexpected consequences signal only that any attempt to control the technology fully are doomed to failure; at best, hospitality involves intelligent servicing of the new technology.
- Different cultures prescribe different codes, norms, and rituals for hospitality, which the guest must accept. When systems development is conceived of as hosting the new technology, methodologies are seen to constitute today's rituals imposed by humans on the technology.
- Following Kant's (1972) discussion of the universal right to hospitality, humans should grant a set of rights to technology, such as the right to visit—but not necessarily the right to stay. This also recalls Heidegger's (1992: 28) idea of being able 'to say yes and no' to modern technology, or the injunction by Weick (1993: 632) to 'be able to drop our tools' in an emergency.
- If the guest is perceived as hostile, the host will treat the guest as an enemy (as during the Industrial Revolution, when the Luddites attacked the machines that threatened their work).

Technology as a guest presents itself to the host as endowed with affordances. Affordances trigger a network of commitments by the host: they define the contours of the host's role as designer, sponsor, or user (Norman 1988). That is just the beginning of an open-ended process. The guest-technology also possesses its own dynamics, and will begin to align the host according to certain needs and constraints. In this way, hospitality diverges from unilateral command and control as prescribed by structured methodologies because, in order to remain the master of the house, the host must release control and serve the guest. Service is in the first instance compliance with affordances. The consequences for the guest-technology are:

- technology is rendered human through hospitality, and the humanization of systems can be a more intriguing project than the virtualization of reality;
- technology can consider the hosting organization as being at its own service, but it should not dominate the host;
- when technology turns into an enemy, it will exploit the organization and its members, finding allies among certain groups in the organization while, at the same time, dropping them when it no longer needs them (Ciborra 2000).

Hospitality involves the risk of misunderstanding, since it typically has to deal with communication across different languages and cultural modes. The guest is intrinsically ambiguous, and can turn into an enemy. Both can become hostages of each other. It is thus a relationship which has to be based on a certain amount of trust, although this is a trust that can be cultivated only across communities or clans, not within the boundaries of one community or clan. If the host becomes a guest in his own home, then the guest becomes a sort of host. Thus, technology hosts the humans thanks to its own array of affordances, and systems development becomes the intriguing business for humans of finding ways to be hosted by the technology—as in Stanley Kubrick's movie *2001*, where the astronauts slowly discover that they are powerless guests of the spaceship's computer, HAL. Enduring standards, legacy systems, interfaces, and old-fashioned languages can be now looked at as the rituals imposed by the ICT infrastructure on humans as guests.

In summary, seeing hospitality as the main phenomenon of the encounter between technology and organizations shows that systems development methods are just the external appearance of a ritual imposed by the human host. Methodologies may be carefully planned and deployed, but cannot dispose of the unpredictability and ambiguity of the guest. Understanding hospitality as a phenomenon we have to deal with when designing, implementing, and using new technologies is not a purely intellectual exercise. We can transcend the programme and priorities set by the rituals of methodologies through such an understanding. We will then be ready to explore new ways of rearranging our commitments towards ourselves as designers and users, and towards the technology as a non-human, ambiguous guest. This enables a different agenda to be set out to deal with new technologies. Trust and friendliness must be coupled with a release of control, as it must be remembered that the host must play the server. Part of the practical ways of coping with technology should become acceptance of the guest's intrinsic ambiguity and mystery, for which Ciborra and Lanzara (1994) borrow poet John Keats' phrase 'negative capability'. An effective host must certainly be able to exercise various forms of care, depending upon the unpredictable circumstances in the unfolding of hospitality. Finally, host and guest will most probably be engaged in forms of reciprocal cultivation by sharing and enriching the respective culture and practices.

Understanding the domain of application: knowledge management, improvisation, and moods

The final topic in this chapter to be exposed to a phenomenological discussion is the very domain of application of many information systems, that of decision making and procedure, planning, and routine. Consideration is also given to the more general domain of knowledge, covering expert systems, Artificial Intelligence (AI), and knowledge management systems (Winograd and Flores 1987). In most of the relevant literature, a difference is acknowledged between structured and unstructured decisions (Simon 1961); between established routines and emerging ones (Nelson and Winter 1982); between tacit and explicit knowledge (Polany 1962; Nonaka and Takeuchi 1995); and between planning and situated action (Suchman 1987). All these topics, models, and approaches represent the core drive for a cognitive understanding of organizations and their information systems. What can the phenomenological approach deliver in this domain, and what difference does it make?

Differences between cognitive and phenomenological approaches

The study of improvisation is a good candidate to illustrate the difference between cognitive and phenomenological approaches. Improvisation is needed to fill the gaps of planning, cope with unexpected consequences, and, in general, face emergencies. Weick (1993) has shown in his vivid account of the Mann Gulch fire disaster how improvisation can be an antidote to panic and, more generally, to those forces, whether natural or psychological, that can bring about the collapse of sense making and organization.[3] More common activities, like making choices on a market, also show many of the characteristics of improvisation. For instance, Hayek (1945) is unsurpassed in interpreting the price system as a way of coordinating local, highly-situated decisions made on the spur of the moment by market agents. Even within hierarchical work organizations, several studies have shown the highly *ad hoc* character of many actions coping with events and situations that do not immediately fit planned, hierarchical procedures (e.g. Suchman 1987; Brown and Duguid 1991).

Improvisation can be looked at as a special case of situated action: highly contingent upon emerging circumstances; unifying design and action; and quick, sudden, and extemporaneous. Indeed, improvisation, like situated action, is now seen in the management literature from the privileged standpoint of cognition as a way to solve problems in the context of emerging circumstances, on the spur of the moment. Improvising is usually linked to the exploitation of tacit knowledge. Such cognitive study of improvisation has concluded that any entity which can reasonably be thought of as planning or executing action can also be thought of as improvising. But here we encounter a dead end. If improvisation is just quick problem solving that takes into account emerging circumstances by some sort of ongoing feedback on the very design of the action being undertaken, then it isn't clear what is new, or so special, in comparison to the more planned or procedural problem solving (Vera and Simon 1993). To overcome such an impasse, we need to step back and revisit the intellectual roots in phenomenology of situated action, of which we consider improvisation a special case.

Improvisation is currently treated in the management and organization literatures as a form of situated action, where the emphasis is placed on its temporal dimension and its description is based largely on a cognitive perspective. It is seen as an activity in which composition and execution, thinking and doing converge in time, or occur simultaneously. Thus, there is a situation, and it is emergent. The trick of improvisation, as opposed to scripting and planning, is to capture all the latest circumstances in the emergent problem solving. However, with regard to the temporal dimension, it is important to recall that the Latin term for improvisation is *extemporalis actio*. This is still used today, and 'extemporaneous' is one of the attributes most referred to when describing improvisation, both in music (e.g. jazz) and managerial literature. The managerial literature also seems to be concerned with the simultaneity of different activities, such as thinking and doing, and the speed of knowing and acting. But extemporaneous does not mean just fast and simultaneous. Etymologically, it means outside of time, or outside the normal flow of time.

Interestingly, the literature mentioned above generally takes no notice of this. Weick (1998) rightly points out that if speed is the main characteristic of improvisation, then in many circumstances coping with a faster tempo would condemn the agent to using preplanned, repetitive procedures to keep the performance going. In other words, higher speed may not encourage improvisation, but a sudden reversion back to old ideas and routines. This leaves us with a conceptual inconsistency: on the one hand, improvisation being advocated for fast product development, prompt market decision making, and successful organizational performances; on the other, it is promoted for carrying out the very activities that seem to require simplification, accelerated production, and less slack, thereby forcing people back on older ideas and away from adaptive improvisation. In short, gaining speed may actually undermine spontaneity and extemporaneity.

A key limitation of situated action approaches

A way back to finding the required alternative to the cognitive view of improvisation can be opened by looking again at the study of situated action versus planning carried out by Suchman (1987). She defined the situation studied as the full range of resources that the actor has available to convey the significance of his or her action, and to interpret the actions of others. Specifically, in analysing how employees deal with photocopiers—and their expert systems, Suchman (1987: 119) suggests that 'the situation of the user comprises preconceptions about the nature of the machine and the operations required to use it, combined with moment by moment interpretations of evidence found in and through the actual course of its use'.

Note here the ingredients of the theory of improvisation: on the cognitive side, they are preconceptions, interpretations, and evidence; on the temporal dimension, it is the moment-by-moment activity. Suchman (1987: viii) adds: 'Action is contingent on a complex world of objects, artefacts and other actors located in space and time. And this is an essential resource that makes knowledge possible and gives action its sense'. Thus, the situated action paradigm states the importance of the fleeting circumstances on which making sense of the action relies, but which these accounts of action routinely ignore. Plans might provide sense or meaning to an action through a formalized representation of events, resources, and interactions over (clock) time, but do not help to cope with unexpected breakdowns and more generally emerging circumstances.

It is interesting to compare this with the planning approach developed in Artificial Intelligence. In this, a physical AI system interacts with the external environment by receiving sensory stimuli, which it converts into symbol structures in its memory. The system then acts upon the environment in ways determined by symbol structures, in which the memory is an indexed encyclopaedia where representations of external situations are stored. Stimuli coming from the environment invoke the appropriate index entries. According to Vera and Simon (1993: 15):

> Sequences of actions can be executed with constant interchange among (a) receipt of information about the current state of the environment (perception), (b) internal processing of information (thinking), and (c) response to the environment (motor activity). These sequences may or may not be guided by long-term plans (or strategies that adapt to the feedback of perceptual information).

In other words, the proponents of the more traditional AI approaches based on representations and symbol processing argue that one can design and build symbol systems that continually revise their description of the problem space and the alternatives available to them. This mimics one of the key ideas of the situated action perspective: the importance of moment-by-moment capture of the full situation of action. Plans can be seen not just as symbolic representations of fixed sequences of actions, but as strategies that determine each successive action as a function of current information about the situation. Here, we come full circle, with the AI perspective coinciding with the definition given above of improvisation as 'situational decision making'. The two approaches contain the same ingredients: symbols, goals, means and ends, plans and actions. They differ only in terms of high speed and fine adaptability.

But what if the view of situated action and improvisation as cognitive enterprises is actually the source of the impasse in the debate, and also the reason why the descriptions discussed so far are not able to address innovatively the issue of the temporal, or better the extra-temporal, dimension of improvisation? Once again, we need to put into brackets the prevailing cognitive view of organizations, information systems, problem solving, and knowledge. Phenomenology can come to the rescue by pointing out that there are at least four different terms for describing and interpreting a 'situation' (Heidegger 1962, using the original German expressions): *Stelle*, meaning position and place; *Lage*, for condition and disposition; *Situation*, as the culminating limit situation of making a choice; and, last but not least, *Befindlichkeit*, which is the term that can help us in opening new dimensions.

Befindlichkeit is derived from the common expression '*Wie ist Ihre Befindlichkeit?*', a courteous way to ask, 'How are you?'. *Befindlichkeit* is the situation one finds oneself in. But what situation? The loose arrangements of resources in the environment? The emerging physical and social circumstances? The answer is none of these. The expression refers to the existential situation of the actor: 'How do you *feel* today?' Thus, *Befindlichkeit* combines the ideas of situatedness and of feeling and faring, of where and how one finds oneself.

The term *Befindlichkeit* significantly captures the common way of inquiring about the situation of the people we encounter in everyday life. Although this is such a routine and ubiquitous a habit, it is still totally absent not only from the symbolic AI-based representations of human problem solving, as one might expect, but also, fatally, from the situated action approach. Indeed, both AI and situated action seem to consider the actor as a cognitive robot. The discussion between these two approaches is about how the robot solves the problems, learns about circumstances, and plans or reacts to emergent conditions. Later appeals to 'situated and embodied knowledge' still evade a reply to the ordinary question, the how-are-you of the actor, his moods, feelings, affections, and fundamental attunement with the situation. What is missing from this situated action literature is precisely an enquiry into the situation of the actor, specifically her or his moods.

The vital need to take account of actors' moods

The original intuition of phenomenology has been to bring into the picture the situation of the actor—those fleeting personal circumstances captured by the term 'mood', not only the situatededness as defined by emerging environmental circumstances. Unfortunately, concern with moods has been lost even in the alternative accounts of situated action, such as the one put forward by Suchman (1987).

Any actor enters into the situation with a mood that is elusive: fear, anxiety, happiness, panic, boredom. This can hardly be controlled, designed, or represented in symbols to be fed into and analysed by computers. Moods are the uncontrollable changing skies of the otherwise flat world of cognition and action, whether planned or situated. According to Heidegger (1962), it is precisely because moods come and go like the weather that they are very close to who we are in the situation. They are ephemeral, sometimes superficial and unexplained, but they precede or, better, ground any mental representation of the situation and action strategy. But moods are far from being just private states. They also disclose the world. When we encounter the world in a situation, certain things, people, or circumstances matter. This 'mattering' is grounded in how one is affected, and this affectedness discloses the world in an intrinsically social way as a threat, as boring, or as exciting. It sets the stage for shaping the problem's definition, solution, design, and action.

Our openness to encountering the world, our being amidst people and circumstances—and the related intentional projects of planned or situated action—are constituted within a fundamental attunement: the mood. These moods can change, but we are never without one. In this respect, not only is symbolic representation not primary, but neither is cognition. Note that if moods provide the ground in which our encountering the world and defining the situation take place, we can seldom choose such a ground. Rather, we are thrown from it into the situation. Moods colour indelibly our being in the situation; unless I am in a mood, I will not be affected, touched, or interested by anything. It is precisely because a mood is not a mere consequence of our actions that its essence and origin tend to remain concealed. Moods are the fundamental ways in which we are disposed in such and such a way; they are not the direct consequence of our thinking, doing, and acting. Instead, they are the presupposition, the medium within which those activities take place. The most powerful moods attune us with the situation so strongly that it almost seems there are no moods at all. And this is the trap into which even those who write about 'embodied knowledge' seem to fall, by failing to put moods at the centre of attention.

The way we care about the world unfolds according to the passing moods that attune us with the situation. Intentionality, the reading of circumstances, the in-order-tos of projects, the selection of appropriate means and ends, are all rooted in the ground of our basic attunement. The study of situated action in general, and of improvisation in particular, has focused so far only on the later stages of this process: on the encounter between intentions and situations. But it has systematically failed to bring the (moody) situation of the actor into the reckoning.

Looking at improvisation as a special disposition or attunement with the situation, a special way of being amidst the world—and of being thrown into it—opens up 'improvisation as mood' as a different point of access to the phenomenon. However, this complicates our enquiry. The cognitive approaches of symbolic representation and of plans or situated actions show how actions can be studied as the carrying out of projects, plans, or intentions, or as emergent responses to circumstances. Moods, on the other hand, often cannot be forcibly brought about and are not necessarily linked to a plan or an action; they are the ground or the medium for them, but not the other way round. As we slip into moods unaware, it is difficult to make an attunement into an object of analysis.

The best approach might be to compare moods in order to single out the characteristics of one in particular—improvisation—by contrasting it with others germane to it. But deciding which of these should be chosen for this comparative exercise is not easy because psychologists suggest there are hundreds of moods. As already mentioned many times, improvisation strikes us because of its sudden, extemporaneous, and full impact. We should therefore contrast it with affections characterized by their negligible or null effect, and by their being stuck close to clock time. Good candidates for such comparisons are panic and boredom. Both have problematic relationships with any form of effective action and with the passing of clock time, though in different ways and for different reasons.

When the fundamental attunement of panic sets in, the ordinary way of understanding and acting in the world stops. The world overwhelms us. It ceases appearing as a set of tools ready for use and we lack the time to implement our projects. Resources are not at hand; in particular, time is not available. We quickly come to inaction or engage frantically in whatever activity comes to hand, after considering possible options and jumping to the conclusion that none will be successful. Angst for the lack of time in pursuing further exploration of alternatives then blocks decision-making. This is the structure of panic as mood. Care is aimless. The world is unusable. Intentionality has nowhere to go except to consider the supreme alternative of death. Time is lacking.

In contrast to panic, which implies that time and other things matter too much (the world overwhelms us), in boredom nothing really matters: the world is indifferent and time never seems to pass. If in panic we fall victim to the world and time, in boredom we try to kill time while being immersed in a fog of indifference. Starting from the superficial ways in which we deal with boredom in everyday life, we can encounter the inextricable relation of this particular mood with time. At first, it takes the form of whatever pastime we engage in to overcome boredom; we pass the time in order to master it because time becomes long, for example when waiting for a train that is late. When we are bored, our attunement, our way of being-amidst, is characterized by being in a time that is hanging and amidst a world that does not offer many resources to fight the slow pace of time.

Moving back to our phenomenological intuition of improvisation as mood, consider a more profound form of boredom: when one feels empty and wants nothing from the world. The whole situation becomes indifferent. The peculiarity of this form of profound boredom is that there is no pastime in sight. Time does not drag us, neither we make it stand still. Rather, one feels removed from the flow of time, indifferent to time—not time deprived, but timeless. This profound indifference is linked to the complete time horizon that encapsulates the refusal of the world as a whole. What is striking of this form of boredom is that all dimensions of time (clock time, past, future, now, etc.) seem not to matter. There is no determinate time-point when this boredom arises. We do not worry at all about the clock, as we do not worry about beings and the world; we are not annoyed by any 'stretched now', that is the time span during which this boredom holds us. Actually, the profound boredom can take hold of us in an instant like 'a flash of lightning'.

Our analysis here suggests that the opposite of improvisation is not planned action, but profound boredom. In such a form of boredom, unarticulated time and the refusal of beings lead to inaction, while in improvisation there is the 'moment of vision': the look of resolute decision in which the full situation of action opens itself, and keeps itself open to our initiative of re-registering, recombination, and intervention. The cognitive perspective focuses on how smart improvisers can quickly re-register the world and recombine resources. This is possible only because, suddenly, the world and its resources and people matter differently, so that they can be singled out and recombined differently. Here, quickness is far from implying rigidity and the opposite to the already-known. Improvisation is that particular mood where reconfiguration takes place of the ground in which we encounter the world, and in which we devise and carry out those projects and actions, which form the objects of the cognitive perspective.

Conclusions

When encountering information systems, before action and before design, there is our being thrown in a world of possibilities (Introna 1997). The mood situates one in respect to these possibilities, discloses some and conceals others. Those which are revealed will matter and will be objects of the simultaneous planning and actions of the improviser. Improvisation is that moment of vision and self-revelation where all the possibilities linked to the being-in-the-situation emerge out of the fog of boredom. Improvisation is the antidote to panic and boredom because it is 'extemporaneous', that is it ruptures the way time entrances us in both situations, either by being completely missing or totally undifferentiated. Thus, once again, the phenomenological understanding reaches a different conclusion to the prevailing cognitive and managerial views, according to which any entity capable of planning or executing can improvise. In the new perspective evoked here, we can rephrase that cognitive science statement as follows: any entity existing, being able to reflect on its existence and endowed with moods, feelings, and emotions, is able to improvise.

The original phenomenological understanding of improvisation as the moment of vision suggests that key decisions are much rarer than perhaps appreciated so far. Many forms of improvisation celebrated in the American Academy of Management sessions dedicated to it appear as pastimes against boredom. Few may be actually antidotes to panic in emergencies. In particular, if we accept Heidegger's (1995) statement, formulated in the late 1920s, according to which profound boredom defines the modern condition, a sombre shadow is cast on the last sixty years of management and organization science. Preoccupation with programming, planning, and rational choice possibly conceals what goes on most of the time, in most organizations, for most people: boredom. Emphasis on rational decision making entrances and distracts us from appreciating that important decisions occur very seldom, while what prevails are all sort of pastimes (programmes, methods, models) that punctuate colourfully and irreversibly the life of workers and managers alike.

Hopefully, phenomenology can still offer us today a series of metaphors, such as bricolage, hospitality and moods, which may lead to a new aesthetic appreciation of information systems, very different both from the positivist management concepts and those anthropological alternatives still suffering from a disguised cognitivism.

References

Abbate, J. (1999). *Inventing the Internet*. Cambridge, MA: MIT Press.

Brown, J. S., and Duguid, P. (1991). 'Organizational Learning and Communities-of-practice: Towards a Unified View of Working, Learning and Innovation'. *Organization Science*, 2/1: 40–57.

Carr, N. (2003). 'IT Doesn't Matter'. *Harvard Business Review*, 81/5: 41–9.

Ciborra, C. (1992). 'From Thinking to Tinkering: The Grassroots of Strategic Information Systems'. *Information Society*, 8: 297–309.

Ciborra, C. (1996) (ed.). *Groupware and Teamwork*. Chichester, UK: John Wiley.
—— (2000) (ed.). *From Control to Drift: The Dynamics of Corporate Information Infrastructures*. Oxford: Oxford University Press.
—— (2002). *The Labyrinths of Information: Challenging the Wisdom of Systems*. Oxford: Oxford University Press.
—— and Lanzara, G. F. (1994). 'Formative Contexts and Information Technology: Understanding the Dynamics of Innovation in Organizations'. *Accounting, Management and Information Technologies*, 4/2: 61–86.
De Certeau, M. (1998). *The Practice of Everyday Life*. Berkley, CA: University of California Press.
Derrida, J. (1997). *De L'Hospitalité*. Paris: Calmann-Levy.
Hayek, F. A. (1945). 'The Use of Knowledge in Society'. *American Economic Review*, September: 519–30.
Heidegger, M. (1962). *Being and Time*. New York: Harper and Row.
—— (1992). *Gelassenheit*. Stuttgart: Neske.
—— (1995). *The Basic Concepts of Metaphysics: World, Finitude, Solitude*. Bloomington, IN: Indiana University Press.
Husserl, E. (1970). *The Crisis of the European Sciences and Transcendental Phenomenology*. Evanston, IL: Northwestern University Press.
Introna, L. D. (1997). *Management, Information and Power: A Narrative of the Involved Manager*. Basingstoke, UK: Macmillan.
Kant, I. (1972). *Perpetual Peace: A Philosophical Essay*. London: Garland.
Latour, B. (1999). *Pandora's Hope: Essays on the Reality of Science Studies*. Cambridge, MA: Harvard University Press.
Nelson, R. R., and Winter, S. G. (1982). *An Evolutionary Theory of Economic Change*. Cambridge, MA: Belknap.
Nonaka, I., and Takeuchi, H. (1995). *The Knowledge-creating Company*. Oxford: Oxford University Press.
Norman, D. A. (1988). *The Design of Everyday Things*. New York: Basic Books.
Polany, M. (1962). *Personal Knowledge: Towards a Post-critical Philosophy*. New York: Harper and Row.
Schön, D. A. (1983). *The Reflective Practitioner*. New York: Basic Books.
Scribner, S. (1984). 'Studying Working Intelligence', in B. Rogoff and J. Lave (eds.), *Everyday Cognition*. Cambridge, MA: Harvard University Press, 13–31.
Simon, H. (1961). *Administrative Behaviour*. New York: Macmillan.
Suchman, L. (1987). *Plans and Situated Actions: The Problem of Human–Machine Communication*. New York: Cambridge University Press.
Vera, A. H., and Simon, H. A. (1993). 'Situated Action: A Symbolic Representation'. *Cognitive Science*, 17: 7–48.
von Hofmannstahl, (1995), *The Lord Chandos Letter*. London: Syrens.
Weick, K. E. (1993). 'The Collapse of Sensemaking in Organizations: The Mann Gulch Disaster'. *Administrative Science Quarterly*, 38: 628–52.
—— (1998). 'Improvisation as a Mindset for Organizational Analysis'. *Organization Science*, 9/5: 543–55.
Winograd, T., and Flores, F. (1987). *Understanding Computers and Cognition: A New Foundation for Design*. Norwood, NJ: Ablex.

Notes

1. A more comprehensive discussion of these and other metaphors can be found in Ciborra (2002).
2. This point, originally made in Ciborra (1992), has been recently debated anew in the management journals (see Carr 2003).
3. The Mann Gulch fire disaster, in which several smokejumpers died, took place in Montana in August 1949. Weick (1993) provides a rich interpretation of why a highly select group can fall apart in an emergency. In particular, by analysing the practices of the few survivors of the accident, the author is able to identify potential sources of resilience that make groups less vulnerable to disruption: improvisation is one of these survival factors.

2

Solution is the problem: a story of transitions and opportunities

IAN O. ANGELL AND FERNANDO M. ILHARCO

Introduction

It is a well-known maxim that solutions create more problems than they solve. On balance, after more than three hundred years of positivism, we are left with a world marked not by solutions but by problems. In this chapter, we confront this state of affairs head on, arguing that the real problem of our times is not problems as such, but solutions. We claim that many tenets of current mainstream thinking about organizations, and about IT/IS, cannot stand up to rigorous scrutiny.

In these times of technological innovation and globalization, the world is revealed within dichotomies: subject/object, man/world, problem/solution, thinking/action, planning/executing, strategizing/structuring, manager/workers, data/information, and so forth. This inheritance of the industrial age, now gone, relies on the Cartesian dualist approach to being, which has invaded a great many aspects of human activities. Even the Earth, as such, is nowadays an object in space, that is, it is an object of the scientific method, while man is of course the Cartesian thinking subject. 'It is not the victory of science that distinguishes our nineteenth [, twentieth and twenty-first] centur[ies], but the victory of scientific method over science' (Nietzsche 1968: 261, §466).

In this chapter, we challenge you to engage with a novel approach to our world, and our 'self-evident' need to adapt and thrive. The chapter aims to point out a world we already know, where things—people, notions, feelings, involvements, events—happen, change, surprise us, disappear, and come again into being. We intend to tell you, the reader, a story that makes sense to you as you already are in this world: a narrative not typified by problems and solutions, but instead relying on an intuitive, open-minded, and enlightened grasping of transitions and chance, learning and change, action and opportunities. We argue that the current and prevalent techno-functionalist thinking is a source of ambiguity and non-sense

The research of Angell described here was undertaken as part of the project at the London School of Economics called *Enabling the Integration of Diverse Socio-cultural and Technical Systems within a Turbulent Social Ecosystem*, which was funded by the UK Engineering and Physical Sciences Research Council (EPSRC) under the Systems Integration Initiative (project GR/R37753).

The research of Ilharco presented in this chapter, developed during his Ph.D. work at the London School of Economics, was supported by fellowships from the Calouste Gulbenkian Foundation and from Luso-American Foundation for Development, both from Lisbon, Portugal.

in many organizations. Given this understanding, we neither infer nor imply that the state of affairs pointed out above is in any way a problem. For this techno-functionalist position is part of an environment that, in itself and on account of the turbulence and complexity it generates, should be taken as an opportunity.

Theoretically, we support our position with an ontological approach to human action based on Nietzsche's post-nihilist thought and on Heidegger's (1962) phenomenology of humanness in *Being and Time*. From this standpoint, we assume that action is itself the ground for all our actions, thoughts, intentions, and interpretations, as well as the usefulness thereof. Within this fundamental context, we aim at presenting a narrative that makes sense to you, the reader, as you already are in-the-world. We claim that the way to thrive in the current ambiguous, complex, and chaotic world is not to search for solutions, but to recognize that what is there, is—and will inevitably change, leading to new opportunities. The more this chapter makes sense to you, we argue, the better it will help you thrive and profit in contemporary business, political, and social environments.

We start with a critique of some of the key current IS and management mainstream fundamental positions. Following that, we develop our grounding approach to the management of IT and organizations. We conclude the chapter by pointing out ways in which this narrative might enable you, the reader, to cope better in our complex, changing, and surprising world.

The problem/solution world

Over the years, many attempts have been developed that see common factors within the variability that is the human condition, in order to categorize human behaviour into a small number of substantially different labels. By pigeonholing the complexity of a given individual's set of distinctions under a single heading, that individual can be contrasted with his fellows, and his actions 'explained'. The business world is awash with convincing categorizations and schemas. They vary from the practical and highly influential 'portfolio grid', developed in the 1970s by the Boston Consulting Group, to the more esoteric-like taxonomy of Burrell and Morgan.

Burrell and Morgan (1979) proposed an arrangement of all social theory into four paradigms, and this has had some influence in the IS field. They claim that there are just four ways in which an individual will look at, and act in, human society. These ways form a two-by-two matrix resulting from the intersection of an ontological with an epistemological dimension. The first dimension contrasts the 'sociology of regulation' (advocating a view of society in terms of integration and cohesion) with the 'sociology of radical change' (claiming that society is in a continual state of crisis). The second dimension identifies an interpretation or knowledge as being either 'objective' or 'subjective' in nature.

Using this matrix, Burrell and Morgan claim that each individual's view of society adheres to just one of four quite distinct and mutually exclusive paradigms, because each paradigm contradicts the views of the other three. Thus, each sociological theory must be based in just one paradigm, since each is examining social phenomena in different ways. Consequently, arguments between proponents of different paradigms cannot be easily resolved, because they are arguing not about points of fact, but from different perspectives about different things—even the words used have subtly different meanings for each group. Professor Geoff Walsham sums up this dilemma very succinctly: 'theory is both a way of seeing and a way of not seeing'.

The reason we are labouring the point of the Burrell and Morgan categorization here is that it delivers (a somewhat functionalist) definition of the functionalist approach we reject: they label the quadrant that concerns objective knowledge in an integrated and stable society as the Functionalist Paradigm. Functionalism is unquestionably the dominant theory within today's business community: the way it both sees itself and what it doesn't see. From the functionalist standpoint, organizations and people, information, technology, and systems are taken as the same for all of us, and as objects for observation, analysis, reduction, and categorization. We, however, assert that this schema, appropriate for physics (and appropriated from physics), is destabilizing and highly dangerous in the management context.

Functionalists say that social reality is governed by immutable laws of causality that can be discovered through the scientific method, based on unbiased observation, supported by empirical evidence, and then used to advantage to explain and build cohesive and stable structures in society: the fundamental tenets of social engineering. This is Sherlock Holmes' position: '... when you have eliminated the impossible, whatever remains, *however improbable*, must be the truth'. This too is Mr. Spock's claim that rational observers come to the same conclusions about a situation because of an objective reality and logic inherent in that situation. Such a normative approach tends to dehumanize the situation, just as the *Star Trek* hero and the 'junkie' of Baker Street do, with their insistence that there are such things as right answers, which can be prescribed through logic.

Functionalists respond to our objections to categorization by asking: but surely there are categorical absolutes in Mathematical Logic? Are its laws wrong? Our response to these questions is, as always, subjectively yes and no. 'Only very naive people are capable of believing that the nature of man could be transformed into a purely logical one' (Nietzsche 1996: 28, §31). 'The world seems logical to us because we have made it logical' (Nietzsche 1968: 283, §521). 'Logic was intended as facilitation; as a means of expression and not as truth; only later did it acquire the effect of truth' (ibid.: 291, §538). Of course, the logical operators 'equality', AND, OR, NOT are all valid for the manipulation of categorical abstractions. But can we be so sure that the operands in such logical operations fit into the chosen category, once and for all time? Can we be sure that human behaviour fits into boxes with neat and discrete labels? Can we ignore the 'debris of detail' that is the difference between a continuously varying similarity of the objects we are considering, and the idealized sameness that identifies a category? For that is where the trouble with categories lies: we see opposites instead of transitions. We see 'false opposites' (Nietzsche 1968: 371, §699), unaware of the 'excluded middle' (Watzlawick 1993) of human value systems, unaware that there is a continuum, and that we humans attempt to split that continuum between uniquely different and discrete objects. Yet these categorical schemas, our value systems, do have a purpose:

. . . and our fundamental tendency is to assert that the falsest judgements [. . .] are the most indispensable to us, that without granting as true the fictions of logic, without measuring reality against the purely invented world of the unconditional and self-identical, without a continual falsification of the world by means of numbers, mankind could not live—that to renounce false judgements would be to renounce life, would be to deny life' (Nietzsche 1990: 35–6, §4).

To us, human schemas are devised for action not for truth, aiming at profiting from reality. So too are numbers. However, category, as a particular kind of schema, and one with such appeal in the contemporary world, has two main problems. Everything that is given the label must be treated 'as if' it is identical to the idealized entity that defines the category. Anything that is not in the category, is un-similar in all respects to the category, and must be treated as totally different. As long as this is beneficial, or at the very least not detrimental to our actions, then that is fine. But this is not always the case. 'All ideals are dangerous: because they debase and brand the actual; all are poisons, but indispensable as temporary cures' (Nietzsche 1968: 130, §223). We must recognize that the logic of mathematics is just idiosyncrasy, and in its use for analysing societal problems it has become mere self-indulgent over-sophistication. Mathematics is not universally appropriate. Was it ever?

What about statistics? Mark Twain was being deluged by a torrent of statistical data, culminating with the rebuff 'the figures don't lie.' His devastating retort: 'it's not the figures lying that worry me, it's the liars figuring.' Twain ably demonstrated that, when it comes to winning the argument, rhetoric proves far more powerful than numerical logic. But, we go further. To us the figures are 'instrumental fictions' (Nietzsche 1979: xlvi), sometimes useful, but fictions nonetheless. These fictions may be appropriate and are thus appropriated by us if and when they make sense to us, according to our experience and to our expectations, and thereby help us thrive in a world we already know by being in it.

Moreover, numbers are not final in that they always have to make sense to us, according to our previous experiencing of the world. If tortured enough, numbers will end up confessing to anything. We don't need numbers to prove that we have won a particular argument. The champions of numbers get very annoyed when they can prove themselves 'right' by measuring 'reality' with their numbers, and yet somehow they still lose the day. They fail to understand that 'what convinces us is not necessarily true—it is merely convincing' (Nietzsche 1968: 15, §17). After all, and it is worth repeating, the use of numbers for measurement (merely comparison against an arithmeticized ideal) is itself dependent on the intellectual trick of 'sameness' that we play on ourselves. Measurement may make sense for the development of technology and of physical artefacts, but to use it as a tool for social engineering is absurd:

The rigor of mathematical physical science is exactitude. Here all events, if they are to enter at all into representation as events of nature, must be defined beforehand as spatiotemporal magnitudes of motion. Such defining is accomplished through measuring, with the help of number and calculation. But mathematical research into nature is not exact because it calculates with precision; rather it must calculate in this way because its adherence to its object-sphere has the character of exactitude. The humanistic sciences, in contrast, indeed all the sciences concerned with life, must necessarily be inexact just in order to remain rigorous. A living being can indeed also be grasped as a spatiotemporal magnitude of motion, but then it is no longer apprehended as living. The inexactitude of the historical humanistic sciences is not a deficiency, but is only the fulfilment of a demand essential to this type of research (Heidegger 1977: 119–20)

Fragmenting and then detailing and measuring the parts of the whole that is each one of us do not describe, let alone explain, that whole. It is naïve to assume that the complexity of our lives can be captured as a mere collection of numerical and categorical data—so-called 'absolute facts'. The Human Genome project has categorized humanity as a long list, as if the number of genes matters. What splendid irony to discover that humble rice has far more genes than humans, and the mouse has almost as many (IHGSC 2001; Venter and Associates 2001). The 'gene' is just a category, a human schema, a model, which as such might be useful, but which evidently always falls short of that which we are pointing to. The important issue is how the genes evolve, relate to each other, and establish themselves in a way that came to generate what we as humans are today—something far more complex and mysterious. Nothing up to now has ever shown itself to be *knowable*.

However, the functionalist mode of thinking is everywhere in the folly of forecasting techniques that merely assign categories to the future—in principle, little different from using the twelve signs of the zodiac. Such forecasts are a belief that categories are meaningful in relation to the future, and controlling that future becomes a matter of labelling it (with numbers in the case of our technical/scientific society), rather than by continually re-evaluating the uncertain situation, and by depending on ourselves and on the trusted people around us. The future is not some smooth trajectory of the past; the discontinuities implicit in change are forever pulling the future away from past trends. Discontinuities mean that the ride of life just isn't that smooth. The current untidy and messy world was not planned by anyone—no country singly or in coalition, no world commission of geniuses, in fact no one at all has at anytime decided the conditions in which we now find ourselves. Even the past is not that objective, clear, and obvious time which has happened; our understanding of the past—what it 'is', or rather what it 'is becoming' (not what it was)—changes with our understanding of the future, with our changing projections, expectations, and ambitions (Heidegger 1962; Merleau-Ponty 1962). With Nietzsche (1968: 377, §708) we see the past perpetually 'becoming' something else, and not an uninterpreted snapshot frozen in time.

Yet few of the so-called experts in the business world doubt the validity of statistical methods for predicting the behaviour of their complex systems. Everywhere we see them misusing the term 'system', when all they actually mean is 'installation'. For a system is what that installation becomes, what it will become, and not what it was intended to be. When you use the tools of technology to solve a particular problem, you may or may not succeed, but what is certain is that completely unexpected phenomena will happen. For example, as little as twenty years ago experts were claiming that the mean time between failures of nuclear power stations was in the order of tens if not hundreds of thousands of years. The meltdown at Chernobyl in April 1986, which was just the worst in a series of over 150 serious nuclear mishaps occurring over a ten year period, has clearly shown that position to be wrong. This is Gall's First Law of Systemantics in action: 'All Systems have Antics' (Gall 1988). Sometimes the antics lead to a system's inversion, where the exact opposite occurs to what was intended. Worried that adverse weather conditions at Cape Canaveral would hold up the Apollo project, NASA constructed a giant awning over the launching pad. It was so large that a micro-climate formed beneath it, so that even though the sun was shining on Florida, it was raining on the rocket.

(Optimistic) rationality

Numerical methods of science spouted by a 'management culture' still dominate management. They seem so attractive to all business sectors. Yet these methods are not going to help managers walk the tightrope stretched above the unknowable abyss that is the future. These methods and methodologies are mere practices, rituals that evolved to be optimal in the modern Machine Age; but will they help us cope in the post-modern Information Age? No! It is now quite clear that our society's over-use of measurement is misplaced ritual, as is computerization, as is 'auditing' (mere output measurement), as are command and control systems. All are 'rain-dances with pseudo science' (Angell and Straub 1999), pretences that we understand our world and are in control of it. That, even after more than three hundred years of positivism, we are still looking for 'the' solution, should tell us something.

Many professionals themselves have sensed this inadequacy in much of ritual management by method and measurement. Yet still, from all sides, there are loud calls for 'rational solutions' to the problems facing us, as if there exists a single unifying rationality. Everywhere we observe the belief in an 'optimistic rationality': being in control of a better world through rational thinking; expressed as science and technology: tidy thoughts implemented by tidy minds on that icon of tidiness, a computer. This current course of management action, which belongs to the wider context that Heidegger (1977) points out as the unfolding of 'enframing', the essence of modern technology, insists on an orthodoxy grounded on engineering principles and on the Tayloristic factory metaphor, which treats the business environment as if it is a machine or a production line. The standard argument is that management must confront the 'post-modern hysteria' brought on by a loss of faith both in the instrumental efficacy of science and technology, and in the high priesthood of experts who proselytize that faith. It must regain that faith, and then everything will be back on track.

Champions of this discourse want our world to be tidy, and try to banish all the 'evil spirits' by forcing tidiness on it with the magic wand of systematic, yet arbitrary use of measurement, systems analysis, opinion polls, market research, socio-economic classifications, performance measures, efficiency audits, and cost-benefit analyses. They call for 'best practice', when best it certainly isn't. Such best practice would be a mistake, for it is becoming increasingly apparent that the ritual usage of conventional methods simply doesn't work in the broader social arena where professionals must operate.

Searching for the right numerical label to represent the future is no different to the mysticism of numerology or astrology. The world does not have to conform to method (Feyerabend 1993). Rather, the method has to be useful in this world, by conforming to it. There is nothing to tell us that important things can always be measured. Experience shows that much that can be (and so is) measured is irrelevant. Maybe the most important things cannot be measured. The only reason that what is being measured is measured at all is simply because it can be measured, and only ritual makes that important.

Yet, the basis of decision-taking has undoubtedly shifted to a numerical justification. And the lust for numerical solutions and other theoretical abstractions is spreading. As Edmund Burke (1999*a*) wrote: 'Not men but measures: a sort of charm by which many people get loose from every honourable engagement'. Numbers, particularly when manipulated by computers, seem to purify otherwise unpleasant acts. Those operating the technology are encouraged to carry out impersonal actions that affect real people in a detached and insensitive manner. People become numbers upon which operations that can adversely affect them are mechanistically performed. The urge to classify, count, measure, and then computerize places key human aspects, such as staff morale and customer goodwill, on the sidelines.

A British Building Society (Savings and Loans) provides a good example of a failure caused by optimizing short-term performance measurement. The Society was counting customer throughput, in an attempt to measure customer satisfaction and staff efficiency. But, when asked a simple specific question about mortgages, clerks in three different branches provided three different answers (all of them wrong!). Each clerk appeared to give the first answer that came to mind, in an effort to improve throughput and hence his or her personal automated performance measure.

Computerized tidiness and chaos

'Madness is something rare in individuals—but in groups, parties, peoples, ages it is the rule' (Nietzsche 1990: 103, §156). The madness of our particular Age is the arrogance that our world can be managed by using the excessively tidy methods of science and technology. The current collective madness believes that, particularly through computerization, the world will become 'the way it ought to be'—an 'obsessive compulsive neurosis' of constantly trying to tidy up the world. This madness increasingly appears as a corporate disposition, particularly in the way management try to 'Manage Change' or 'Engineer Knowledge' with tidy quantitative methods.

The literature is full of cure-alls that sell such tidiness to the gullible manager in the form of ritual application of computerized methods. Consultants charge onto the scene, promising a world neatly described in networks of boxes, triangles, hexagons, circles, and arrows. It is well to remember that, as Chekhov wrote in *The Cherry Orchard*: 'When a lot of different remedies are proposed for a disease, that means the disease can't be cured.'

The unholy twins of management-by-'bubbleware' and management-by-numbers projects the world as a deterministic black box—linking preordained input parameters to a predictable output. Neurotic companies attract tidy minds, which reinforce organizational neuroses. However, intrinsic ambiguity cannot be resolved into some tidy pattern, and jumping onto a bandwagon of methodologies is merely impulsive stress-relief. The result is a business world full of insecure managers who feel that the remedies are successful for everyone else, but not for them.

Putting numbers on a computer just amplifies the error. Computers can deal with objective well-structured problems with amazing speed, with detail, but they cannot cope with subjective subtlety, ambiguity, and complexity—computers are not in-the-world, but we are! There is a basic management problem with integrating computerized information systems with human activity systems, because they are fundamentally different. Computers do not work in physical space or within a biological existence, but in a mathematical dimension, a subtle fact too easily forgotten until it is too late.

The checks and balances fundamental to error tolerance and correction in any human activity system can become invalid in the rule-based world of computerized information systems. We simply cannot act fast enough to keep up with the machines, and so instead we often surrender responsibility. But because it is only in hindsight that we can analyse the subtle differences between the dimensions of man and computer, there is an enormous potential for misunderstanding. The sheer speed of positive feedback from computers means they are just not in tune with the human mode of error-correction. By our failure to deal with this different time scale, errors are amplified and this feeds back into chaotic situations.

All facts are context sensitive. Contexts are personal. Nuances of detail, as well as deliberate, accidental, and arbitrary actions feed back and continuously modify and amplify elements, processes, and sub-systems within that organization. In most cases, this feedback has minimal effect on the systems (negative feedback); it simply reinforces organizational systems. But every so often positive feedback—the seed of chaos—explodes the stability. The devil is in the detail. An initially marginal event can, through positive feedback, be the cause of the long-term dramatic events of the Lorenz butterfly effect (Gleick 1987). The trouble is that there are innumerable marginal events out there, but very few precipitate the butterfly effect—and those can be identified only in hindsight. Even physicists no longer accept the universality of 'similar systems in similar environments will undergo similar changes', and physicists, after all, work in a reasonably well-behaved universe of discourse where the meaning, measurement, and choice of a particular 'similarity' can be justified.

This highlights the fallacy of creating a machine/model to simulate an organization. The implied ambiguity and incorrect emphasis, the arbitrary extra components that, even if perceived, are taken to be non-systemic and irrelevant, the uncertainty and emergence, mean that the model ends with just one of an infinity of possible universes. Unfortunately, too often management chooses to restrict its understanding to mechanisms and to functionality, and do not deal with the systemic nature of business. It ignores the magnifying effect of positive feedback in the ill-structured and turbulent business environment, and overlooks the 'butterfly effect', where the marginal and insignificant trends of today become the major business opportunities/risks of tomorrow.

The trouble is that by squeezing complex experience into tidy categories, innumerable marginal details shear off, and it takes only one to precipitate the butterfly effect. The prevalent form of linear thinking denies all but negative (and thus tidy) feedback. It totally rejects the idea that the act of imposing tidiness on an untidy world, or on a world that is ordered, but in a different way, can itself lead to chain reaction of chaotic positive feedback which generates hazards that refute any prior analysis. We may know the general form of a hazard, but we may not recognize a particular occurrence until we are wise after the event.

And yet, neurotic management still projects the world as a deterministic black box—linking preordained input parameters to a predictable output. No expert can appear on television without the ubiquitous microcomputer peeping over his shoulder. Each business program displays wall-to-wall, knowledge-giving computerized charts. We really do seem to believe that computers increase business potency. The history of human problem solving is littered with examples of this 'first step' fallacy—they think they are reaching for the moon, but all they've done is climb the nearest tree (Dreyfus and Dreyfus 1986). It is only after the nonsense stops that a technology can be used propitiously to its full potential.

But the nonsense goes on. In light of our dominant scientific mindset, it was inevitable that computerized information systems would be thrust upon organizations, all with the highest motives. But information technology is just the latest technological adventure whereby man feels he can subjugate nature by mere will. However, as Emerson wrote 'New technology, owes ecology, an apology'. A history of technological achievements, including the development of the computer itself, sustains a scientific optimism that is apparent in the huge telephone-number sums being thrown into the bottomless money-pit of computing. There is no need to give examples here, because stories of where computer implementation is a shambles are legion. Far too often, information technology is considered culturally neutral, when in reality it is value-ridden (e.g. Spengler 1926; Ellul 1964; Heidegger 1969, 1977; Habermas 1970, 1987; Borgmann 1984, 1999; Ihde 1990; Zimmerman 1990; McLuhan 1994; Winner 1995). IT hides a powerful intellectual imperialism. This predicament is nothing new: it is the culmination of the trend that is the unquestioned application of scientific method, which was recognized over two hundred years ago by Edmund Burke (1999*b*): 'But the Age of Chivalry is gone. That of economists, sophisters, and calculators has succeeded.' For, behind much application of information technology is the belief that human thought is mere calculation; that we are no more than biological analogue computers. This state-of-affairs was referred to by Heidegger (1969, 1977) as the hidden meaning of the essence of modern technology: the *danger* that we become what we essentially are not.

Computer simulations maintain a pretence that they represent reality within their limited models. Yet unquestioning acceptance of simulations of open-ended human experience is non-sense. The application of IT obscures the difference between correlation and causality, and confuses superficial process with substance. Models can only ever be a pale shadow of what actually happens, and can never emulate the subtle, and not so subtle, checks and balances and the feedback of unknown and unknowable interactions. They really cannot hope to emulate the infinity of parameters implicit in being there (Heidegger 1962). Computerization is a prisoner of societal consequences which cannot be controlled, no matter what the management regime. Theoretically valid ideas can cause failure, while blatantly incorrect facts can induce success.

The self-serving logic of methodically applying technology takes on a significance far greater than the original problem. The use of technology becomes *compulsive*. Because of a predominant functionalist 'information ideology', initial success with IT can induce *obsessive* managers to introduce yet more and more layers of technology. To justify the extra effort and expense, managers look around for partners in search of 'synergy'. The implications of each new layer of technology always feed back in the form of new questions about the appropriateness of solutions, and of new problems: *neurosis*. Any tidy demarcation will always collapse in a confusion of multiple overlapping systems. For example, the vast majority of software simply deals with other software, and not the management problems at hand. Complexity increases to a point where utility turns into reliance, reliance becomes dependence, and the Law of Diminishing Returns precipitates a galloping descent into a management nightmare.

It is ludicrous to think we can control uncertainty. According to Goodhart's Law, a well-known maxim of the distinguished Economics professor at the London School of Economics: 'Any observed statistical regularity will tend to collapse once pressure is placed on it for control purposes.' This law is just as apposite for management in general. What Charles Goodhart is saying is that you cannot mix up cause and effect. The dominant belief is that with 'proper control procedures' we can impose order. This is a complete misunderstanding of the human condition. Control doesn't create order, quite the contrary. Order is systemic and may have come about in the complexity of human actions (specifically the feedback of interactions with the environments), but not necessarily from human intent. Order must emerge first, and this order tolerates control. Only by the concession of that order does the consequent control impose structure and stability. Don't confuse order with structure and stability, don't confuse cause with effect.

Any observed regularity in society is an effect, but the moment it is measured and that measure used within an IT control system, a false assumption is being made that the regularity is the cause of the observed order. However, all order is transitory regularity; order allows controls to work, and then order fails; consequently the certainty of control and structure collapses. Uncertainty precedes the birth of a new order, followed by new controls. Managers must act as midwives at this birth, by standing out against the urge that devours company resources in trying to control what is not controllable. They must not try to create stability and order, but rather focus on temporary equilibrium, surfing the waves of transitions and change, and profiting from them.

Today's management theory is simply not coping. It is apparent that business problems can no longer be treated as mechanistic problem/solution models, predominantly as technical tasks. Managers must forget the promises made in the methods of the exact sciences, and learn to live with that uncertainty, and love it—enjoying the sheer wonder of it. It is not sensible to think that we can use technology to control the uncertainty implicit in the real world. There is a lesson to be learned in the actions of Long Term Capital Management (LTCM) that precipitated the chaos that almost brought down western capitalism in 1998. The 'Masters of the Universe' from Wall Street thought that their mathematical models, developed by the likes of Nobel Laureates Robert Merton and Myron Scholes, delivered a mathematical guarantee of a 'risk-less risk' that could beat the system. Accordingly, hedge-funds leveraged already huge amounts of money into astronomical sums, which were then placed as bets. With the vast sums involved, even very small percentage gains turn into a tidy profit. Indeed, they did win big in the 1980s and 1990s, which is why the banks were happy to 'lend' them ever increasing sums of money. Then LTCM, which had leveraged its $4.5 billion into a $1.25 trillion bet, suddenly lost 44% of its capital. This is serious money, and only swift action by the US Federal Reserve Bank avoided global financial Armageddon. Banks were forced to write off hundreds of millions of dollars. Even the award of a Nobel Prize is no guarantee that a particular economic schema will work.

Too many companies think they can force tidiness (risk-less risk) on their world with the computerized, yet arbitrary, use of measurement. However, by automating deterministic and mechanistic processes, management may introduce further levels of complexity that can obscure their lack of understanding, and serve to deny the uncertainty in those processes. The unquestioning acceptance of their models and simulations of open-ended human experience is unreasonable. Uncertainty is not a matter of incomplete information, because in-the-world, in uniqueness, it is self-evident that there is no such thing as *complete* information. But neurotic companies still want their world to be well behaved; to be tidy. They believe it is virtuous to collect numerical data—the so-called absolute facts. This only shows that many managers have no idea of what uncertainty is. It is a foreboding, where the surprise of imminent change is outside of our control. It certainly does not conform to a neat computational logic. Surely, deep down, we must all know that profound uncertainty contradicts the smug certainty of formal methods. Every mechanism and measure we use has limitations, and a hidden agenda of choice. Their interaction with components in a system reflects an act of faith in the existence of the element, the property, and the system. This existence has to be perceived, a boundary imposed, and no emergent property can be measured or manipulated before it has emerged from the unknown.

Coercing our untidy world into a straitjacket of categories is forcing square pegs into round holes, and this sheds a debris of detail that conspires with the context to form a critical mass, and a subsequent explosion. Something has gone wrong. Cartesian dualism, and its mindset, the problem/solution dichotomy, has invaded a great many aspects of human activities. Is this a problem? Paradoxically, our answer is no! This chapter is about opportunity—an 'always and already there' opportunity, on account of a changing world in which we are immersed. To us, the fact that Cartesian dualism dominates much of management and IS/IT discourses is an opportunity. Belief in the contrary would not have escaped the very same limited approach we are criticizing. The current status quo of shortcomings, non-sense, of partial-solutions, and of complete problems, should be grasped as an opportunity—an opportunity for those who willingly view things differently, and who thrive and benefit from the situation in which we all are immersed.

Thriving in transitions

Heidegger's (1962) phenomenological investigations into humanness and Nietzsche's post-nihilist thought, underpinned with some insights from complexity theory, lead us to reject this present state of affairs. It is self-evident that similar causes can generate different consequences, and that different causes can produce similar consequences. Human unpredictability, chance, uncertainty, bias, prejudice, opportunities, agendas, and—above all—humanity's necessarily crude and simple ignorance of what is going on, support our claim.

Let us move ahead by decisively grounding our approach in some down-to-earth self-evidence about who we all are, that is, about who we, ourselves, *already* are:

> I am not a 'living creature' nor even a 'man', nor again even 'a consciousness' endowed with all the characteristics, which zoology, social anatomy, or inductive psychology recognize in these various products of the natural or historical process—I am the absolute source, my existence does not stem from my antecedents, from my physical and social environment; instead it moves out towards them and sustains them, for I alone bring into being for myself (and therefore into being in the only sense that the word can have for me) the tradition, which I elect to carry on, or the horizon whose distance from me would be abolished—since that distance is not one of its properties—if I were not there to scan it with my gaze (Merleau-Ponty 1962: viii–ix).

'I' am the absolute source, which is there before reflection, even before any kind of awareness has begun. That which the world is, is not what we put into words. The world is that which those words already assume, and refer to precisely as the world. There can be no humans, no genes, no systems, no organizations, no tradition, no horizons or ambitions, no anything if 'I' am not there, in-the-world to make sense of it all (Heidegger 1962), to distinguish beings as such, to capture this or that as something. This 'as something' (ibid.) is itself the meaning that opens up, for each of us separately, a world as a world, our own being as human, an organization as an organization, etc. This necessary and grounding account is the primary schema of all the theories and methods in which each of us dwells and is entangled, and with which we each act 'as if' we ourselves are in a world of certainty.

Bearing this in mind, we state explicitly the two fundamental assumptions of this chapter, and which we claim as self-evident. A 'world that is' (and is becoming) is that which is most evident to us. Kant (1985) considered it a scandal that a proof of existence of the external world had not yet been produced. Contrariwise, Heidegger (1962) regarded it as a scandal that such a proof was searched for at all. For who we ourselves are cannot be stripped out of the world, and so we come to be revealed as 'being-in-the-world' (Heidegger 1962; Merleau-Ponty 1962). We further assume that the way in which the world always already is for us, is in our 'be-ing' and 'be-coming'. It is always and already in our actions. Therefore, action is the ground of groundings; it is that which is primary, in that it is always already happening.

Besides Nietzsche's thought and Heidegger's *Being and Time* (1962), this chapter is highly influenced by the analysis of the Frankfurt School of social theorists (although not by their Marxist ideology), particularly by Herbert Marcuse. He argued that there is a vicious circle of cognitive tyranny within technology that has crushed humanity down to a 'one dimensional man' (Marcuse 1964), who is subservient to the technological structures of his own making. Although his work was written down before the rise of the all-powerful computer, Marcuse's insights are very convincing, even more so in the Information Age. On account of these theoretical grounds, we will argue in our narrative that there is a need for breaking free from the limits of this one-dimensional man and, in particular in our case, of freeing the organizational account of IT/IS from schemas heavily embedded by the techno-functionalist paradigm. Our non-functionalist position draws heavily on Heidegger's (1962, 1977) account of being-in-the-world, of his phenomenology of technology, and on that strand of IS research that has been developing for some time around this foundation (e.g. Winograd and Flores 1986; Dreyfus and Dreyfus 1986; Introna 1997; Ciborra 1998).

Instrumental fictions

We recall that a human is a 'who', not a thing or an object (Heidegger 1962). What most distinguishes we humans from other beings is precisely the 'who' that we are. Each human lives his or her life in its being-ness, bringing forth a world that matters to him or to her. A person is a performer of acts, which are something non-physical—'a person exists only in the performance of intentional acts, and is therefore essentially *not* an object' (Heidegger 1962: 73).

Each of us is unique, with his own past, intentions, aims, objectives, and experiences, his own biological, physical, and mental structures. Each of us is a private and personal perspective of it all. As a 'who', 'I' select according to my 'instrumental fictions', my schemas from among whatever I might distinguish within the niche in which I dwell. 'Not to know but to schematize—to impose upon chaos as much regularity and form as our practical needs require' (Nietzsche 1968: 278, §515). In-the-world we are experts in acting. Intuitively, we repeat what worked—this is what we *know* best. 'We are not databases stocked with trillions of propositions that orient us in life. Oriented living comes first' (Polt 1999: 69).

In-the-world, some actions immediately show up as do-able, as making sense, and others as not. Whatever doesn't seem to work, either because it did not work before, or because it counters what has worked, shows up as non-feasible ways of acting (Dreyfus 1991: 185). On the other hand, anything that has worked has shaped our structures, moulded our disposition, affected our attunement—as such, it has opened up specific possibilities for us to act in the future. The structural congruence that leads the manager to repeat what has worked is the instinctive behaviour to maintain himself as what he is for himself: projecting and articulating possibilities into the future.

As an observer (or rather a self-observer if we are to be rigorous), I am necessarily partial, and hence I disregard much. In doing so I simplify, and I then convince myself that I have clarified the situation. Instrumental fictions (schemas, theories, models) enter, and everything irrelevant just disappears. 'Whether you see a thing or not depends on the theory which you use. It is theory which decides what can be observed', as Einstein commented to Heisenberg. 'Even our intellect is a consequence of the conditions of existence' (Nietzsche 1968: 273, §498), a culmination of feedback in this species' successful (to date) quest for survival. Human intelligence cannot be separated from what is the human total. These schemas are fundamental, impelled by the human will to categorize. 'The categories are "truths" only in the sense that they are conditions of life for us' (ibid.: 278, §515). 'Just as certain human organs recall the stage of evolution of the fish, so there must also be in our brain grooves and convolutions that correspond to that cast of mind: but these grooves and convolutions are no longer the riverbed along which the stream of our sensibility runs' (Nietzsche 1996: 36, §43).

My schemas filter out the potential overload of sense data by limiting the flow of information, thus enabling my thoughts to crystallize on a course of action. Note the plural is used here: schemas. No single schema can be a complete description of the terrain that is the niche I inhabit and where I am already acting. Each schema is simplified and partial, in the sense that the need for action delimits what data are included, and what is excluded. The course of action in which I am already engaged is what supports the usually and traditional distinction between data and information; information is the data that is useful for me in my already on-going course of action. The etymology of the word *in-form-ation* highlights this aspect. Its Latin origins point to the joining of the meanings of interior, inside, personal (*in*), of forming, shaping, modelling (*formare*), and of action, in action, for action (*ation*) (Crane 2002). Data are absorbed, interiorized, becoming so to speak *in-form-ation*, as it informs and is useful to my already going or devised course of action.

We, as individuals, develop numerous schemas corresponding to different perspectives, different resolutions, different metaphors, and different purposes, each limited from its very creation by the appropriateness of its application. They are all superimposed onto a totality from which we can each choose appropriately to deal with any particular situation. We are *thrown throwers* (Heidegger 1999)—we must choose. That is, as a *who* I find myself *thrown* into the world, always and already with a past I cannot leave, *throwing* projections, ambitions, and my will into my own future. I have myself to be, I have to choose what to be, what to do or not to do at each and every particular moment of my life. Furthermore, in action I choose already, transparently, intuitively, and instinctively, most of the time. 'One responds on the basis of a vast past experience of what has happened in previous situations, or more exactly, one's comportment manifests dispositions that have been shaped by a vast amount of previous dealings, so that in most cases when we exercise these dispositions everything works the way it should' (Dreyfus 1991: 68).

Each instrumental fiction is a partial guess that focuses the senses. Each schema is both biased by previous experience of the unknowable 'reality' around me, and incomplete. All my uniquely individual interpretations of sense data start with provisional approximations. The schemas enter a dynamic feedback loop where I, the individual, collect and synthesize perceived regularities in the sense data. These regularities then progressively and recursively adjust the schemas. This ongoing and irreducible feedback loop is not only part of my individual experience of the world as it is, but also the culmination of the natural selection of both the species and all the sub-groups to which I as an individual belong. Mostly, this sampling of regularities delivers useful data that enables me to make choices—although sometimes I can be tricked, as happens with optical illusions. This should be a reminder that feedback does not deliver an explanation of our world, merely a description. 'We call it "explanation", but it is "description" which distinguishes us from earlier stages of knowledge and science. We describe better—we explain just as little as any who came before us' (Nietzsche 1974: 174, §112).

We describe? In creating their individual schemas, most people vaguely follow the agglomerated structures laid down in the past, either directly by interpreting writings and other 'knowledge' artefacts acquired first-hand, or second-hand via family, friends, fellows, and teachers. Mostly the 'education' we get from these people is in essence the development of socially acceptable schemas. We have no choice in the way we create these stable beliefs. We are predisposed to a form of atomism in which we create cognitively convincing categories within our schemas. To do this we separate things from other things—entities are separated by a void, designated, and hence differentiated. This is the fiction underpinning the edifice of classification, which itself drives the thinking of man, the Aristotelian and Cartesian 'rational animal', who 'places his behaviour under the control of abstractions' (Nietzsche 1979: 84).

The same schema being apparently used by different people further complicates this situation. That is, they may think they are using the same schema, but at the very least there will be slightly different interpretations of that schema. Even absolute units of measurement have a unique relevance to the individual. Therefore, there will be mistaken applications of the schema. Some mistakes will be very subtle, some blindingly obvious, some minor, others catastrophic. In September 1999, much to the embarrassment of NASA, the Mars Climate Orbiter burnt up in the atmosphere of the Red Planet. It turned out that the engineers who worked on the project used

Imperial Units of measurement (feet and inches), whereas the scientists whose job it was to get the spacecraft into orbit around Mars were using metric units. When the data needed to control the flight-path were transferred between the two teams, nobody thought to convert from inches to centimetres. The multimillion dollar craft went up (or was that down) in a cloud of smoke.

However, most differences are not that clear-cut. Any distinction, necessarily private to an individual, may be very similar to those of others. Be very clear, the distinction is unique to that individual. What is more, there is no single classification for each individual, rather a whole raft, gleaned from personal experience and communicated from others, each giving that individual a whole multitude of perspectives on the world, not all of them mutually consistent.

These categorical schemas are evidently not truth, but are merely an act of choice; an atomism, imposed by the observer. Once we get away from any notion of absolute validity, then we can see those choices as arising from the purposes and priorities of the observer that are grounded in action. Hence, it is impossible to separate the observer from the observed. As observers of ourselves we cannot get it all: no system can totally describe itself (Gödel 1992). We are each our own issue, and that is what is most essential for us (Heidegger 1962). It is impossible to separate the structures of a personal schema from the interpretations made from it. Both Searle (1983: 156–7) and Maturana and Varela (1992: 17–32) address this issue using the same metaphor: an eye cannot see itself. By the same token we cannot expect the brain to understand itself. 'How should a tool [our intellect] be able to criticize itself when it has only itself for the critique?' (Nietzsche 1968: 269, §486).

Similarity and sameness

Each of us experiences a different world, albeit with 'similarities'. I can consider any one of my schemas as similar to yours; and furthermore, within my personal schemas I can choose labels that group 'things' together, and which, according to my experiencing, are similar. However, what makes those things in themselves 'similar' to one another is totally another matter. Although we can know 'similarity', Kant (1985) is clear that we can never know 'sameness'. Things appear similar to me because my sampling of them with my personal schema inevitably throws up tiny differences—differences that I choose to ignore for pragmatic reasons. However, similarity is not sameness, although it is the human condition to treat similarity 'as if' it is sameness, and then assume that all comparisons between such sense-data choices are absolute facts.

Every schema, or 'closure' to use a popular technical term (e.g. Lawson 2001), rests on the will and capacity to treat that which is different but similar as being the same. However, as the playwright Pirandello wrote, '[a] fact is like a sack. It won't stand up until you put something in it'. A fact is merely an approved communal judgement, positioned within a context. What something is, is the context in which it appears (Heidegger 1962). The references that a thing holds in the referential whole (ibid.) in which we always already are, is the meaning of the thing in itself. The context provides the meaning—only within a context can something come forth as something. The aspect we want to stress here becomes clear if we acknowledge that all contexts are individual, which is in accordance with the theoretical foundations of this chapter, and we claim as self-evident. Thus, meaning—what something is perceived to be—is also individual.

A fact does not exist for me until I have grasped it, necessarily according to what I am, as having been (Heidegger 1962), and to what I have become and will become (Nietzsche 1968: 377, §708), according to my own becoming, among my personal schemas. Everything occurs in a context. I am trapped because the taken-for-granted categories that I choose are both the fundamental building blocks of my schemas, and my way of differentiating meaning and identifying such facts. As the context changes, so does category, and so does meaning. Such ever-changing meaning can only ever be uniquely individual—similar, but never the same as the meaning inferred by others. Even individual meaning is inconsistent over time because my always and already acting in the world is a permanent experiencing of new relations, phenomena, and possibilities that change both my future and my past.

Therefore, even those who share the same (whatever that means) social systems necessarily do not share exactly the 'same' schemas or the 'same' perspectives on those systems. Shared schemas become private schemas because of unique interpretations. Although there is a general cultural commonality of meaning within social groupings, each schema is truly individual, reflecting how that individual became what he or she is, not only because of all the groups to which he or she belongs, but also despite of them. Individually, I am what I have become, and what I will become, in the dynamic feedback of influences from a lifetime of personal experiences. However, each individual has to cope with multiple overlapping perspectives that need not even be mutually consistent, and that certainly aren't rigidly fixed. Naturally the events that 'I' sense in the world are in flux, but more importantly 'I', the sensor, am also in flux.

Category is not truth, but merely an act of choice (not necessarily a conscious choice) underpinning a cognitive fiction. That choice says it is acceptable, appropriate even, to treat similarity as though it is sameness, and then to assume that all comparisons between such data-choices are absolute facts; it is 'the presupposition that there are identical things, that the same thing is identical at different points of time' (Nietzsche 1996: 16, §11). Why do I ignore this fundamental error? Because the assumption of 'sameness' throws up temporary regularities, but nonetheless useful regularities, that I can use as the basis of maps that guide me through the turbulence of existence. As Nietzsche (1968: 13, §12) commented: 'We have measured the value of the world according to categories *that refer to a purely fictitious world.*' And:

> Man has for long ages believed in the concepts and names of things as *in aeternae veritates* he has appropriated to himself that pride by which he raised himself above the animal: he really thought that in language he possessed knowledge of the world. The sculptor of language was

not so modest as to believe that he was only giving things designations, he conceived rather that with words he was expressing supreme knowledge of things (Nietzsche 1996: 16, §11).

Yet the use of category, in names, words, and numbers, is our way of differentiating meaning. However, as time moves on, the environment changes, as does category, and so does meaning. The first lines of the *Tao Te Ching* (Lao Tzu 1963) sum up the human dilemma: 'the way that can be spoken of is not the constant way; the name that can be named is not the constant way'. Every time I name a thing I conspire with the error of sameness. 'Every word is a prejudice' (Nietzsche 1996: 322, §55). Even Lewis Carroll's Humpty Dumpty knew that a word 'means just what I choose it to mean—neither more nor less'. Every time I compare one thing with another, and in particular measure it by comparing it with an artificially designated standard unit (an ideal), I conspire with this error of sameness. Such categories are where confusion begins, in that a debris of detail on the edge of uncertainty—namely the difference between ideal/classification and the individual thing being categorized—will be strewn around and about.

No categorical solutions are applied in a vacuum, and this includes technological solutions. We have to experience the problem; there has to be situated opportunistic management. However, at the same time management must deal with any implications of the application of particular methods, avoiding any emergent hazards, all the while profiting from any unexpected opportunities that may arise. Management techniques, based on rigid deterministic thinking, can precipitate problems. The tension between rigid control mechanisms and the dynamic emergence of systemic phenomena is itself a source of uncertainty. Modern management must learn to live with this, and not deny the inevitable ambiguity and vagueness. What is commonly referred to as information technology actually only relates to the production and application of tools and artefacts. Managing, both the functionality of the technology and the phenomena of complex systems that spontaneously emerge around information technology, is the real challenge for managers today.

Conclusions: usefulness as criterion

Something has gone wrong, and that something is the cure, not the disease. We do not need another medicine to cure what cannot be cured. Being-in-the-world is not a disease. We accept the world as it is: the one you, yourself as you are, have already experienced in all its complexities, ambiguities, contradictions, and uncertainties. It is time to reject the unreal reality of much of modern business thinking. Change, uncertainty, ambiguity, and complexity must not be sensed as sources of problems to be subsequently solved, but as transitions, potential sources of opportunities, innovation, and profit. Of course there is no final solution, no definitive theory, no solve-all methodology. We argue that it is precisely because of this that the road is open for those managers who are willing to get involved, to commit themselves to change, to innovation, and to the acceptance of the transitory and imperfect nature of all our actions, solutions, and successes.

As we take action, some feedback will oscillate and amplify the loop of processes, eventually totally destabilizing the organization. Problems increase exponentially, as do the technological solutions. We are always only ever one step away from chaos. Anyone who runs a computer system will recognize this scenario. Some loops will imply risk, and these must be intercepted and curtailed. Even the feedback loops that enable commercial opportunities must be monitored, so that they don't oscillate into positive feedback. The manager's job is to cope with the feedback loop, not just tinker with the tools.

The issue is not about management; it is about 'thinking management', which emphasizes flexibility and avoids tying in the organization to the whims of a single supplier, or a single market, a single department, a single individual, a single culture, a single product, a single IT unit, a single perspective, or of a single schema. Emphasizing variety and flexibility provides further justification for strategies based on the development of human resources, and tactics based on informed decisions. Of all organizational resources, the workforce is arguably the most flexible, and that flexibility can be encouraged and exploited through the provision of education and training schemes.

Only education—that is, questioning and thinking, developing and changing—might enable one to see opportunities where others see threat, or to consider possible what everyone else assumes as impossible. Only education can change you as much, opening up new and newer roads of meaning and possibilities for the organization. A proactive policy of engaging professionals in regular educational schemas is one of the most obvious corollaries of our position.

Managers should see the *status quo*, the present one, or any other for that matter, as an opportunity. Where does that leave us? Individually and in groups, we must make sense of the world, by categorizing it with all-encompassing labels synthesized from the previously potent schemas that have come down to us as individuals. What makes sense? Any situation where input from the senses conforms to some preconceived notion, a schema individually built up by each of us in becoming what we have become. We can do nothing about the schemas in our hardware makeup, but the schemas in our software, that is, of our culture and personality, are there for the taking, for the making. For these schemas, the products of man's ambition, imagination and *Will to Power* (Nietzsche 1968), are the ways in which we deal with so-called 'reality' in becoming what we are.

This is not nihilism. Quite the contrary, it transcends nihilism. This is a post-nihilist attitude which acknowledges that it is self-evident that something 'universal' is there. However, in asserting the universal is not necessarily a knowable terrain, it pragmatically assumes that it just doesn't matter whether we 'correctly' and in truthfulness clarify it or not. Trained-for-action soldiers sing: 'We're here because we're here'. We are alive, and we just have to get on with it. In this world, *Dasein* (Heidegger 1962) has himself/herself to be, he or she is responsible for who he or she wants to be.

Too many theories, methodologies, and so-called explanations mean we are not getting the whole picture—that there is no right answer. Action was introduced above as the principal schema that grounds it all. For it is self-evident that we are in a world to make something out of ourselves, in order to make sense of IT and organizations, and to profit from our involvement in the world. The pragmatic 'solution' is to go on with the situation that has no solution. Paradoxically, a genuine and full realization of this fact, actually opens up the world, and enables us to take advantage of emergent, transitory, and surprising would-be solutions, partial would-be solutions, and no-solutions: what we call schemas.

In my action in-the-world, I am the centre of it all because it is evident that I am the absolute source of everything I perceive. 'Objective' features of things, their present-at-hand characteristics (Heidegger 1962), that is, their observed, de-contextualized, analysed, and fragmented presence, reveal themselves to me only within a larger, meaningful context, which cannot itself be explained in terms of its presentness-at-hand. 'The astronomer determines that a certain star is millions of kilometres away from the sun. That is correct, but it means something to the astronomer and to the rest of us only if we can relate it back to the lifeworld in which three kilometres are a gentle afternoon stroll, and thirty kilometres are a good day's hike' (Polt 1999: 59). A computer technician measures the surface of a screen to be 27 centimetres wide and 21 centimetres high. So what? What do these facts mean? What is their relevance? These measurements would be meaningless, and the technician would never have bothered finding them out, if they were not already meaningful in the world in which he exists, lives, and works as a technician—life comes first. Trauger, quoted in Polt (ibid.), highlights this point: 'Scientists do what fascinates them, and what fascinates them is not something you can discover with science. They are interested in investigating where planets come from, say, not because science tells them to do that, but because as human beings they find that interesting'.

As beings-in-the-world (Heidegger 1962), we have always and already understood things and events on the basis of their already-established places in the whole, that is our involvement with that world (ibid.). 'The Objective distances of Things present-at-hand do not coincide with the remoteness and closedness of what is ready-to-hand within-the-world' (ibid.: 141), that is, of what we non-thematically, transparently, and intuitively are already using and relying on. Present-at-hand distances are understood, interpreted, and assumed on the basis of my *having been* (ibid.). A present-at-hand distance *per se* has no meaning. Its meaning only appears on the basis of my individual life as a unique *having been*.

Consequently, there can be no right theory, for that would be a contradiction in terms. If there were such a thing as a correct theory then there would be no theory at all, just that which is, because all theories would add up to nothing at all. We build up theories and methods, not for the sake of those theories and methods, but because we might find them useful and appropriate for our already on-going involvement and action in the world. What we must uncover is not the correctness or truthfulness of theories. 'He who considers more deeply knows that, whatever his acts and judgements may be, he is always wrong' (Nietzsche 1996: 182, §518). What we must uncover is rather the appropriateness of theories and methods. Einsteinian physics is inconsistent with the prior development and discoveries of Newtonian mechanics. Yet, nowadays either approach is used as and when it is appropriate for particular but different situations. Furthermore, the different models of the physics of Einstein and Newton are themselves incompatible with quantum physics; but again, any one of those theories and their corresponding models will be used as and when they give an advantage when dealing with the researcher's involvement in the world.

Is being forever wrong a problem? No! For '[a]ll models are wrong, [but] some models are useful' (Burnes 2000: 1). 'The falseness of a judgement is to us not necessarily an objection to a judgement... The question is to what extent it is life-advancing, life-preserving, species-preserving, perhaps even species-breeding' (Nietzsche 1990: 35, §4). Perhaps falseness is too harsh a word here. Far better to use Nietzsche's term 'instrumental fiction'; we repeat: a possibly useful fiction, but a fiction nonetheless. Why useful? Because it overcomes rigid inertia and helps us make decisions and spurs us to action. We possess a convincing criterion of reality 'in order to misunderstand reality in a shrewd [and advantageous] manner' (Nietzsche 1968: 314, §584).

Did you find this critique on the state-of-affairs of current IT management, of its practice and theory, disconcerting? Did it offer you new distinctions and nuances in your world? Did it make sense to you and make you a little or even decidedly uncomfortable? It is our assertion that answering 'yes' to any of these questions is beneficial to you as an individual, for it demonstrates your genuine capacity to thrive in a world where the main problem is perceived as the main solution. The issue, as any enlightened manager would recognize, is just how to explore and profit from the current 'status quo' and from the natural propensity of things.

We aimed at telling you, the reader, a story that makes sense *to you*. In so doing we claim that this chapter would have much more influence than the traditional positivist recipes. According to the ontology on which we base our narrative, the most important impact of our text lies in its potential of being ready-to-hand (Heidegger 1962), that is, of making sense to you, of being absorbed, embodied (Merleau-Ponty 1962; Dreyfus 1991; Introna 1997), and relied upon transparently in your future actions. This chapter has its main empirical consequences on the kind of effects it might bring to action as, having been *arrested* within you—that is, 'understood' as defined in the *Oxford Paperback Dictionary and Thesaurus* (1997: 32)—its articulations and insights might unfold and adjust intuitively and instinctively within your ongoing action in-the-world. Our patterns of action tend to repeat what we genuinely and intuitively understand to have worked and what is useful (Dreyfus 1991). So, describing, thinking, and reflecting all influence action in the degree to which they are understood and absorbed by you. The chief empirical implication of our narrative is the way in which, in a particular situation, you might take advantage of that narrative to assist you in coping in-the-world, that is, the readiness-to-hand of our chapter.

Our narrative does not present a tidy picture. Why should it? In fact, to be consistent with its thesis, the chapter cannot promise to deliver a tidy and full articulation of the phenomena addressed. We reject such claims of the supporters of Cartesian epistemologies. Cartwright (1983: 53) noted that '[t]here is no reason to think that the principles that best organize will be true, nor that the principles that are true will organize much'; and Nietzsche (1968: 273, §497) commented that '[t]he most strongly believed a priori "truths" are for me—*provisional assumptions*; e.g., the law of causality, a very well acquired habit of belief, so much a part of us that not to believe in it would destroy the race. But are they for that reason truth? What a conclusion!'

The way to thrive in the current ambiguous, complex world, always and already in transition, is not to search for solutions, but to recognize how that which is there, is—and will inevitably change, leading to new opportunities, just as 'you could not step into the same river twice' (Heraclitus quoted in Plato 1998, §402A). The more this chapter make sense to you, we argue, the better it will help you thrive and profit in contemporary business, political, and social environments.

Good Luck! We really do mean Good Luck, for according to the Latin and Sanskrit etymological roots of 'luck', a deeper meaning of the expression is *make the future your own!* (Crane 2002).[1]

References

Angell, I., and Straub, B. (1999). 'Rain-Dancing with Pseudo-Science', *Cognition, Technology & Work*, 1: 179–96.

Borgmann, A. (1984). *Technology and the Character of Contemporary Life: A Philosophical Inquiry*. Chicago and London: University of Chicago Press.

—— (1999). *Holding On to Reality: The Nature of Information at the Turn of the Millennium*. Chicago and London: University of Chicago Press.

Burke, E. (1999*a*). *Selected Works of Edmund Burke, Vol. 1*. Indianapolis, IN: Liberty Fund Inc.

—— (1999*b*). *Reflections on the Revolution in France*. Oxford: Oxford University Press.

Burnes, B. (2000). *Managing Change*. London: Financial Times Prentice Hall.

Burrell, G, and Morgan, G. (1979). *Sociological Paradigms and Organisational Analysis*. London: Heinemann.

Cartwright., N. (1983). *How The Laws of Physics Lie*. Oxford: Oxford University Press.

Ciborra, C. U. (1998). 'From Tool to *Gestell*'. *Information Technology and People*, 11: 305–27.

Crane, G. R. (2002) (ed.). *The Perseus Project*. www.perseus.tufts.edu, November 2001–April 2002.

Dreyfus, H. (1991). *Being-in-the-world: A Commentary on Heidegger's Being and Time, Division 1*. Cambridge, MA: MIT Press.

—— and Dreyfus, S. (1986). *Mind over Machine*. Oxford: Basil Blackwell.

Ellul, J. (1964). *The Technological Society*. New York: Vintage Books.

Feyerabend, P. (1993). *Against Method*. London: Verso.

Gall, J. (1988). *Systemantics: The Underground Text of Systems Lore—How Systems Really Work and How they Fail*. Ann Arbor, MI: General Systemantics Press.

Gleick, J. (1987). *Chaos: Making a New Science*. New York: Viking Press.

Gödel, K. (1992). *On Formally Undecidable Propositions of Principia Mathematica and Related Systems*. New York: Dover.

Habermas, J. (1970). 'Technology and Science as Ideology', in J. Habermas, *Toward a Rational Society*. Boston: Beacon Press.

——(1987). *The Theory of Communicative Action: Lifeworld and System, a Critique of Functionalist Reason, Vol. 2*. Cambridge: Polity Press.

Heidegger, M. (1962). *Being and Time*. Oxford: Blackwell.

——(1969). *Identity and Difference*. New York and London: Harper and Row.

——(1977). *The Question Concerning Technology and Other Essays*. New York: Harper Torchbooks.

——(1999). *Contributions to Philosophy (On Endowment)*. Bloomington, IN: Indiana University Press.

Ihde, D. (1990). *Technology and the Lifeworld*. Bloomington, IN: Indiana University Press.

IHGSC (2001), International Human Genome Sequencing Consortium. 'Initial Sequencing and Analysis of the Human Genome'. *Nature*, 409: 860–921, 15 February.

Introna, L. (1997). *Management, Information and Power*. London: Macmillan.

Kant, I. (1985). *Crítica da Razão Pura*. Lisbon: Fundação Gulbenkian.

Lao Tzu (1963). *Tao Te Ching*. London: Penguin.

Lawson, H. (2001). *Closure: A Story of Everything*. London and New York: Routledge.

McLuhan, M. (1994). *Understanding Media*. Cambridge, MA: MIT Press.

Marcuse, H. (1964). *One-Dimensional Man*. Boston: Beacon Press.

Maturana, H. R., and Varela, F. J. (1992). *The Tree of Knowledge: The Biological Roots of Human Understanding*. Boston and London: Shambala.

Merleau-Ponty, M. (1962). *Phenomenology of Perception*. London and New York: Routledge.

Nietzsche, F. (1968). *The Will to Power*. New York: Vintage Books.

——(1974). *The Gay Science*. New York: Vintage Books.

——(1979). *Philosophy and Truth; Selections from Nietzsche's Notebooks from the Early 1870's*. London: Humanities Press.

——(1990). *Beyond Good and Evil*. London: Penguin Books.

——(1996). *Human, All Too Human*. Cambridge: Cambridge University Press.

Oxford Paperback Dictionary and Thesaurus (1997). Oxford: Oxford University Press.

Plato (1998). *Cratylus*. Indianapolis, IN: Hackett Publishing.

Polt, R. (1999). *Heidegger: An Introduction*. London: UCL Press.

Searle, J. (1983). *Intentionality: An Essay in the Philosophy of Mind*. Cambridge: Cambridge University Press.

Spengler, O. (1926). *The Decline of the West*. London: George Allen and Unwin.

Venter, J. C. and Associates (2001). 'The Sequence of the Human Genome'. *Science*, 291: 5507, 14 February.

Watzlawick, P. (1993). *The Situation is Hopeless, but not Serious*. New York: W.W. Norton.

Winner, L. (1995). 'Citizen Virtues in a Technological Order', in A. Feenberg and A. Hannay (eds.), *Technology and the Politics of Knowledge*. Bloomington, IN: Indiana University Press.

Winograd, T., and Flores, F. (1986). *Understanding Computers and Cognition: A New Foundation for Design*. Reading, MA: Addison Wesley.

Zimmerman, M. E. (1990). *Heidegger's Confrontation with Modernity: Technology, Politics, Art*. Bloomington, IN: Indiana University Press.

Note

1. *Suus* (*sua, suum*) is the word most used in ancient Latin texts to mean luck and opportunity. Literally, *suus* might be translated as *yours* (Crane 2002). It meant also *of* or *belonging to himself, herself, his own, her own, his, her, its, their, one's, hers, theirs*. This Latin word *suus*, in its turn, is akin to the Sanskrit word *svá*, which meant 'own' (ibid.).

3

On using ANT for studying information systems: a (somewhat) Socratic dialogue

Bruno Latour

Prologue

For this chapter on the use of Actor Network Theory (ANT), I have conflated several real conversations into one imaginary one.[1] Since English was not the native tongue of any of the participants, some improprieties have been kept.[2]

The scene in which the dialogue takes place is an office at the LSE on a dark February Tuesday. It is the end of the afternoon, close to the time for moving to the nearby Beaver pub for a pint of beer. A quiet but insistent knock is heard on the door of the *Professor.* An IS doctoral *student* peers into the office.

Introduction: on negative and positive theories

Student: Am I bothering you?

Professor: Not at all; these are my office hours anyway. Come in, have a seat.

S: Thank you.

P: So . . . I take it that you are a bit lost?

S: Well, yes. I am finding it difficult, I have to say, to apply Actor Network Theory to my case studies.

P: No wonder—it isn't applicable to anything!

S: But we were taught . . . I mean . . . it seems like hot stuff around here. Are you saying it's really *useless*?

P: It might be useful, but only if it does not 'apply' to something.

S: Sorry: are you playing some sort of Zen trick here? I have to warn you, I'm just an Information Systems doctoral student, so don't expect . . . I'm not too into French stuff either, just read a bit of *Thousand Plateaus* but couldn't make much sense of it . . .

P: I'm sorry, I wasn't trying to say anything cute. Just that ANT is first of all a *negative* argument. It does not say anything positive on any state of affairs.

S: So what can it do for me?

P: The best it can do for you is to say something like: 'When your informants mix up organization and hardware and psychology and politics in one sentence, don't break it down first into neat little pots; try to follow the link they make among those elements that would have looked completely incommensurable if you had followed normal academic categories.' That's all. ANT can't tell you positively what the link is.

S: So, why is it called a 'theory' then, if it says nothing about the things we study?

P: It's a theory... and a strong one I think... but its about *how* to study things, or rather how not to study them. Or, rather, how to let the actors have some room to express themselves.

S: Do you mean that other social theories don't allow that?

P: Yes, and because of their very strengths: they are good at saying *positive* things about what the social world is made of. In most cases that's fine. The ingredients are known. But not when things are changing fast and—I would add—not in information studies, where boundaries are so terribly fuzzy. *New* topics, that's what you need ANT for.

On networks and the importance of description

S: But my agents, actors... I mean the people I'm studying at IBM, are actor networks. They are connected to a lot of other things, they are all over the place...

P: But see, that's the problem. You don't need Actor Network Theory to say that: any available social theory would do the same. It's useless to pick this very bizarre argument to show that your informants are in a network.

S: But they are. They form a network. Look, I have been tracing their connections: computer chips, standards, schooling, money, rewards, countries, cultures, corporate board rooms... everything. Haven't I described a network in your sense?

P: Not necessarily. I agree this is terribly confusing, and it's our fault: or rather it's entirely Callon's fault! He's the one who came up with this horrible word... But you should not confuse the network that is drawn by the description and the network that is used to make the description.

S: ...?

P: Really! Surely you'd agree that drawing with a pencil is not the same thing as drawing the shape of a pencil. It's the same with this ambiguous word, network. With Actor Network you may describe something that doesn't at all look like a network; conversely, you may describe a network which is not all drawn in an 'Actor Networky' way. You are simply confusing the object with the method. ANT is a method, and mostly a negative one at that; it says nothing about the *shape* of what is being described with it.

S: This is confusing... my IBM folks, are they not forming a nice, revealing, significant network?

P: Maybe yes... I mean, surely, yes—but so what?

S: Then, I can study them with Actor Network Theory!

P: Again, maybe yes, but maybe not. It depends entirely on what you make your actors, or actants, do. Being connected, being interconnected, being heterogeneous is not enough. It all depends on the sort of action that is flowing from one to the other, hence the words 'net' and 'work'.

S: Do you mean to say that once I have shown that my actors are related in the shape of a network, I have not yet done an ANT study?

P: That's exactly what I mean; ANT is more like the name of a pencil or a brush than the name of an object to be drawn or painted.

S: But when I said ANT was a tool and asked you if it could be applied, you objected.

P: Because it's not a tool. Or, rather, because tools are never 'mere' tools ready to be applied: they always change the goals as well. Actor Network . . . I agree the name is silly . . . allows you to produce some *effects* that you would have never obtained by any other social theory. That's all that I can vouch for. It's a very common experience: drawing with a lead pencil or with charcoal is not the same either; cooking tarts with a gas oven is not the same as with an electric one.

S: But that's not what my supervisor wants. He wants a frame in which to put my data.

P: If you want to store more data, buy a bigger hard disk . . .

S: He always says to me: 'You need a framework.'

P: Ah? So your supervisor is in the business of selling pictures? It's true that frames are nice for that: gilded, white, carved, baroque, aluminium, etc. But have you ever met a painter who began her masterpiece by first choosing the frame? That would be a bit odd, wouldn't it?

S: You're playing with words. By 'frame' I mean a theory, an argument, a general point, a concept—something for making sense of the data. You always need one.

P: No you don't! Tell me, if some X is a mere 'case of' Y, what is more important to study: X which is the special case, or Y which is the rule?

S: Probably Y . . . but X too, just to see if it's really an application of . . . well, both I guess.

P: I would bet on Y myself, since X will not teach you anything new. If something is simply an instance of some other state of affairs, go study this state of affairs instead. A case study that needs a frame in addition is a case study that was badly chosen to begin with!

S: But you always need to put things into a context, don't you?

P: I have no patience for context, no. A frame makes a picture look nicer, it may direct the gaze better, increase the value, but it doesn't add anything to the picture. The frame, or the context, is precisely what makes no difference to the data, what is common knowledge about it. If I were you I would abstain from frameworks altogether. Just describe.

S: 'Just describe'. Sorry to ask, but is this not terribly naïve? Is this not exactly the sort of empiricism, or realism, that we have been warned against? I thought your argument was more sophisticated than that.

P: Because you think description is easy? You must be confusing it, I guess, with strings of clichés. For every hundred books of commentaries, arguments, glosses, there is only one of description. To describe, to be attentive to the concrete states of affairs, to find the uniquely adequate account of a given situation—I have, myself, always found this incredibly demanding. Ever heard of Harold Garfinkel?

On interpretative and objectivist perspectives

S: I'm lost here, I have to say. We have been told that there are two types of sociology, the interpretive and the objectivist. Surely you don't want to say you are of the objectivist type?

P: You bet I am! Yes, by all means.

S: You? But we have been told you were something of a relativist. You have been quoted as saying that even the natural sciences are not objective . . . So, surely you are for interpretive sociology, viewpoints, multiplicity of standpoints, all that.

P: I have no patience for interpretive sociologies, whatever you may call by that name. No. On the contrary, I firmly believe that sciences are objective: what else could they be? They're all about objects, no? I simply say that objects might look a bit more complicated, folded, multiple, complex, entangled than what the 'objectivist', as you say, would like them to be.

S: But isn't that exactly what 'interpretive' sociologies argue?

P: Oh no, not all. They would say that *human* desires, *human* meanings, *human* intentions, etc. introduce some 'interpretive flexibility' into a world of inflexible objects, of pure causal relations. That's not at all what I am saying. I would say that this computer, this screen, this keyboard—as objects—are made of multiple layers, exactly as much as you, sitting here are: your body, your language, your questions. It's the object itself that adds multiplicity, or rather the thing, to speak like Heidegger, the 'gathering'. When you speak of hermeneutics, whatever you do you always expect the second shoe to drop: someone inevitably will add 'but of course there also exist "natural", "objective" things that are not interpreted'.

S: That's just what I was going to say! There are not only objective realities, but also subjective ones. This is why we need both types of social theories.

P: See? That's inevitable, a trap: 'Not only but also'. Either you extend the argument to everything, but then it becomes useless as 'interpretation' becomes another synonym for 'objectivity'—or else you limit it to one aspect of reality, the human, and then you are stuck, since objectivity is always on the other side of the fence. And it makes no difference if the other side is greener or more rotten, it's beyond reach anyway.

S: But you wouldn't deny that you too possess a standpoint, that ANT is situated too, that you too add another layer of interpretation, a perspective?

P: No, why would I 'deny' it? But so what? The great thing about a standpoint is, precisely, that you can change it. Why would I be stuck with it? Astronomers in Greenwich, down the river from here, have a limited perspective from where they are on earth. And yet, they have been pretty good at shifting this perspective, through instruments, telescopes, satellites. They can now draw a map of the distribution of galaxies in the whole universe. Pretty good, no? Show me one standpoint, and I will show you two dozen ways to shift out of it. Listen, all this opposition between 'standpoint' and 'view from nowhere', you can safely forget. And also this difference between 'interpretive' and 'objectivist'. Leave hermeneutics aside and go back to the object: or rather to the thing.

S: But I am always limited to my situated viewpoint, to my perspective, to my own subjectivity.

P: You are very obstinate! What makes you think that having a viewpoint means 'being limited', or especially 'subjective'? When you travel abroad and you follow the sign 'belvedere', 'panorama', 'bella vista', when you finally reach the breathtaking site, in what way is this a proof of your 'subjective limits'? It's the thing itself, the valley, the peaks, the roads that offer you this grasp, this handle. The proof is that two metres lower you see nothing because of the trees, and two metres higher nothing because of a parking lot. And yet you have the same 'subjectivity', and have exactly your very same 'standpoint'. If something supports many viewpoints, it's just that it's highly complex, intricately folded, nicely organized, and beautiful—yes, *objectively* beautiful. If you can have many viewpoints on a statue, it's because the statue itself is in three dimensions and allows you, yes, *allows* you to turn around it.

S: But certainly, nothing is objectively beautiful: beauty has to be subjective . . . taste and colour, relative . . . I am lost again. Why would we spend so much time here fighting objectivity then? What you say can't be right.

P: Because the things people call 'objective' are most of the time a series of clichés. We don't have a good description of anything: of what a computer, a piece of software, a formal system, a theorem is. We know next to nothing of what this thing you are studying—*information*—is. How would we be able to distinguish it from subjectivity? In other words, there are two ways to criticize objectivity: one is by going *away* from the object to the subjective human view point. But the other direction is the one I am talking about: back to the object. Don't leave objects to be described only by the idiots. As we saw in class yesterday, a computer described by Alan Turing is quite a bit richer and more interesting than the ones described by *Wired Magazine*. The name of the game is to get back to empiricism.

S: Still, I am limited to my own view.

P: Of course you are, but again, so what? Don't believe all that rubbish about being 'limited' to one's perspective. All of the sciences have been inventing ways to *move* from one standpoint to the next, from one frame of reference to the next, for heaven's sake: that's called relativity.

On relativism, ANT and context

S: Ah! So you confess you are a relativist.

P: But of course, what else could I be? If I want to be a scientist and reach objectivity, I have to be able to travel from one frame to the next, from one standpoint to the next.

S: So you associate objectivity with relativism?

P: 'Relativity', yes, of course. All the sciences do the same. Our sciences too.

S: But what is *our* way to change our standpoints?

P: I told you, we are in the business of descriptions. Everyone else is trading on clichés. Enquiries, polls, whatever—we go, we listen, we learn, we practice, we become competent, we change our views. Very simple really: it's called field work. Good field work always produces a lot of descriptions.

S: But I have lots of descriptions already. I'm drowning in them. That's just my problem. That's why I'm lost and that's why I thought it would be useful to come to you. Can't ANT help me with this mass of data? I need a framework.

P: 'My Kingdom for a frame!'. Very moving; I think I understand your desperation. But no, ANT is pretty useless for that. Its main tenet is that actors themselves make everything, including their own frames, their own theories, their own contexts, their own metaphysics, even their own ontologies... So the direction to follow would be more descriptions, I am afraid.

S: But descriptions are too long. I have to *explain* instead.

P: See? This is where I disagree with most of the training in the social sciences.

S: You would disagree with the need for social sciences to provide an explanation for the data they accumulate? And you call yourself a social *scientist* and an objectivist!

P: I'd say that if your description needs an explanation, it's not a good description, that's all. Only bad descriptions need an explanation. It's quite simple really. In any case, what is meant by an 'explanation' most of the time? Adding another actor to give to those already described the energy necessary to act. But if you have to add one, then the network was not complete, and if the actors already assembled do not have enough energy to act, then they are not 'actors', but mere intermediaries, dopes, puppets. They do nothing, so they should not be in the description anyhow. I have never seen a good description in need, then, of an explanation. But I have read countless bad descriptions to which nothing was added by a massive addition of 'explanations'.

S: This is very distressing. I should have known: the other students warned me not to touch ANT stuff... Now you are telling me that I shouldn't even try to explain anything.

P: I did not say that, but just that either your explanation is relevant and, in practice, you are adding a new agent to the description as the network is simply longer than you thought—or it's not an actor that makes any difference and you are simply adding something irrelevant, which helps neither the description nor the explanation. In that case, throw it away.

S: But my colleagues use a lot of them: 'IBM corporate culture', for instance, or 'British isolationism', or 'market pressure', or 'self-interest'. Why should I deprive myself of those contextual explanations?

P: You can keep them if this amuses you, but don't believe they explain anything—they are mere ornaments. At best they apply equally to all your actors, which means they are absolutely irrelevant since they are unable to introduce a difference among them. As a rule, context stinks. It's simply a way of stopping the description when you are tired or too lazy to go on.

On written descriptions, stories, and theses

S: But that's exactly my problem: to stop. I have to complete this Ph.D.; I have just eight more months. You always say 'more descriptions', but this is like Freud, indefinite analysis. When do you stop? My actors are all over the place. Where should I go? What is a complete description?

P: Now that's a good question because it's a practical one. As I always say: a good thesis is a thesis that is done. But there is another way to stop than by 'adding an explanation' or 'putting it into a frame'.

S: Tell me it then.

P: You stop when you have written your 80,000 words or whatever is the format here, I always forget.

S: Oh! That's really great, so helpful, many thanks! I feel so relieved . . . So my thesis is finished when it's finished . . . But that's a textual limit, it has nothing to do with method.

P: See? That's again why I totally disagree with the ways doctoral students are trained. Writing texts has everything to do with method. You write a text of so many words, in so many months, for so much grant money, based on so many interviews, so many hours of observation, so many documents. That's all. You do nothing more.

S: Of course, I do: I learn, I study, I explain, I criticize, I . . .

P: But all those great goals: you achieve them through a text, don't you?

S: Of course, but it's a tool, a medium, a way of expressing myself.

P: There is no tool, no medium, only mediators. A text is thick. That's an ANT tenet, if any.

S: Sorry, Professor, I told you, I have never been into French stuff. I can write in programming languages like C and even C++, but I don't do Derrida, semiotics, any of that. I don't believe the world is made of words and all of that . . .

P: Don't try to be ironic, it doesn't suit the engineer in you. And anyway I don't believe that either. You ask me how to stop. I am just telling you that the best you will be able to do, as a Ph.D. student, is to *add* a text read by your advisors and maybe a few of your informants—three or four fellow doctoral students—to a given state of affairs. Nothing fancy in that. Just plain realism. One solution for how to stop is to 'add a framework', an 'explanation'; the other is to put the last word to the last chapter of your damned thesis.

S: I have been trained in the sciences. I am a systems engineer—I am not coming to Information Systems to abandon that. I am willing to add organizations, institutions, people, mythologies, psychology to what I already know. I am even prepared to be 'symmetric', as you said, about those various factors. But don't tell me that science is about telling nice stories. This is the difficulty with you. One moment you are a completely objectivist—naïve realist even: 'just describe'—and the other moment a completely relativist: 'tell some nice stories and run'. Is this not so terribly French?

P: And that makes you so terribly what? Don't be silly. Who talked about 'nice stories'? Not me. I said you were *writing* a Ph.D. thesis. Can you deny that? And then I said that this so-many-words-long Ph.D. thesis—which will be the only lasting result of your stay among us—is thick.

S: Meaning?

P: Meaning that it's not just a transparent window pane, transporting without deformation, the information about your study. Can you deny that? I assume that you agree with this slogan of my class: 'There is no in-formation, only trans-formation', or translation if you want. Well, then, this is surely also true of your Ph.D. thesis, no?

S: Maybe, but in what sense does it help me be more scientific, that's what I want to know. I don't want to abandon the ethos of science.

P: Because this text, depending on the way it's written, will or will not capture the actor network you wish to study. The text, in our discipline, is not a story, not a nice story, it's the functional equivalent of a laboratory. It's a place of trials and experiments. Depending on what happens in it, there is or there is not an actor—and there is or there is not a network being traced. And that depends entirely on the precise ways in which it is written. Most texts are just plain dead.

S: But no one mentions 'text' in our programme. We talk about 'studying information', not 'writing' about it.

P: That's what I am telling you: that you are being badly trained! Not teaching social science doctoral students to *write* their theses is like not teaching chemists to do laboratory experiments. That's why I am teaching nothing nowadays but writing. I feel like an old bore always repeating the same thing: 'describe, write, describe, write, exercise . . .'

S: The problem is that that's not at all what my supervisor wants. He wants my case studies to be generalizable. He does not want 'mere description'. So, even if I do what you want, I will have one nice description of one state of affairs, and then what? Then, I still have to put it into a frame, find a typology, compare, explain, generalize. That's why I'm starting to panic.

P: You should panic only if your actors were not doing that constantly as well, actively, reflexively, obsessively: they too compare, they too produce typologies, they too design standards, they too spread their machines as well as their organizations, their ideologies, their states of mind. Why would you be the one doing the intelligent stuff while they act like a bunch of morons? What they do to expand, to relate, to compare is what you have to describe as well. It's not another layer that you would have to add to the 'mere description'. Don't try to shift from description to explanation: simply continue the description. What your own ideas are about IBM is of no interest compared to how this bit of IBM itself has managed to spread.

On actors who leave no trace and don't need to learn from a study

S: But if my people don't act, if they don't actively compare, standardize, organize, generalize, what do I do? I will be stuck! I won't be able to add any other explanations.

P: You are really extraordinary. If your actors don't act, they will leave no trace whatsoever either. So you will have no information at all: you will have nothing to say.

S: You mean I should remain silent when there is no trace?

P: Incredible! Would you raise this question in any of the natural sciences? It would sound totally silly. It takes a social scientist to claim that they can go on explaining even in the absence of any information. Are you really prepared to make up data?

S: No, of course not, but still I want . . .

P: Good, at least you are more reasonable than most of your colleagues. No trace left, thus no description, then no talk. Don't fill it in. It's like a map of a country in the sixteenth century: no one went there, nor did anyone came back—so leave it blank! *Terra incognita.*

S: But what about invisible entities acting in a hidden way?

P: If they act, they leave some trace, then you have some information, then you can talk about them. If not, just shut up.

S: But if they are repressed, denied, silenced?

P: Nothing on earth allows you to say they are there without bringing in the proof of their presence. That proof might be indirect, exacting, complicated, but you need it. Invisible things are invisible. Period.

S: 'Proof'. Isn't that terribly positivistic?

P: I hope so, yes. What's so great about saying that things are acting whose existence you can't prove? I am afraid you are confusing social theory with conspiracy theory: although, these days, I agree, most of social science comes down to that.

S: But if I add nothing, I simply repeat what actors say.

P: What would be the use of adding invisible entities that act without leaving any trace and make no difference to any state of affairs?

On reflexivity and explanations

S: But I have to teach the actors something they didn't know; if not, why would I study them?

P: You social scientists, you always baffle me. If you were studying ants, instead of ANT, would you expect ants to *learn* something from your study? Of course not. They know, you don't. They are the teachers, you learn from them. You explain what they do to yourself, for your own benefit, not for them, who don't care a bit. What makes you think that a study was supposed to teach things to the people being studied?

S: But that's the whole idea of the social sciences! That's why I'm here: to criticize the ideology of information, to debunk the many myths of information technologies, to gain a critical edge over all the technical hype. If not, believe me, I would still be in Silicon Valley, and I would be making a lot more money . . . well, maybe not since the bubble burst . . . But anyway, I have to provide some reflexive understanding to the people . . .

P: . . . who of course, before you came, were unreflexive!

S: In a way, yes. No? They did things but did not know why . . . What's wrong with that?

P: What's wrong is that it's much too cheap. Most of what social scientists call 'reflexivity' is just asking totally irrelevant questions to people who ask other questions for which the analyst does not have the slightest beginning of an answer. Reflexivity is not a birthright you transport with you, just because you are at the LSE. You and your informants have different concerns—when they intersect it's a miracle, and miracles are rare . . .

S: But if I have nothing to add to what actors say, I won't be able to be critical.

P: See, one moment you want to explain and play the scientist, while the next moment you want to debunk and criticize and play the militant . . .

S: I was going to say that one moment you are a naïve realist: back to the object; and the next you say that you just write a text that adds nothing but simply trails behind your famous 'actors themselves'. This is totally apolitical. No critical edge that I can see.

P: Tell me, Master Debunker, how are you going to gain a 'critical edge' over your actors? I am eager to hear.

S: Only if I have a framework. That's what I was looking for in coming here, but obviously ANT is unable to give me any.

P: And I am pretty glad it doesn't . . . This framework of yours, I assume, is hidden to the eyes of your informants, and revealed by your study?

S: Yes, of course. That should be the value of my work—I hope so at least. Not the description, since everyone knows that anyway, but the explanation, the context they have no time to see, the typology. That's what I can deliver, and they are interested, and ready to give me access to their files, and willing to pay for it!

P: Good for you. What you are telling me is that in your six months of field work you can, by yourself, just by writing a few hundred pages, produce more knowledge than those 340 engineers and staff you have been studying?

S: Not 'more' knowledge maybe, but different, yes, I hope. Shouldn't I strive exactly for that? Is this not why I am in this business?

P: I am not sure what business you are in, but how *different* is the knowledge you produce? That's the whole question.

S: It's the same kind of knowledge as all the sciences, the same way of explaining things: by going from the case at hand to the cause, and once I know the cause—then I can generate the effect as a consequence. What's wrong with that? It's like a pendulum that has been moved far from equilibrium. If I know Galileo's law, I don't even need to look at any concrete pendulum anymore, I know exactly what will happen—provided I forget the perturbations, of course.

P: Of course! So what you are hoping for is that your explanatory framework will be to your case study what Galileo's law is to the fall of the pendulum—minus the perturbations.

S: Yes, I guess so, sort of, though less scientific naturally. Why? What's wrong with that?

P: Nothing. It would be great, but is it feasible? It means that, whatever a given concrete pendulum does, it will add no new information to the law of falling bodies. The law holds *in potentia* everything there is to know about the state of affairs. The concrete case is simply the 'realization of a potential', to be technical, which was already there.

S: Is not that an ideal explanation?

P: That's just the problem: it's an ideal, and it's the ideal of an ideal explanation. I doubt somewhat that your IBM subsidiary behaves that way. And I am pretty confident you can't produce the law of its behaviour that will allow you to deduce everything as the realization *in concreto* of what was already there potentially.

S: Minus the perturbations . . .

P: Yes, yes, yes, this goes without saying . . . admirable modesty . . .

S: Are you making fun of me here? Striving for that sort of framework seems worthwhile to me . . .

P: But is it desirable? See, what you are really telling me is that the actors in your description make *no difference whatsoever.* They have simply realized a potential, apart from minor deviations. Which means they are not actors at all: they simply carry the force that comes through them. So, you have been wasting your time describing people, objects, sites that are nothing, in effect, but passive intermediaries since they do nothing on their own. Your fieldwork is simply wasted. You should have gone directly to the potential.

On the abyss between structuralism and ANT

S: But that's what a science is for! Just that: finding the hidden structure that explains the behaviour of those agents you thought were doing something, but in fact are simply place-holders for something else.

P: So you are a structuralist! Out of the closet, finally. Place-holders, that's what you call actors? And you want to do Actor Network Theory at the same time! That's stretching the limits of eclecticism pretty far.

S: Why can't I do both? Certainly if ANT has any scientific content, it has to be structuralist.

P: Have you realized that there is the word 'actor' in actor network? Can you tell me what sort of action a place-holder does in a structuralist explanation?

S: It fulfils a function, that's what is so great about structuralism. Any other agent in the same position would be forced to do the same . . .

P: So, a place-holder, by definition, is entirely substitutable to any other?

S: Yes, that's what I am telling you is so great.

P: But that's also what is so silly and what makes it radically incompatible with ANT: an actor that makes no difference, in my vocabulary, is not an actor at all. An actor, if words have any meaning, is exactly what is *not* substitutable for anyone else; it's a unique event, totally irreducible to any other. Except, that is, if you *make* one commensurable with another one by some sort of standardization, but even that requires a third actor, a third event.

S: So you are telling me that ANT is not a science.

P: Not a structuralist science, that's for sure.

S: That's the same, any information system science . . .

P: No! If information is transformation, translation . . . IS cannot rely, by definition, on any structuralist explanation.

S: 'Systems of transformation', that's exactly what structuralism is!

P: No way, my friend, since in structuralism nothing happens. You don't seem to fathom the abyss that exists between structuralism and ANT: if I want to have actors in my account, they have to *do* things, not to be place-holders; if they do something they have to make a difference. If they make no difference, drop them, start the description anew. You want a science in which there is no event.

S: Eventful stories, that's what you want! Always stories. I am talking about explanation, knowledge, critical edge, not writing scripts for soap operas on TV.

P: I was getting to that. You want your bundle of a few hundred pages to make a difference, no? Well then, you have to be able to prove that your description of what people do, when it comes back to them, *does* make a difference to the way they were doing things. Is this what you call having a 'critical edge'?

S: I guess so, yes.

P: But then you would agree that it wouldn't do to provide them with irrelevant causes that make no difference at all to what they do, because they are too general?

S: Of course not; I was talking about *real* causes.

P: But those won't do either because, if they existed—which I doubt very much—they would have no other effect than transforming your informants into the place-holders of other actors, which you call function, structure, etc. So, in effect, they would not be actors anymore but dopes, puppets: and when I say puppet . . . I have shown that even puppets are making you do quite a lot of unexpected things . . . Well, anyway, you are making them out to be nothing; the best they could do would be to add some minor perturbations like the concrete pendulum which only adds slight wobbles.

S: ...

P: Now you have to tell me what is so politically great about transforming those you have studied into hapless, 'actless' place-holders for hidden functions that you, and you only, can see and detect?

S: Hmm, you have a way of turning things upside down ... I am not so sure now. If actors become aware of the determinations imposed on them ... more conscious ... more reflexive ... is their consciousness not raised somewhat? They can now take their fate in their own hands. They become more enlightened, no? If so, I would say that now, at last—in part thanks to me—yes, there are more actors now, fully.

On science, authority, and relevance

P: Bravo, bravissimo! So an actor for you is some fully determined agent, a place-holder for a function, plus a bit of perturbation, plus consciousness. Horrible, simply horrible ... and he wants to do ANT! After you have reduced them from actors to place-holders you want to add insult to injury and generously bring them the reflexivity they had before and that you have ignored by treating them in a structuralist way. Magnificent! They were actors before you came in with your 'explanation'—don't tell me that it's your study that might make them so. Great job. Bourdieu could not have done better ...

S: You might not like Bourdieu, but at least he is a real scientist, and even better, he is politically relevant. As far as I can tell, you are neither ...

P: Thanks. I have been studying the links between science and politics for about thirty years, so I am hard to intimidate with talk of what is 'politically relevant'.

S: Arguments of authority don't intimidate me either, so thirty years makes no difference to me.

P: *Touché* ... But your question was: 'What can I do with ANT'? I answered it: no structuralist explanation. The two are completely incompatible. Either you have actors who realize potentialities and they are not actors at all, or you describe actors who are making virtualities actual (this is Deleuze by the way), and that requires very specific texts, and your connection with those who you study requires very specific protocols to work. I guess this is what you would call 'critical edge' and 'political relevance'.

S: So, where do we differ? You too want to have a critical edge.

P: Yes, maybe, but I am sure of one thing: it's not automatic, and most of the time, it will fail. Two hundred pages of interviews, observations, etc. will not make any difference whatsoever. To be relevant requires another set of extraordinary circumstances. It's an event. It requires an incredibly imaginative protocol. As great, as rare, as surprising as Galileo with his pendulum.

S: What should I do then? Pray for a miracle?

P: But why do you want your tiny little text to be more automatically relevant—or not—to those who might be concerned than, say, a huge laboratory of natural sciences? Look at how much it takes for Intel to be relevant for mobile phones! And you want everyone to have a label 'LSE inside' at no cost at all? To become relevant, you need extra work.

S: Just what I need: the prospect of even more work.

P: But that's the whole point: if it's scientific, then it can fail; if it's automatic, across the board, all purpose, then it can't possibly be scientific. It's simply irrelevant.

S: Great reassurance, nice of you to remind me that I can fail my thesis!

P: You are confusing science and mastery. Tell me, can you imagine one single topic to which, for instance, Bourdieu's critical sociology which you are so fond of, could possibly *not* apply?

S: But I can't imagine one single topic to which ANT would apply automatically!!

P: Beautiful, you are so right, that's exactly what I think . . .

S: That was not meant as a compliment.

P: But I take it as a true one! An application of anything is as rare as a good text of social science.

S: May I politely remark that, for all your exceedingly subtle philosophy of science, you have yet to tell me how to write one . . .

P: You were so eager to add frames, context, structure, to your mere descriptions, how would you have listened to me?

S: But what's the difference between a good and a bad ANT text?

P: Now, that's a good question! Answer: The same as between a good and a bad laboratory. No more, no less.

S: Well, okay, um, thanks . . . It was nice of you to talk to me. But I think after all, instead of ANT, I will use Luhmann's autopoiesis as an underlying framework—that seems to hold a lot of promise. Or maybe I will use a bit of both. Don't you like Luhmann?

P: I would leave aside all 'underlying frameworks', if I were you.

S: But, your sort of 'science', it seems to me, means breaking all the rules of social science training.

P: I prefer to break them and follow my actors . . . As you said, I am, in the end, a naïve realist.

S: But see, I'm just a Ph.D. student. You're a professor. You have published, you can afford to do things that I can't. I have to listen to my supervisor. I simply can't follow your advice too far.

P: Why come to me then?

S: For the last half hour, I have to confess, I've been wondering the same thing . . .

Background reading

Callon, M., and Latour, B. (1981). 'Unscrewing the Big Leviathans: How Do Actors Macrostructure Reality? Advances in Social Theory and Methodology', in K. Knorr and

A. Cicourel (eds.), *Toward an Integration of Micro and Macro Sociologies*. London: Routledge, 277–303.

Ciborra, C. U. (2000) (ed.). *From Control to Drift: The Dynamics of Corporate Information Infrastructures*. Oxford: Oxford University Press.

Law, J., and Hassard, J. (1999) (eds.). *Actor Network and After*. Oxford: Blackwell.

Notes

1. In this volume, ANT is also discussed in Monteiro (Chapter 7), Kallinikos (Chapter 8), and Smithson and Tsiavos (Chapter 11).
2. I would like to thank John Tresch for his help in correcting the use of English in the dialogue.

4

Towards a sociology of information technology

SASKIA SASSEN

Introduction

The technical attributes of new information and communication technologies increasingly dominate explanations of contemporary change and development. As Wajcman (2002) points out, many sociologists see technology as the impetus for the most fundamental social trends and transformations.[1] To this, I would add a tendency to understand or conceptualize these technologies in terms of technical properties and to construct the relation to the sociological world as one of applications and impacts. The challenge for sociology is not so much to deny the weight of technology, but rather to develop analytic categories that allow us to capture the complex imbrications of technology and society.

I want to develop two particular aspects of this challenge in this chapter, focusing especially on digital technologies. I will argue that understanding the place of these new technologies from a sociological perspective requires avoiding a purely technological interpretation and recognizing the embeddedness and the variable outcomes of these technologies for different social orders. They can indeed be constitutive of new social dynamics, but they can also be derivative or merely reproduce older conditions. Secondly, such an effort will, in turn, call for categories that capture what are now often conceived of as contradictory, or mutually exclusive, attributes. I will examine these two aspects by focusing on three analytic issues for sociology: the embeddedness of the new technologies, the complex interactions between the digital and the material world, and the mediating cultures that organize the relation between these technologies and users. In the second half of the chapter, I examine these analytic issues as they get instantiated in substantive sociological arenas, each based on prior research: the interactions between capital fixity and capital mobility; the gendering of access to and use of electronic space; and the emergence of a new politics of places on global networks (see, respectively, Sassen 2000*a*, 2001, 2002).

The difficulty that analysts and commentators have had in specifying or understanding the impact of digitization on multiple conditions results essentially from two analytic flaws. One of these, which is especially evident in the USA, confines interpretation to a technological reading of the technical capabilities of

A version of this chapter first appeared in *Current Sociology*, 50/3: 365–88.

digital technology. This is crucial for the engineering side, but is problematic for a sociological understanding. Such a purely technological reading of technical capabilities inevitably neutralizes, or renders invisible, the material conditions and practices, place-boundedness, and thick social environments within and through which these technologies operate. Another consequence of this type of reading is to assume that a new technology will *ipso facto* replace all older technologies that are less efficient, or slower, at executing the tasks the new technology is best at. We know that historically this is not the case.

A second tendency is the continuing reliance on analytical categorizations that were developed under other spatial and historical conditions, that is, conditions preceding the current digital era. Thus, the tendency is to conceive of the digital as simply and exclusively digital, and the non-digital (whether represented in terms of the physical/material or the actual, which are all problematic, though common, conceptions) as simply and exclusively that: non-digital. These either/or categorizations filter out alternative conceptualizations, thereby precluding a more complex reading of the impact of digitization on material and place-bound conditions.

The embeddedness of digital technologies

Digital networks are embedded in both the technical features and standards of the hardware and software, and in actual societal structures and power dynamics (Latour 1991; MacKenzie and Wajcman 1999; Lovink and Riemens 2002).[2] There is no purely digital economy and no completely virtual corporation or community. This means that power, contestation, inequality, and hierarchy inscribe electronic space and shape the production of software.

The fact that electronic space is embedded and cannot be read as a purely technological condition, or merely in terms of its technical features, is illuminated by the nature of segmentations evident inside electronic space. One instance is captured in the differences between private and public-access digital networks, such as the dedicated networks of financial services firms and wholesale financial markets.[3] The Internet is a different type of space from the private networks of the financial industry; and the firewalled corporate sites on the web are different from the public-access portion of the web. The financial markets operating largely through private dedicated digital networks are a good instance of private electronic space. The three properties of digital networks—decentralized access, simultaneity, and interconnectivity—have produced strikingly different outcomes in the private digital space of global finance from the distributed power of the public-access portion of the Internet. Although the power of these financial electronic networks rests on a kind of distributed power, with millions of investors and their millions of decisions, it ends up as concentrated power. The trajectory followed by what begins as the distributed power we associate with the public-access Internet may assume many forms; the case of financial networks shows this form can be radically different from that of the Internet.

This difference points to the possibility that network power is not inherently distributive. Intervening mechanisms which may have little to do with the technology *per se* can reshape its organization. To keep it as a form of distributed power requires that it be embedded in a particular kind of structure. We cannot take the distributed power and, hence, the democratizing potential of digital networks as an inevitable feature of this technology, as is so often the case in Utopian readings, perhaps most prominently exemplified by John Perry Barlow's 1996 'Declaration of Independence of Cyberspace' (see www.eff.org/~barlow/Declaration-Final.html).

Beyond these issues of intentionality and use, lies the question of infrastructure and access (e.g. Thomas 1995; NTIA 1998; Sassen 1998; Shade 1998; Petrazzini and Kibati 1999; Darkwa and Mazibuko 2000). Electronic space is going to be far more present in highly-industrialized countries than in the less-developed world; and far more present for middle-class households in developed countries than for poor households in those same countries (Hoffman and Novak 1998; Jensen 1998; Harvey and Macnab 2000). However, what needs emphasizing here is that there are very cheap ways of delivering access to the Internet, far cheaper than the standard telephone system, and hence that once such access is secured, the opportunities for low-income households and communities, especially in the global south, can increase enormously (e.g. Nadeau *et al.* 1998; ITU 1998; Mele 1999).

An additional issue is the privatization of infrastructure that has also taken place since the mid-1990s. The backbone has been privatized where it was previously financed by the US government, that is to say, taxpayers. This, in turn, changes the normative issues about private appropriations of Internet space as a public space. But it does so only partly, since it does not override the new distinction between privatized Internet space and public-access space, even if for a fee (for a resource to be public it need not necessarily be free). Internet space can remain public even if there is a fee to be paid for access, but privatized Internet space is not accessible, no matter what the fee.

Recognizing the embeddedness of electronic space, in my research I have come to regard the Internet as a space produced and marked through the software that shapes its use and the particular aspects of the hardware mobilized by the software (Sassen 1999). These features can also function as an indicator of transformations in the articulations between electronic space and larger institutional orders. There are significant implications attached to the fact that one of the leading Internet software design focuses since the turn of the century has been on firewalled intranets for firms and encrypted tunnels for firm-to-firm transactions. This saves companies the cost of private computer networks, the associated staffing and servicing, and the costs of frame-relay connections or of using intermediaries for firm-to-firm transactions. Peer-to-peer software is a significant development in this direction.

Both firewalling and encrypted tunnels represent, in some sense, private appropriations of a 'public' space. The growing interest in e-commerce has also stimulated the development of software linked to identity verification, trademark protection, and billing. The rapid growth of this type of software and its use in the Internet does not necessarily strengthen the publicness of electronic space (e.g. Elkin-Koren 1996). This is especially significant if there is less production of software aimed at strengthening the openness and decentralization of the Net, as was the case in the earlier phases of the Internet. Far from strengthening the Internet's democratic potential, as many liberal and neo-liberal commentators maintain, this type of commercialization can threaten it. It also carries major implications for the impact of democratizing initiatives.

However, electronic space remains a crucial force for new forms of civic participation, especially in its public-access portion. Non-commercial uses still dominate the Internet, even though the race is on to invent ways of expanding electronic commerce and ensuring the safety of payment transactions. But, at the same time, there has been a proliferation of non-commercial uses and users. Civil society, whether it be individuals or non-governmental organizations (NGOs), is an energetic presence in electronic space. From struggles around human rights, the environment, and workers strikes around the world to genuinely trivial pursuits, the Net has emerged as a powerful medium for non-elites to communicate, support each other's struggles, and create the equivalent of insider groups at scales going from the local to the global (e.g. Frederick 1993; Kobrin 1998; Ronfeldt *et al.* 1998).

Looking at electronic space as embedded allows us to go beyond the common duality between Utopian and dystopian understandings of the Internet and electronic space generally. For instance, even as it reproduces masculine cultures and hierarchies of power, electronic space also enables women to engage in new forms of contestation and in proactive endeavours in multiple different realms, from political to economic. Further, in the context of globalization these initiatives can go global and bypass national states and major national economic actors, thereby opening a whole new terrain for initiatives of historically disadvantaged peoples and groups (e.g. Correll 1995; Cleaver 1998; Ronfeldt *et al.* 1998; Mele 1999).

Three analytic issues that capture various features of the embeddedness are the complex imbrications between the digital and material conditions, the destabilizing of existing hierarchies of scale made possible by the new technologies, and the mediating cultures between these technologies and their users. The next three sections develop these issues.

Digital/material imbrications

Hypermobility or dematerialization are usually seen as mere functions of the new technologies. This understanding erases the fact that it takes multiple material conditions to achieve this outcome. Once we recognize that the hypermobility of the instrument, or the dematerialization of the actual piece of real estate, had to be actually produced, we introduce non-digital variables in our analysis of the digital. On the other hand, much of what happens in electronic space is deeply inflected by the cultures, the material practices, the imaginaries that take place outside electronic space. Much of what we think of when it comes to cyberspace would lack any meaning or referents if we were to exclude the world outside cyberspace. In brief, digital space and digitization are not exclusive conditions that stand outside the non-digital. Digital space is embedded in the larger societal, cultural, subjective, economic, imaginary structurations of lived experience and the systems within which we exist and operate.

For instance, producing capital mobility takes capital fixity: state-of-the-art built environments, well-housed talent, and conventional infrastructure—from highways to airports and railways. These are all partly place-bound conditions, even when the nature of their place-boundedness differs from what it may have been a hundred years ago when place-boundedness was far more likely to be a form of immobility. Today it is a place-boundedness that is inflected or inscribed by the hypermobility of some of its components, products, and outcomes. Both capital fixity and mobility are located in a temporal frame where speed is ascendant and consequential. This type of capital fixity cannot be fully captured through a description confined to its material and locational features, that is, through a topographical description (Sassen 2001).

Conceptualizing digitization along these lines allows us to recognize the ongoing importance of the material world even in the case of some of the most de-materialized digitized activities. This can be illustrated by the case of finance, one of the most digitized activities and one that involves a dematerialized instrument. Yet it cannot simply be thought of as exclusively digital. To have electronic financial markets and digitized financial instruments requires enormous amounts of *materiel*, not to mention human talent (which has its own type of physicality). This *materiel* includes conventional infrastructure, buildings, airports, and so on. Much of this is inflected by the digital insofar as it is a function of financial markets. And much of the digital composition of financial markets is inflected by the agendas that drive global finance.

Digitization brings with it an amplification of those capacities that make possible the liquefying of what is not liquid. Thereby, digitization raises the mobility of what we have customarily thought of as not mobile, or barely mobile. At its most extreme, this liquefying dematerializes its object. Once dematerialized, it gains hypermobility—instantaneous circulation through digital networks with global span. In my view, it is important to underline that the hypermobility gained by an object through dematerialization is but one moment of a more complex condition. Representing such an object as hypermobile is, then, a partial representation since it includes only some of the components of that object, that is, those that can be dematerialized. Much of what is liquefied, and circulates in digital networks and is marked by hypermobility, remains physical in some of its components. A great deal of my work on global cities (e.g. Sassen 2001) has been an effort to conceptualize and document the fact that the global digital economy requires massive concentrations of material conditions in order to be what it is. Finance is an important intermediary in this regard: it represents a capability for liquefying various forms of non-liquid wealth and for raising the mobility (i.e. hypermobility) of that which is already liquid.

The real estate industry further illustrates some of these issues. Financial services firms have invented instruments that liquefy real estate, thereby facilitating investment and circulation of these instruments in global markets. Yet, part of what constitutes real estate remains very physical. At the same time, however, that which remains physical has been transformed by the fact that it is represented by highly liquid instruments that can circulate in global markets. It may look the same, it may involve the same bricks and mortar, it may be new or old, but it is a transformed entity.

We have difficulty capturing this multivalence through our conventional categories: if it is physical, it *is* physical; and if it is liquid, it *is* liquid. In fact, the partial representation of real estate through liquid financial instruments produces a complex imbrication of the material and the de-materialized moments of that which we continue to call real estate. And so does the partial endogeneity of physical infrastructure in electronic financial markets.

Mediating practices

There are multiple ways of examining the interactions between the new digital technologies and their users. There is a strong tendency in the literature to conceptualize the matter of use as an unmediated event, as unproblematized activity. In contrast, a long-standing concern with what I have called 'analytic borderlands' has led me to try to detect the mediations in the act of using the technologies. In my research, I find that use is constructed or constituted in terms of specific cultures and practices, through and within which users articulate the experience/utility of electronic space. Thus, my concern here is not with the purely technical features of digital networks and what these might mean for users, nor is it simply with its impact on users. The concern is, rather, with this in-between zone that constructs the articulations of cyberspace and users.

This conceptualization clearly rests on the earlier proposition that electronic space is embedded, and not a purely technological event. Electronic space is therefore inflected by the values, cultures, power systems, and institutional orders within which it is embedded. If we were to explore these issues in terms of gendering, or specifically the condition of the female subject, we would then posit that—insofar as these various realms are marked by gendering—this embeddedness of cyberspace is also gendered at least in some of its components and, further, that so is cyberspace itself.[4] This is so even though there is enormous variability in this gendering by place, age, class, race, nationality, and issue orientation; at the same time, there are likely to be various situations, sites, and individuals not marked by gendering, or marked by hybrid or queered genderings.[5]

The second consequence of this embeddedness is that the articulations between cyberspace and individuals, whether as social, political, or economic actors, are constituted in terms of mediating cultures; it is not simply a question of access and understanding how to use the hardware and the software. To some extent, these mediating cultures are likely to be shaped by gendering as well as other marking conditions.

The destabilizing of older hierarchies of scale

The complex imbrications between the digital (as well as the global) and the non-digital bring a destabilizing of older hierarchies of scale, and often dramatic rescalings. As the national scale loses significance along with the loss of key components of the state's formal authority, other scales gain strategic importance. Most especially among these are subnational scales, such as the global city, and supranational scales such as global markets. Older hierarchies of scale, dating from the period that saw the ascendance of the nation state, continue to operate; they are typically organized in terms of institutional size and territorial scope: from the international, down to the national, the regional, the urban, to the local. But today's rescaling dynamics cut across institutional size and across the institutional encasements of territory produced by the formation of national states (Sassen 2000*b*). This does not mean that the old hierarchies disappear, but rather that rescalings emerge alongside the old ones which can often trump the latter.

Existing theory is not enough to map today's multiplication of non-state actors and forms of crossborder cooperation and conflict, such as global business networks, NGOs, diasporas, global cities, transboundary public spheres, and the new cosmopolitanism. International Relations theory is the field which to date has had the most to say about crossborder relations. But current developments associated with various mixes of globalization and the new information and communications technologies point to the limits of International Relations theory and data. Its models and theories remain focused on the logic of relations between states and the scale of the state, at a time when we see a proliferation of non-state actors, crossborder processes, and associated changes in the scope, exclusivity, and competence of state authority over its territory, all partly enabled by these new technologies. Theoretical developments in other disciplines may prove important; especially relevant in the case of sociology's contribution is the type of network theory developed in economic sociology.

These transformations in the components of international relations and the destabilization of older hierarchies of scale can be captured in a variety of instances. For example, much of what we might still experience as the 'local' (an office building or a house or an institution right there in our neighbourhood or downtown) actually is something I would rather think of as a micro-environment with global span, insofar as it is deeply inter-networked. Such a micro-environment is in many senses a localized entity, but it is also part of global digital networks which give it immediate far-flung span. To continue to think of this as simply local is not very useful. More importantly, the juxtaposition between the condition of being a sited materiality and having global span captures the imbrication of the digital and the non-digital. It also illustrates the inadequacy of a purely technological reading of the technical properties of digitization, which would lead us to posit the neutralization of the place-boundedness of that which precisely makes possible the condition of being an entity with global span.

A second example is the bundle of conditions and dynamics that marks the model of the global city. Just to single out one key dynamic: the more globalized and digitized the operations of firms and markets, the more their central management and coordination functions (and the requisite material structures) become strategic. It is precisely because of digitization that simultaneous worldwide dispersal of operations (whether factories, offices, or service outlets) and system integration can be achieved. And it is precisely this combination, that raises the importance of central functions. Global cities are, among other components, strategic sites for the combination of resources necessary for the production of these central functions. The crossborder network of global cities emerges as one of the key components in the architecture of 'international relations'.

New interactions between capital fixity and hypermobility

Information technologies have not eliminated the importance of massive concentrations of material resources but have, rather, reconfigured the interaction of capital fixity and hypermobility. The complex management of this interaction has given some cities a new competitive advantage at a time when the properties of the new ICTs could have been expected to eliminate the advantages of agglomeration, particularly for leading and globalized economic sectors. We have now come to understand that the vast new economic topography implemented through electronic space is one moment, one fragment, of an even vaster economic chain that is in good part embedded in non-electronic spaces. There is not yet a fully virtualized firm or economic sector. Even finance, the most digitized, dematerialized, and globalized of all activities has a topography that weaves back and forth between actual and digital space. To different extents, in different types of sectors and different types of firms, a firm's tasks are now distributed across these two kinds of spaces; further, the actual configurations are subject to considerable transformation as tasks are computerized or standardized, markets are further globalized, and so on.

To illustrate these issues, I focus on three particular aspects of the interaction of capital mobility and fixity: the ongoing importance of social connectivity and central functions for global digitized economic sectors; the variety of locational options available to firms in partly digitized economic sectors; and the multiplication of the possible spatial correlates of centrality made possible by the new ICTs (for a detailed presentation of the subject, see Sassen 2001).

The importance of social connectivity and central functions

While the new technologies do indeed facilitate geographic dispersal of economic activities without losing system integration, they have also had the effect of strengthening the importance of central coordination and control functions for firms and for markets. Major centres have massive concentrations of state-of-the-art resources that allow them to 'produce' the most strategic and complex of these central functions. It is not enough to have the ICT infrastructure. It also takes a mix of other resources: state-of-the-art material and human resources, and the social networks that maximize connectivity. Much of the added value these technologies can produce for advanced service firms and advanced markets represents, then, a new type of urbanization economy insofar as achieving this value-added potential depends on conditions external to the firms and markets themselves, and to the technologies as such.

This new type of urbanization economy allows firms to maximize the benefits they can derive from the new technologies and to maximize their capabilities for operating globally. Even electronic markets rely on traders and banks which are located somewhere; for instance, Frankfurt's electronic futures market is actually embedded in a global network of financial centres, each of which concentrates resources that are necessary for Frankfurt's market to thrive.

A second fact that is emerging with greater clarity concerns the meaning of 'information'. There are two types of information that matter to advanced services firms. One is the datum, which may be complex but is standardized and easily available to these firms (e.g. the details of a privatization in a particular country). The second type of information is far more difficult to obtain because it is not standardized. It requires interpretation, evaluation, judgement. It entails negotiating a series of data sets and a series of interpretations in the hope of producing a higher-order type of information. Access to the first kind of information is now global and immediate thanks to the digital revolution. But it is the second type of information that requires a complicated mixture of elements, not only technical but also social—what we could think of as the social infrastructure for global technical connectivity. It is this type of social infrastructure which gives major financial centres a strategic role. In principle, the technical infrastructure for connectivity can be reproduced anywhere, but not the social connectivity (e.g. Garcia 2002).

When the more complex forms of information needed to execute major international deals cannot be obtained from existing data bases, no matter what one can pay, then one needs the social information loop and the associated *de facto* interpretations and inferences that come with bouncing off information among talented, informed people. It is the importance for firms and markets of this complex type of 'information' that has given a whole new importance to credit rating agencies, for instance. Part of the rating has to do with interpreting and inferring. When this interpreting is 'authoritative' because it originates with an established entity for its 'production', it becomes 'information' for the rest of us.

The process of making inferences/interpretations into 'information' takes quite a mix of talents and resources. Risk management, for example, which has become crucial with globalization due to the growing complexity and uncertainty that comes with operating in multiple countries and markets, requires enormous fine tuning of central operations. We now know that many, if not most, major trading losses in the 1990s have involved human error or fraud. The quality of risk management will depend heavily on the top people in a firm rather than simply on technical conditions, such as electronic surveillance. Consolidating risk management operations in one site, usually a central one for the firm, is now seen generally as more effective. This has been the case for several major banks, including Chase and Morgan Stanley DeanWitter in the USA and Deutsche Bank and Credit Suisse in Europe.

In brief, urban centres provide the mix of resources and the social connectivity which allow a firm or market to maximize the benefits of its technical connectivity.

Locational patterns

Information technologies have had a sharp effect on the spatial organization of economic activity. But this effect is not uniform: the locational options of firms vary considerably. It is not simply a matter of reducing the weight of place. The scattered evidence for the last decade, which saw the widespread use of information technologies by firms in a broad range of sectors, allows us to identify three types of globally-operating firms in terms of their locational patterns.

First, firms with highly standardized products/services see an increase in their locational options insofar as they can maintain system integration no matter where they are located. This might also hold for firms with specialized products/services that do not require elaborate contracting and subcontracting or suppliers networks, all conditions which tend to make an urban location more efficient. Data entry and simple manufacturing work can be moved to wherever labour and other costs might be lowest. Headquarters can move out of large cities and to suburban locations or small towns.

A second locational pattern is that represented by firms deeply involved in the global economy and, hence, have increasingly complex headquarter functions. Perhaps ironically, the complexity of headquarters functions is such that they get outsourced to highly specialized service firms. This frees up the headquarter to locate anywhere, as long as they can access a highly specialized networked service sector somewhere, most likely in a city.

The third locational pattern is that evident in highly specialized networked service sectors. It is these sectors, rather than the headquarters, that benefit from spatial agglomeration at the point of production; in this regard, it is these firms rather than large corporate headquarters which are at the core of economic global city functions. These economic global city functions are to be distinguished from political global city functions, which might include the politics of contestation by formal and informal political actors enabled by these economic functions. This particular form of political global city function is, then, in a dialectical relation—both enabled and in opposition—to the economic functions (see Sassen 1998; Bartlett 2001).

Firms in these networked service sectors are embedded in intense transactions with other such firms in kindred specializations and are subject to time pressures and the constraints of imperfect information discussed in the preceding section. Along with some of the features contributing to agglomeration advantages in financial services firms, this has the effect of rendering the network of specialized service firms more place-bound than the hypermobility of their products and of their professionals would indicate.

The spatialities of the centre

The combination of the new capabilities for mobility along with the advantages of urbanization economies for leading globalized economic sectors suggests that spatial concentration remains a key feature even in a global digital economy. But it is not simply a continuation of older patterns of spatial concentration. There is no longer a simple straightforward relation between centrality and such geographic entities as the downtown, or the central business district (CBD). Up to quite recently, however, centrality was synonymous with the downtown or the CBD. The new technologies and organizational forms have altered the spatial correlates of centrality. Several of the organizing hypotheses in the global-city model concern the conditions for the continuity of centrality in advanced economic systems in the face of major new organizational forms and technologies that maximize the possibility for geographic dispersal (see Introduction to Sassen 2001 and, for a variety of other perspectives, Salomon 1996; Landrieu *et al.* 1998; Fainstein 2001; Orum and Chen 2002).

Given the differential impacts of the capabilities of the new information technologies on specific types of firms and of sectors of the economy, the spatial correlates of the 'centre' can assume several geographic forms, likely to be operating simultaneously at the macro level.

First, the centre can be the CBD—as it still is largely for some of the leading sectors, notably finance—or an alternative form of CBD, such as Silicon Valley. Yet, even as the CBD in major international business centres remains a strategic site for the leading industries, it is one profoundly reconfigured by technological and economic change (Graham and Aurigi 1997; Fainstein 2001; Cicollela and Mignaqui 2002). Further, there are often sharp differences in the patterns assumed by this reconfiguring of the central city in different parts of the world (e.g. Marcuse and van Kempen 2000).

Second, the centre can extend into a metropolitan area in the form of a grid of nodes of intense business activity. One might ask whether a spatial organization characterized by dense strategic nodes spread over a broader region does in fact constitute a new form of organizing the territory of the 'centre', rather than, as in the more conventional view, an instance of suburbanization or geographic dispersal. Insofar as these various nodes are articulated through digital networks, they represent a new geographic correlate of the most advanced type of 'centre' (Veltz 1996; Yeung 2000). This is a partly de-territorialized space of centrality.

This regional grid of nodes represents, in my analysis, a reconstitution of the concept of region. Further, it should not be confused with the suburbanization of economic activity. I conceive of it as a space of centrality partly located in older socio-economic geographies, such as that of the suburb or the larger metropolitan region, yet as being distinct precisely because it is a space of centrality. Far from neutralizing geography, the regional grid is likely to be embedded in conventional forms of communication infrastructure, notably rapid rail and highways connecting to airports. Ironically perhaps, conventional infrastructure is likely to maximize the economic benefits derived from telematics. I think this is an important issue that has been lost somewhat in discussions about the neutralization of geography through telematics (for exceptions to this trend, see Peraldi and Perrin 1996; Veltz 1996; Landrieu *et al.* 1998).

A third form of centre we are seeing is the formation of a transterritorial 'centre' constituted via intense economic transactions in the network of global cities. These transactions take place partly in digital space and partly through conventional transport and travel. The result is a multiplication of often highly-specialized circuits connecting sets of cities. These networks of major international business centres constitute new geographies of centrality. The most powerful of these at the global level bind the major international financial and business centres, including: New York, London, Tokyo, Paris, Frankfurt, Zurich, Amsterdam, Los Angeles, Sydney, and Hong Kong. But this geography now also includes cities such as Bangkok, Seoul, Taipei, São Paulo, and Mexico City (see Yeung 2000; Sassen 2001; Schiffer Ramos 2002). In the case of a complex landscape such as Europe's, we see in fact several geographies of centrality: one global, others continental and regional.[6]

Fourth, new forms of centrality are being constituted in electronically generated spaces. For instance, strategic components of the financial industry operate in such spaces.

These various trends point to a profound transformation, but it is not one characterized by the neutralization of capital fixity, or of the built environment, or, in the end, the city. Rather than being neutralized, these emerge with renewed and strategic importance in some of their features, that is to say, not as a generalized condition but as a very specific condition. It is a particular type of built environment, of conventional communication system, of city—in brief, a particular type of spatiality—that accommodates and furthers the new digital dynamics.

Mediating cultures: women's cyberpresence and cyberopportunities

The embeddedness of cyberspace and the larger social reflexivity this entails are evident in the facts about the presence of women in cyberspace, where they remain under-represented even as their Internet usage is growing sharply. This combination captures the contradictory features of women's conditions in the larger social world today. In addition, where the specificity of cyberspace enables the emergence of new cultures of interaction between cyberspace and the larger social order, there is also specificity in the opportunities and forms of presence of women (for a variety of viewpoints, see Grint and Gill 1995; Cherny and Weise 1996; Marcelle 1998; Bastani 2000; Holloway, Valentine, and Bingham 2000).

Aggregate-level data show clearly that, in country after country, women still account for less than half of all Internet usage but are rapidly raising their share, often at faster rates than men. By 2000, women were half of all people online in the USA, about 46–47 per cent in New Zealand and Sweden, and between 42 and 45 per cent in Singapore, Ireland, Australia, France, and Korea. These aggregate figures contain highly specific trends. In the United States, for instance, among first-time users in the year 2000, women exceeded men slightly, and usage increased by 125 per cent among girls twelve to seventeen years of age. Internet usage is also growing faster among women in a country as diverse from the USA as South Korea.

The specificity of cyberspace and the new cultures of interaction it entails are made evident by the presence of women in e-businesses started by women and in the proliferation of new women-oriented websites. Table 4.1 shows a sampling of women start-ups as of 2000; these are firms owned and operated by women.[7]

Table 4.1. *Selection of dotcom start-ups owned and operated by women, 2000*

Category	Website
Portal, content, and community ventures	Astronet
	AudioBasket
	eSampo
	Ivillage
	ThirdAge Media
Web-based service ventures	Desktop.com
	EDGAR Online
	E-Loan
E-commerce ventures	Della.com (renamed weddingchannel.com)
	oneNest
	Sparks.com
	SuperVerticals
E-business applications and web-technology ventures	CoVia
	eCommerce Industries
	Marimba
	RightWorks

Source: Carlassare (2001).

Clearly, given the dynamism in these sectors, we can expect turnover in ownership, not to mention cessation of operations. We included samples of firms in four distinct categories:

(*a*) portal, content and community ventures;
(*b*) web-based services;
(*c*) e-commerce; and
(*d*) e-business applications and web-technology ventures.

In addition, Table 4.2 lists women-oriented websites created for and/or run by women, and e-mail electronic forums ('listserves') oriented towards issues related to technology and women. Beyond the fact that the sites in Tables 4.1 and 4.2 are largely instrumental, they tell us a wider story about the Internet and women (e.g. Boneva, Kraut, and Frohlich 2001; Wellman and Haythornthwaite 2002). They become a collective refutation of a very common representation of cyberspace as reducing sociability and engagement with one's community. On the contrary, it can build local community. At the same time, being located in cyberspace makes it far more possible that the networks connecting each of these types of local sites might become transnational, probably an unplanned trajectory for many of them. In the next section, I return to this subject through a discussion of new types of women-activists networks.

The limits of electronic space in bringing about changes in existing hierarchies of power and privilege may also be inferred from the fact that existing cyber segmentations can trump women-oriented agendas. There is no doubt that cyberspace brings new opportunities for women both in business domains and in larger civic as well as home settings. For instance, in highly digitized sectors, women as professionals have experienced new opportunities and they may fight for greater equality with men in these economic sectors. But they do so largely within the confines of existing hierarchies of economic power. In this regard, it may be naïve to overestimate the emancipatory power of cyberspace in terms of its capacity to neutralize gender distinctions (e.g. Adam and Green 1998; Shapiro 1998; Hampton and Wellman 1999).

Of central importance for gauging the socio-political implications for women of their presence in, and use of, cyberspace is the potential transformation into micro-environments with global span of a whole range of 'local' conditions or institutional domains where women remain the key actors. Among these domains are the household, the community, the neighbourhood, the local school, and health care provider. What I mean by their transformation into micro-environments with global span is that technical connectivity will create a variety of links with other similar local entities in other neighbourhoods in the same city, in other cities, and in neighbourhoods and cities in other countries. A community of practice can emerge that creates multiple lateral, horizontal communications, collaborations, solidarities, supports. It can enable women (or feminine 'subjects' generally) to pursue projects not easily accommodated in their local, often limiting and oppressive, situation.

Table 4.2. *Selection of technology and women-oriented e-mail listserves*

Name	Function
AFRO-TECHIES	Discussion group for technically-inclined women of African descent seeking to expand the experience and knowledge of black women in technology
ASIA-WOMEN-IT	Discussion of issues and concerns relating to women in Asia and the new global information and communication technology.
A-WIA (American Women's Internet Association)	Organization and list for American women and their supporters 'actively involved within the Internet environment'.
CYBORG-L	Run by 'Women on the Net', a UNESCO-SID project to provide a multicultural gender perspective on international communication systems.
FACES	Cyber-resource and international mailing list for women interested in the media and communication arts to share projects, exhibits, critical opinions, and text.
FEMINAMAIL	Affiliated with women-oriented search engine Femina; weekly update to Femina database.
FEMINANET	List to ask for help finding online sites for personal, work-related, and academic interests.
GK97-GENDER	Connected to Global Knowledge 97 Conference; focuses on gender issues related to knowledge and information technologies.
GRANITE	Platform for discussion to stimulate research from a feminist/women's perspective of gender and new information technologies.
GRRLTALK	Discussion of GNU/Linux operating system, Open Source Software movement, Free Software Foundation, etc.
ISOC WOMEN-DISCUSS	Discussion of access to the Internet and information.
MAC-WOMEN	Help forum for users of Macintosh computers.
MAIDEN-L	For women new to the Internet who need help.

Table 4.2. *Continued*

Name	Function
NOWA.INTERNATIONAL	For women who provide computer training to women, dealing with gaining access to new technology, women-specific training, and a general networking medium.
SPIDERWOMAN	Forum for women who manage and/or design websites.
UHURA	Collaborative research project online, for women researching some aspect of the net.
VS-ONLINE STRAT	Forum for issues related to women's organizations' utilization of electronic communication and publishing technologies.
WEBWOMEN-CHAT	Non-technical list for women on the web, 'to keep the chatter away from the focused, technical lists'.
WEBWOMEN-GRAPHICS	Anything related to the creation and/or manipulation of computer graphics.
WEBWOMEN-HTML	For women web-content developers.
WEBWOMEN-TECH	For women who manage the technical aspects of websites.
WISDOM	Women's Internet Site Development and Online Mentoring for Australian women and others interested in Internet literacy.
WOMEN-L	Discussing women's issues with a focus on the Internet and technology.
WOMEN IN TECHNOLOGY	Both local and national coverage, through the list organization Tropica.
WOMEN OF KALI	Moderated list for discussions of feminist politics especially concerning misogyny in the media and on the net.
WOMEN SPACE	How women and women's organizations are using the Internet.

This brings with it a number of significant possibilities. Where previously women's engagement in these domestic or family-related institutions reproduced their isolation from larger public spheres and crossborder social initiatives, that engagement now can emerge as the anchor for participation (e.g. Henshall 2000 and Bastani 2000). Returning to the information in Tables 4.1 and 4.2, several of the websites are centred in female-typed domains, yet by being online open themselves to women from many other communities beyond their own physical neighbourhood or city and country. Secondly, in a context where globalization has opened up the world of international transactions to non-state actors of all sorts, women, especially through NGOs have gained a whole new ascendancy.[8] Where before women interested in international relations were typically confined to what was at the time a fairly invisible and hardly influential world of NGOs, today NGOs are emerging as key players and therewith are propelling women into situations they rarely had access to in the past. Cyberspace makes it possible for even small and resource-poor NGOs to connect with other such NGOs and engage in global social efforts. This is an enormous advance for women engaged in certain types of struggles, particularly those concerning women's issues, whether these are fought through women's organizations or through more general NGOs, such as human rights organizations (e.g. Cherny and Weise 1996; Adam and Green 1998; Espinoza 1999).

Conclusions: citizen networks in a global digital age

Digital networks are contributing to the production of counter-geographies of globalization. These can be constituted at multiple scales. Digital networks can be used by political activists for global or non-local transactions. But they can also be used for strengthening local communications and transactions inside a city. Highlighting how the new digital technology can serve to support local initiatives and alliances across a city's neighbourhoods is extremely important in an age where the notion of the local is often seen as losing ground to global dynamics and actors.

We can conceptualize these 'alternative' networks as counter-geographies of globalization because they are deeply imbricated with some of the major dynamics constitutive of globalization yet are not part of the formal apparatus or of the objectives of this apparatus, notably in relation to the formation of global markets, the intensifying of transnational and translocal networks, and the development of communication technologies which easily escape conventional surveillance practices. The strengthening and, in some of these cases, the formation of new global circuits are embedded or made possible by the existence of a global economic system and its associated development of various institutional supports for crossborder money flows and markets. These counter-geographies are dynamic and changing in their locational features. And they include a very broad range of activities, including a proliferation of criminal activities.

Through the Internet, local initiatives become part of a global network of activism without losing the focus on specific local struggles (e.g. Cleaver 1998; Ronfeldt *et al.* 1998; Espinoza 1999; Mele 1999). This enables a new type of crossborder political activism, one centred in multiple localities yet intensely connected digitally. Activists can develop networks for circulating not only information (about environmental, housing, political issues, etc.) but also political work and strategies. There are many examples of such a new type of crossborder political work. For instance

SPARC (Society for the Promotion of Area Resource Centres), started by and centred on women, began as an effort to organize slum dwellers in Bombay to get housing. Now it has a network of such groups throughout Asia, and some cities in Latin America and Africa. This is one of the key forms of critical politics that the Internet can make possible: a politics of the local with a big difference, as these are localities that are connected with each other across a region, a country or the world. However, because the network is global does not mean that it all has to happen at the global level.

The turn of the century marked a particular moment in the history of digital networks, one when powerful corporate actors and high-performance networks have been strengthening the role of private digital space and altering the structure of public-access digital space (Sassen 2000*a*). Digital space has emerged not simply as a means for communicating, but as a major new theatre for capital accumulation and the operations of global capital. Nevertheless, civil society—in all its various incarnations—is also an increasingly energetic presence in cyberspace (for a variety of viewpoints, see Frederick 1993; Poster 1997; Rimmer and Morris-Suzuki 1999; Miller and Slater 2000). The greater the diversity of cultures and groups, the better it is for this larger political and civic potential of the Internet and the more effective the resistance to the risk that the corporate world might set the standards.[9] This is made even more significant by the possibility of using the Internet for the non-elites to communicate transnationally at a time when a growing set of issues, such as human rights and the environment, are seen as escaping the bounds of nation states.

This is not the cosmopolitan route to the global. This is about the global as a multiplication of the local. These are types of sociability and struggle deeply embedded in people's actions and activities. They are also forms of institution-building work that can come from localities and networks of localities with limited resources, and from informal social actors. We see here the potential transformation of women, 'confined' to domestic roles, who can emerge as key actors in global networks without having to leave their work and roles in their communities. From being experienced as purely domestic, these 'domestic' settings are transformed into micro-environments located on global circuits. They do not have to become cosmopolitan in this process, they may well remain domestic in their orientation and remain engaged with their households and local community struggles, and yet they are participating in emergent global social circuits. A community of practice can emerge that creates the kind of multiple lateral, horizontal communications collaborations, solidarities, and supports identified in the previous section, which can enable local political or non-political actors to enter into crossborder politics.

The architecture of digital networks, primed to span the world, can actually serve to intensify transactions among residents of a city or region, it can serve to make them aware of neighbouring communities, gain an understanding of local issues that resonate positively or negatively with communities that are right there—in the same city—rather than with those that are at the other end of the world. Or it can serve to intensify transactions around the local issues of communities that are actually at opposite ends of the world. It is a peculiar mix of intense engagement with the local, with place, and an awareness of other 'local' engagements across the globe. In brief, social activists can use digital networks for global or non-local transactions, as well as using them to strengthen local communications and transactions inside a city or rural community.

Cyberspace is, perhaps ironically, a far more concrete space for social struggles than that of the national political system. It becomes a place where non-formal political actors can be part of the political scene in a way that is much more difficult in national institutional channels. Nationally, politics needs to run through existing formal systems: whether the electoral political system or the judiciary (taking state agencies to court). Non-formal political actors are rendered invisible in the space of national politics.

Cyberspace can accommodate a broad range of social struggles and facilitate the emergence of new types of political subjects that do not have to go through the formal political system.[10] Much of this becomes visible on the street, where a great deal of urban politics is concrete—enacted by people rather than dependent on massive media technologies. Street-level politics makes possible the formation of new types of political subjects that do not have to go through the formal political system in order to practice their politics. Individuals and groups that have historically been excluded from formal political systems, and whose struggles can be partly enacted outside those systems, can find in cyberspace an enabling environment both for their emergence as non-formal political actors and for their struggles.

References

Adam, A., and Green, E. (1998). 'Gender, Agency, Location and the New Information Society', in B. Loader (ed.), *The Cyberspace Divide*. London: Routledge, 83–97.

Bartlett, A. (2001). 'Politics Remade: Modernization and the New Political Culture in England'. Unpublished thesis. Chicago: Department of Sociology, University of Chicago.

Bastani, S. (2000). 'Muslim Women On-line'. *Arab World Geographer*, 3/1: 40–59.

Boneva B., Kraut, R., and Frohlich, D. (2001). 'Using E-mail for Personal Relationships: The Difference Gender Makes'. *American Behavioral Scientist*, 45/3: 530–49.

Calabrese, A., and Burgelman, J.-C. (1999). *Communication, Citizenship and Social Policy: Re-thinking the Limits of the Welfare State*. Lanham, MD: Rowman and Littlefield.

Carlassare, E. (2001). *Dotcom Divas: E-business Insights from the Visionary Women Founders of 20 Net Ventures*. New York: McGraw-Hill.

Cherny, L., and Weise R. (1996) (eds.). *Wired Women: Gender and New Realities in Cyberspace*. Seattle, WA: Seal Press.

Cicollela, P., and Mignaqui, I. (2002). 'The Spatial Reorganization of Buenos Aires', in Sassen, S. (ed.), *Global Networks/Linked Cities*. New York and London: Routledge.

Cleaver, H. (1998). 'The Zapatista Effect: The Internet and the Rise of an Alternative Political Fabric'. *Journal of International Affairs*, 51/2: 621–40.

Correll, S. (1995). 'The Ethnography of an Electronic Bar: The Lesbian Cafe'. *Journal of Contemporary Ethnography*, 24: 270–98.

Darkwa, O., and Mazibuko, F. (2000). 'Creating Virtual Learning Communities in Africa: Challenges and Prospects'. *First Monday* 5.5. http://firstmonday.org/issues5_5/darkwa/index.html

Elkin-Koren, N. (1996). 'Public/Private and Copyright Reform in Cyberspace'. *Journal of Computer Mediated Communication* 2/2. www.ascusc.org/jcmc/vol2/issue2/elkin.html

Espinoza, V. (1999). 'Social Networks Among the Poor: Inequality and Integration in a Latin American City', in B. Wellman (ed.), *Networks in the Global Village*. Boulder, CO: Westview Press, 147–84.

Fainstein, S. (2001). *The City Builders*. Lawrence, KS: Kansas University Press.

Featherstone, M., and Burrows, R. (1995). *Cyberspace/Cyberbodies/Cyberpunk: Cultures of Technological Embodiment*. London: Sage.

Frederick, H. (1993). 'Computer Networks and the Emergence of Global Civil Society', in L. M. Harasim (ed.), *Global Networks: Computers and International Communications*. Cambridge, MA: MIT Press, 283–95.

Garcia, L. (2002). 'The Architecture of Global Networking Technologies', in S. Sassen (ed.), *Global Networks/Linked Cities*. New York and London: Routledge, 39–69.

Graham, S., and Aurigi, A. (1997). 'Virtual Cities, Social Polarization, and the Crisis in Urban Public Space'. *Journal of Urban Technology*, 4/1: 19–52.

Grint, K., and Gill, R. (1995). *The Gender–Technology Relation: Contemporary Theory and Research*. London: Taylor and Francis.

Hampton, K., and Wellman, B. (1999). 'Netville Online and Offline'. *American Behavioral Scientist*, 43/3: 475–92.

Hargittai, E. (1998). 'Holes in the Net: The Internet and International Stratification'. Paper presented at INET '98 Conference: The Internet Summit. Geneva, 21–4 July. www.isoc.org/inet98/proceedings

Harvey, A. S., and Macnab, P. A. (2000). 'Who's Up? Global Interpersonal Temporal Accessibility', in D. Janelle and D. Hodge (eds.), *Information, Place and Cyberspace: Issues in Accessibility*. Berlin: Springer, 147–70.

Harvey, R. M. (2001). 'Global Cities of Gold'. Dissertation Research. Chicago: Department of Sociology, University of Chicago.

Henshall, S. (2000). 'The COMsumer Manifesto: Empowering Communities of Consumers through the Internet'. *First Monday* 5.5. http://firstmonday.org/issues5_5/henshall/index.html

Hoffman, D. L., and Novak, T. P. (1998). 'Bridging the Racial Divide on the Internet'. *Science*, 280, 17 April: 390–1.

Holloway, S. L., Valentine, G., and Bingham, N. (2000). 'Institutionalising Technologies: Masculinities, Femininities and the Heterosexual Economy of the IT Classroom'. *Environment and Planning A*, 32/4: 617–33.

ITU (1998), International Telecommunication Union. *Challenges to the Network: Internet for Development*. Geneva: ITU.

Jensen, M. (1998). *Internet Connectivity in Africa*, January. http://demiurge.wn.apc.org/africa

Knop, K. (1993). 'Re/Statements: Feminism and State Sovereignty in International Law'. *Transnational Law and Contemporary Problems*, 3/2: 293–344.

Kobrin, S. J. (1998). 'The MAI and the Clash of Globalizations'. *Foreign Policy*, 112/Fall: 97–109.

Landrieu, J., May, N., Spector, T., and Veltz, P. (1998) (ed.). *La Ville Eclatee*. La Tour d'Aigues, France: Editiones de l'Aube.

Latour, B. (1991). 'Technology is Society Made Durable', in J. Law (ed.), *A Sociology of Monsters*. London: Routledge, 103–31.

Loader, B. (1998) (ed.). *Cyberspace Divide: Equality, Agency and Policy in the Information Age*. London: Routledge.

Lovink, G., and Riemens, P. (2002). 'Digital City Amsterdam: Local Uses of Global Networks', in S. Sassen (ed.), *Global Networks/Linked Cities*. New York and London: Routledge.

MacKenzie, D. (1999). 'Technological Determinism', in W. H. Dutton (ed.), *Society on the Line: Information Politics in the Digital Age*. Oxford: Oxford University Press, 39–41.

—— and Wajcman, J. (1999). *The Social Shaping of Technology*. Milton Keynes, UK: Open University Press.

Marcelle, G. M. (1998). 'Strategies for Including a Gender Perspective in African Information and Communications Technologies (ICTs) Policy'. Paper presented at ECA International Conference on African Women and Economic Development, Addis Ababa, April.

Marcuse, P., and van Kempen, R. (2000). *Globalizing Cities. A New Spatial Order*. Oxford: Blackwell.

Mele, C. (1999). 'Cyberspace and Disadvantaged Communities: The Internet as a Tool for Collective Action', in M. A. Smith and P. Kollock (eds.), *Communities in Cyberspace*. London: Routledge, 264–89.

Miller, D., and Slater, D. (2000). *The Internet: An Ethnographic Approach*. Oxford: Berg.

Munker, S., and Roesler, A. (1997) (eds.). *Mythos Internet*. Frankfurt: Suhrkamp.

Nadeau, J., Lointier, C., Morin, R., and Descoteaux, M. A. (1998). 'Information Highways and the Francophone World: Current Situation and Strategies for the Future'. Paper presented at INET '98 Conference: The Internet Summit. Geneva, 21–4 July. www.isoc.org/inet98/proceedings

Nettime (1997). *Net Critique*, compiled by G. Lovink and P. Schutz. Berlin: Edition ID-ARchiv.

NTIA (1998), National Telecommunications and Information Administration. *Falling Through the Net II: New Data on the Digital Divide*. Washington, DC: NTIA, July. www.ntia.doc.gov/ntiahome/net2/falling.html

Ong, A. (1996). 'Globalization and Women's Rights: The Asian Debate on Citizenship and Communitarianism'. *Indiana Journal of Global Legal Studies*, 4/1.

Orum, A., and Chen, X. (2002). *Urban Places*. Malden, MA: Blackwell.

Peraldi, M., and Perrin, E. (1996) (eds.). *Reseaux Productifs et Territoires Urbains*. Toulouse: Presses Universitaires du Mirail.

Petrazzini, B., and Kibati, M. (1999). 'The Internet in Developing Countries'. *Communications of the ACM*, 42/6: 31–6.

Poster, M. (1997). 'Cyberdemocracy: Internet and the Public Sphere', in D. Porter (ed.), *Internet Culture*. London: Routledge, 201–18.

Reaume, D. G. (1992). 'The Social Construction of Women and the Possibility of Change: Unmodified Feminism Revisited'. *Canadian Journal of Women and Law*, 5/2: 463–83.

Rimmer, P. J., and Morris-Suzuki, T. (1999). 'The Japanese Internet: Visionaries and Virtual Democracy'. *Environment and Planning A*, 31/7: 1189–206.

Ronfeldt, D., Arquilla, J., Fuller, G., and Fuller, M. (1998). *The Zapatista 'Social Netwar' in Mexico*, MR-994-A. Santa Monica, CA: RAND.

Salomon, I. (1996). 'Telecommunications, Cities and Technological Opportunism'. *The Annals of Regional Science*, 30: 75–90.

Sassen, S. (1998). *Globalization and its Discontents*. New York: New Press.

—— (1999) 'Digital Networks and Power', in M. Featherstone and S. Lash (eds.), *Spaces of Culture: City, Nation, World*. London: Sage, 49–63.

—— (2000*a*). 'Digital Networks and the State: Some Governance Questions'. *Theory Culture & Society*, Special Section on Globalization and Sovereignty. 17/4: 19–33.

—— (2000*b*). 'Territory and Territoriality in the Global Economy'. *International Sociology*, 15/2: 372–93.

—— (2001). *The Global City: New York, London, Tokyo*. Princeton, NJ: Princeton University Press.

—— (2002) (ed.). *Global Networks/Linked Cities*. London: Routledge.

Schiffer Ramos, S. (2002). 'Sao Paulo: Articulating a Cross-border Regional Economy', in Sassen (ed.), *Global Networks/Linked Cities*. London: Routledge, 205–28.

Shade, L. R. (1998). 'A Gendered Perspective on Access to the Information Infrastructure'. *The Information Society*, 14: 33–44.

Shapiro, S. (1998). 'Places and Spaces: The Historical Interaction of Technology, Home, and Privacy'. *The Information Society*, 14: 275–84.

Thomas, R. (1995). 'Access and Inequality', in N. Heap, R. Thomas, G. Einon, R. Mason, and H. MacKay (eds.), *Information Technology and Society: A Reader*. Buckingham, UK: Open University Press, 90–9.

Veltz, P. (1996). *Mondialisation Villes et Territoires: L'Economie d'Archipel*. Paris: Presses Universitaires de France.

Wajcman, J. (1991). *Feminism Confronts Technology*. Cambridge: Polity Press.

—— (2002). 'Addressing Technological Change: The Challenge to Social Theory'. *Current Sociology*, 50/3: 347–63.

Warf, B., and Grimes, J. (1997). 'Counterhegemonic Discourses and the Internet'. *The Geographical Review*, 87: 259–74.

Wellman, B., and Haythornthwaite, C. (2002) (eds.), *The Internet in Everyday Life*. Oxford: Blackwell.

Yeung, Y. (2000). *Globalization and Networked Societies*. Honolulu, HI: University of Hawaii Press.

Notes

1. This literature is discussed in Wajcman (2002). For critical examinations that reveal particular shortcomings of technology-driven explanations, see e.g. Nettime 1997; Hargittai 1998; Loader 1998 and, more generally, Latour 1991; Munker and Roesler 1997; MacKenzie 1999; MacKenzie and Wajcman 1999.
2. Using a different vocabulary, we can see Latour (1991) making a radical statement in this direction. Lovink and Riemens (2002) give us a detailed account of the multiple non-digital conditions (including neighbourhood sub-cultures) that had to come together in order to create the enormously successful citywide digital inter-network called Digital City Amsterdam.
3. In Sassen (1999), I have examined the extent to which our thinking about electronic space and network power has been shaped by the properties of the Internet, disregarding the crucial differences between the public-access digital networks of the Internet and private digital networks to which there is no access, no matter what one is willing to pay.

4. Much of what has been described for cyberspace in the specialized and general literature is explicitly or implicitly far more likely to be about particular groups of men because these have thus far dominated usage and produced many of the cybercultures (e.g. Holloway, Valentine, and Bingham 2000). Thus, we also need more information about men who do not fit those particular groups.
5. The concept of gendering has increasingly become problematic and is used here as shorthand for a complex bundle of issues. There is a vast critical literature on various aspects relating to gendering and feminist categories (e.g. for a broad range of issues see Wajcman 1991; Reaume 1992; Featherstone and Burrows 1995; Ong 1996). The notion of queering gender is, in this context, a powerful re-positioning.
6. Methodologically, I find it useful to unpack these inter-city transactions into the specific, often highly specialized circuits that connect particular sets of cities. For instance, when examining futures markets, the set of cities includes Sao Paulo and Kuala Lumpur. These two cities fall out of the picture when examining the gold market; this market, on the other hand, includes Johannesburg and Sydney (e.g. Harvey 2001).
7. The listed websites for each of these women start-ups provide more detailed information.
8. For a more theorized account of these issues, see Sassen (1998) and Knop (1993).
9. The Internet may continue to be a space for democratic practices, but it will be so partly as a form of resistance against overarching powers of the economy and of hierarchical power (e.g. Warf and Grimes 1997; Calabrese and Burgelman 1999), rather than the space of unlimited freedom which is part of its romantic representation. The images we need to bring into this representation increasingly need to deal with contestation and resistance, rather than simply freedom and interconnectivity.
10. I have made a parallel argument for the city, especially the global city, being a more concrete space for politics. In many ways, the politics of *reivindicación* (restoration of one's rights or honour) being enacted in cyberspace resonates with many of the activisms evident in large cities today: struggles against police brutality and gentrification, struggles for the rights of the homeless and immigrants, and struggles for the rights of gays, lesbians and queers.

PART II

Theories at work

5

Knowledge as infrastructure

OLE HANSETH

Introduction

ICT solutions are often designed and adopted in order to support, facilitate, and rationalize ongoing activities in the way they are already being carried out. In other cases, the new technology is expected to enable ways of working that are radically different and more efficient than existing ways. For instance, the Internet has been seen as opening up—or even causing—such radical changes. Predictions have been made that the increasingly wider adoption of the Internet will bring about paradigm changes in our technological solutions (from local systems to global networks), as well as in organizing business activities (from closed hierarchies to open, flat, global networks). These paradigm changes have often been described as a transformation from the 'old' to the 'new' economy.

To a large extent, such a transformation process can be seen as a learning process because of the need for the construction and adoption of a lot of new knowledge. For example, new knowledge is required about the design of better business processes in various business sectors and organizations, together with the creation of ICT solutions and the processes, strategies, and methodologies for developing such solutions. The ICT solutions to make such visions a reality will certainly be complex, and so will be the learning processes.

The motivation behind this chapter is to enquire into these processes. In it, I argue that the complexity and systemic character of the knowledge constituting a paradigm makes changes from an old to a new one very challenging. These challenges may often be beyond what we are able to control; in many cases, we may be trapped into one paradigm simply because it makes us unable to generate the knowledge needed to constitute the new one.

Knowledge has many aspects, and different communities interested in issues related to organizational learning, innovation, or knowledge management focus on different aspects.[1] The simplest perspective of knowledge is found among members of Artificial Intelligence and Knowledge Management communities, who see knowledge as being built up of different elements that we can store in our brain or our computer. Others, still seeing knowledge as a cognitive material, describe this kind of knowledge as 'explicit' and contrast it with 'implicit' or 'tacit' knowledge. Yet others, adopting a more phenomenological perspective, see knowledge

I would like to thank Chrisanthi Avgerou, Claudio Ciborra, Bo Dahlbom, Jannis Kallinikos, Eric Monteiro, and Sundeep Sahay for their helpful comments on previous versions of this chapter.

as deeply embodied and embedded into our body and our practices. Further, knowledge may be seen as embedded into institutions and material structures like houses and information systems. Knowledge is also shared by members of communities, such as communities-of-practice or other social units.

The perspective on knowledge adopted in this chapter assumes that knowledge has all these characteristics. But one more aspect is added: individual pieces of knowledge are not independent of each other. Knowledge is highly systemic. Individual pieces are linked together into complex structures in various ways. These structural characteristics play important roles in knowledge processes, such as knowledge construction, diffusion, adoption—or whatever we call them. It is these systemic aspects of knowledge that will be explored in this chapter, their nature as well as their implications. Using Actor Network Theory (ANT), I will describe how knowledge can be thought of—at least partly—as a 'non-human actor' that may be rather autonomous and powerful, playing important roles in the emerging knowledge society.

Basic knowledge concepts

Knowledge as a network

I will address the systemic/complexity aspects of knowledge by looking at it as a 'network', including as an actor network. This is underpinned by two fundamental assumptions:

(*a*) individual 'pieces' of knowledge are related and interdependent; and
(*b*) different individuals adopt the same 'piece' of knowledge, and that piece of knowledge is embedded into routines and practices; accordingly, these individuals, routines, and practices are linked together and become interdependent.

I will try to illustrate this by taking ICT as a paradigm example of a network in general. And the paradigm example of ICT these days is the Internet, a network made up of a huge number of computers linked together. Standards are crucial in this kind of network. For example, the computers on the Internet are linked together by means of a number of standardized protocols and formats, using the same standards implemented on all computers on the Internet around the world. We can call these kinds of compatibility standards 'horizontal', to distinguish them from what we may call 'vertical' standards that are crucial in the construction of other kinds of technologies. In computers, we find that such standards related to software modules (file system, operating system, databases, etc.) support each other, often according to a layered model. The interface between operating systems and the applications running on top of them is a paradigm example. Standardized interfaces are important in order to make it as easy as possible to let each specific module support many different other ones.

We can find the same relationships, interdependencies, and kinds of standards in other areas, such as medical knowledge. Modern hospitals have medical doctors whose knowledge belongs to a wide range of different specialities. They collaborate with each other in many different ways, following procedures and practices for diagnosing and treating patients. In addition, the pieces of knowledge upon which such collaboration is based requires a sort of compatibility and 'standardized interface' between the practices. There is also extensive and increasing collaboration and communication between hospitals, mostly—but not exclusively—among members of the same speciality. For instance, doctors ask each other for advice, send patients from one hospital to another, come together to discuss each others' patients, share data for research purposes, compare data quality assurance purposes, and so on. To achieve this objective, the sharing of standardized practices and knowledge is crucial.

Considering knowledge as a network also implies that theories of 'network economics', also known as 'information economics', should apply. Key concepts in these disciplines are: network externalities, increasing returns, path-dependence, and lock-in (Shapiro and Varian 1999). In the following subsections, I briefly explain these ideas and illustrate their relevance for networks of computers, as well as networks of knowledge.

Network externalities. The term 'network externalities' denotes the fact that an economic transaction may have effects for a network of actors external to the transaction itself: those actors not involved in the transaction as such. An example of such a transaction is when someone buys an e-mail system and access to an e-mail service provider. This transaction also has effects for those already having access to this specific e-mail service, as there is one more user with whom they can communicate.

The value of an e-mail system is not primarily linked to the functions it provides its users, for example to maintain a local archive of messages, but rather to the number of people you can communicate with using the system. This means that when a user adopts an e-mail system implementing a specific standard and links it to a network based on that standard, the value of the e-mail systems of the existing users, and of the whole network, increases due to the fact that there is one more user with whom they can communicate. The same holds for knowledge. If a medical doctor adopts some new knowledge, for instance in terms of a new procedure, those following the same—or other compatible and interdependent procedures—will find it easier to collaborate and communicate with that doctor; accordingly, the value of this knowledge has increased.

Increasing returns. Network externalities can also leads to 'increasing returns': one user's adoption of a specific technology increases the value of that technology for those already having adopted it, so the value of the technology as such increases as more users adopt it (Arthur 1994; Shapiro and Varian 1999). When this kind of positive externality leads to increasing returns, standards are important because the value of a specific technology depends on its number of users. In addition, the technology will have its highest value when all potential users adopt a version of this technology following the same standard (e.g. the TCP/IP protocols for the

Internet; Microsoft Windows operating system; VHS for video recording; 220 volt electricity connections; GSM Global System for Mobile cellular communications; and traditional 'QWERTY' typewriter-like keyboard).

Thus, if network externalities and increasing returns are seen to apply to knowledge, then the standardization of knowledge becomes beneficial although—just as in the case of technologies—standardizing knowledge will, of course, also have its drawbacks. We can say that the value of a standard is primarily determined by externalities. A standard cannot be used for anything in itself, as it enables interaction only with other adopters of the same standard. Accordingly, the value of a standard is the number of adopters in its installed base.

Path dependencies and lock-ins. Network externalities also lead to path-dependent processes, high switching costs, and lock-in situations. The notion of path-dependence applies to situations where two or more standards are competing. If the installed base of standard A is becoming larger than that of standard B, standard A's value increases compared to standard B, which will make more users prefer standard A, thereby increasing its relative advantages even more, and so on. When the installed base of one standard has increased significantly beyond that of another, a state of irreversibility is often created. This may cause 'tipping' (Akerlof 1970) into a situation where 'the winner takes it all', which again may lead to lock-in. A common example of this (including for the standards mentioned above) is where one standard becomes so dominant that we are not able to switch to a technically improved alternative, even in situations where the existing standard is definitely inferior. This happens because the 'use value' of the new alternative is zero as long as nobody has adopted it. As everybody then waits for others to adopt it, nothing happens. This indicates how we are normally not locked-in by standards in strict technical terms, but because switching to another standard may involve high costs.

Even though there are these kinds of lock-in situations, standards do change—and path dependence is also involved in the (evolutionary) change of standards. However, it is precisely because of lock-ins that we are restricted in the degree to which we may change standards. To make a standard 'adoptable', it has to enable a significant compatibility with the existing installed base. This backward-compatibility issue is well known from a wide range of technologies and products; Microsoft Windows is an example with which most of us are familiar.

To the extent that knowledge can be considered as a network, knowledge creation and diffusion are also subject to path-dependent processes and lock-in situations. In the case of medical knowledge and practices, for instance, knowledge and practices that are not compatible with existing ones are of low value to potential adopters. Compatible ones will then be preferred.

Knowledge as infrastructure

Having described knowledge as a network, I will now move a step further and consider it as infrastructure. An infrastructure is a standardized network as described above. But, as encountered in our everyday life, it has some additional features. For instance, infrastructures are shared resources for supporting a wide range of activities for a community. Further, infrastructures are often experienced as big and heavy (and accordingly hard to change), invisible (often underneath the surface of the earth), and stable. Even though knowledge may not have the same physical embodiments, it can also be said to have similar features. Accordingly, I see the notion of infrastructure as richer and more appropriate than the more abstract concept of network. In particular, I find the notion of infrastructure appropriate as a way of underscoring the systemic character of knowledge, and the fact that this also makes knowledge big, heavy, and rigid—and not light and flexible as is often wrongly assumed.

Related perspectives

The perspective on knowledge presented here is also found in other research that focuses on knowledge. One such field is innovation theory and related disciplines, which has found that innovation occurs most often in areas where many different companies take part in the same, or related, activities. This is illustrated by phenomena called 'clusters' (Porter 1998) or 'industrial districts', such a Silicon Valley or the Bologna region (e.g. Castells 1985; Piore and Sabel 1986).

The existence and important role of such phenomena is largely explained by the concept of 'knowledge spillovers'. A spillover is simply a form of network externality, created by the need to generate new knowledge for the development of a new product. In most, or perhaps all cases, this knowledge will be partly of a more general character, as it can be used for the development of other products as well. When people move between companies in a region, or through other knowledge sharing/dissemination activities, these spillovers create a shared stock of knowledge. Because of the importance of the collective accumulation of knowledge as a shared resource by a community, innovation theorists talk about knowledge as infrastructure (Steinmueller 1996). In addition, the existence of such spillovers is used by economists to explain more general phenomena, like the existence of cities (Castells 1985).

This perspective on knowledge is also in line with a well-known concept from the theory of science: Thomas Kuhn's (1970 [1962]) 'scientific paradigms'. That is based on a view of knowledge as a heterogeneous object containing the 'pure' knowledge in itself, the methods and practices shared by the members of a paradigm, and the tools and instruments they are using—as well as the institutions (journals, departments, etc.) connected to the production and dissemination of the specific knowledge. Kuhn describes normal scientific work ('normal science') as the extension of the knowledge base of a paradigm: the production of new knowledge that is correct in terms of what is considered as being known within the paradigm. As a result, normal science is a knowledge-production process maintaining backward compatibility. This is important in Kuhn's model of science because solving the problems, or understanding the phenomena motivating the scientists, require ever larger amounts of knowledge. It is this that usually underlies the arguments that research within a field needs to be cumulative.

The existence of network externalities also helps us to explain why scientific revolutions—the changes of paradigms—are so problematic and rare. This is because paradigms are incompatible, or incommensurable in Kuhn's terms. Even though the basic theoretical notions of a paradigm are 'proved' not to be true, major parts of the whole network of knowledge may still be useful for many purposes, as illustrated by the change from a geo-centric to helio-centric world view. Accepting a new paradigm means invalidating all the knowledge belonging to the old paradigm while the new one still is 'empty' and valueless: there are no researchers there to collaborate with, no journals in which to publish, no conferences to attend, no useful knowledge base to which you can contribute. Nobody will join the new paradigm until it already has a significant number of other members—the 'critical mass'. Everyone stays in the old paradigm because all others do. We are locked-in.

Changing paradigms, infrastructures, and standards

Despite all the constraints outlined in the previous section, paradigm changes do occur. And they do so in ways that are strikingly similar to the transformation of infrastructures, or of standards in use. Such infrastructure changes can be split into three categories:

(*a*) Extensions: where the existing infrastructure may be extended while maintaining full backward compatibility, which can happen by either adding new nodes to the network (e.g. new computers to the Internet) or new functionality (e.g. new infrastructure layers on top of existing ones).
(*b*) Moderate: replacing one standard with another by means of gateways translating between the standards.
(*c*) Radical: replacing the old standard with a radically new and incompatible one.

In the radical case, an infrastructure usually starts to grow to solve specific problems for a smaller group. As the technology is improved and matures, it becomes more useful for other groups, which again generates new ideas about how the technology may support other activities; in turn, this generates new user requirements that again lead to further improvement of the technology. And as the infrastructure grows in terms of users, it becomes increasingly attractive for more and more users.

This is the way the Internet has grown and evolved since it was initiated as ARPANET in the late 1960s by the US military's Advanced Research Projects Agency (ARPA). In its early days, this network was not an infrastructure in the rich way it is today. It was a simple technology connecting a few users, primarily for experimental applications. But through this experimental use, new ways of employing the technology were discovered, and new users were attracted. Some of the new ways of using the Internet required new technological features and functions. This is how the Internet continued to grow, taking over more and more of the role and function played by other communication infrastructures, such as telecommunications and postal services. The technology was changing according to the 'extensions' model mentioned above, with the emerging infrastructure being extended both in terms of computers and links between them, as well as in terms of technological functions. Gradually, it was turned into an infrastructure supporting a wide range of shared or collaborative activities for an increasingly growing user community across the world.

The Internet's functionality has been improved by maintaining 100 percent backward compatibility: adding new services—such as e-mail, file transfer, and the web—onto existing ones, without changing what was previously offered. At some points, however, the Internet arrived at states where parts of the existing technology have become inferior, and replacement with improved alternatives has been required.[2] Such developments follow the model for moderate change mentioned above.

Examples of changes in paradigms

The similarity between changes in scientific paradigms and technological infrastructures can be illustrated through two examples: the transformation in cancer research that came to view the cause of cancer as genetic diseases or injuries rather than viruses; and the role of Pasteur's discovery of microbes in the establishment of modern bacteriological medicine.

Fujimura's (1996) presentation of the cancer 'viruses to genes' story identifies 'standardized packages' as the key concept for understanding the transformation, because the successful design of such a package is the main factor explaining how this transformation was made possible. The term 'standard packages' points to the two crucial aspects: it was a package that contained a number of different interdependent elements, and both the individual elements and the package as a whole were standardized.

The elements in the 'genetics for cancer researchers' package were a set of techniques, tools, and instruments required to do this kind of research, which included a standardized breed of mouse (a real biological one—trademarked OncoMouse). Further, the package also included the genetic knowledge required to use the tools and techniques correctly. The standardization of such tools, instruments, techniques, materials (including the mouse), and knowledge was a precondition for the comparison and accumulation of results and knowledge among cancer researchers, which are also part of the requirements needed to make a paradigm work. For such a package to be effective, more requirements than just being accepted as a standard have to be satisfied: the package also has to 'scale', in that the tools, instruments, and materials have to be low cost and the knowledge required should be easy to learn. Those involved succeeded in designing the package so that these requirements were fulfilled.

This cancer research transformation process from viruses to genes was indeed a radical change, so should follow the model for such changes mentioned above. And it did. Genetics research has been a part of biology for a long period. Gradually, some of the molecular biologists started looking into issues related to cancer, which led to the proposal of the hypothesis that cancer is caused by our genes. Starting in biology and moving towards cancer, a bandwagon was created—just as for infrastructures.

Pasteur's discovery (or construction) of microbes is usually considered the development of a totally new paradigm, completely inconsistent with the incorrect assumptions, beliefs (or theories) held at the time.[3] However, Latour (1998) shows how the construction of the new paradigm happened as a transformation process made up of a long series of steps, each maintaining a strong degree of backward compatibility. Latour argues that this transformation could actually happen only as such a piecemeal, evolutionary process. Two elements are central in this process. First, Pasteur's new ideas about microbes could become accepted only if they could help to explain the problems and challenges that the members (the so-called hygienists) of the old paradigm were concerned about. Second, Pasteur's ideas could be developed only on the basis of other bodies of knowledge, including the knowledge it aimed at invalidating.

Implications of seeing knowledge as network and infrastructure

If the assumptions about the network character of knowledge are valid, we would expect learning to follow a path-dependent process, as argued by Arthur (1994). Radical change would then be limited, because the knowledge required by new and radically-different practices may be incompatible with the existing stock of knowledge. The new knowledge would be valueless until there has been an accumulation of a significantly huge stock—a critical mass—of additional knowledge that is compatible with it. This accumulation cannot start if there is not already a network with which the new knowledge is compatible. This hypothesis is in line with Kuhn's outline of incommensurable scientific paradigms, and the difficulties in changing from one to another.

I suggest that it is appropriate to consider knowledge as being highly embedded in a complex web of relationships and dependencies, where it is difficult to change one part without the other. Material practices, institutions, taken-for-granted notions, communities, etc. contribute to the kinds of knowledge produced, and the ways in which it is produced and used. Knowledge could be said to be related to a materially heterogeneous 'installed base' that has a certain transparency, a taken-for-grantedness, which works as a fundamental, shared resource for the community relating to it. Such a view of knowledge provides striking similarities with common definitions of information infrastructures (Star and Ruhleder 1996; Hanseth 2000). This perspective could suggest a sense of the complexity of knowledge, its entanglement and intertwining with many other entities within a large and heterogeneous network. Acknowledging this would have implications for how we view knowledge, learning, and innovation.

The heterogeneity of reality

So far, I have portrayed knowledge in its purest form, isolated and independent from other parts and aspects of reality. Of course, this is not how things are. By adopting the perspective of communities-of-practice, we can see that knowledge is inseparable from our working (and other) practices; more strongly, we can see that knowledge *is* practice. Further, practices are based on specific tools and technologies, organizing structures, various kinds of institutions, and other important factors. Knowledge is also dependent upon, and tied to, the context within which its epistemic culture and machinery is produced, in the terms formulated by Knorr-Cetina (1999). Most of these 'layers' of reality often constitute large networks which have themselves obtained infrastructural characteristics.

Working practices in large organizations, such as hospitals, are very complex. In order to make a hospital work as a whole, each task within it needs to fit with most other tasks. The same is true for the different tools and instruments being used, including patient-related information systems. Each of these networks draws upon a wide range of standards. The different elements are interdependent in a way that makes them very hard to change, beyond maintaining full backward compatibility by adding new elements that are compatible with the existing network. In addition, all these networks (practices, knowledge, technology, organizational structures, etc.) are interdependent and linked together into one huge network—or, expressed perhaps in an even better way, into a network of interdependent and interconnected networks having an enormous inertia. Radical paradigm changes, for instance from the old to the new economy, may then often require a redesign of all these networks, which is a task of enormous challenge.

An IS-based example of an organizational paradigm change

I will now examine the above ideas and perspectives in more detail by exploring a paradigm example of what the transformation of an organization from the old to the new economy is about, and what kind of challenges that involves. This case is about the introduction of a global information system in a maritime classification company, which is here called MCC.[4] The company has about 4,000 employees in more than a hundred countries, working in offices located mainly in cities having a major harbour or shipyard, or both. Computer technology has been applied to support the classification work since the 1960s.

Planning for a big-bang change

Up to the mid-1990s, computer support of the classification work at MCC had been dealt with by centralized mainframe-based systems. All communication between the headquarters and the company's different offices had taken place by exchange of paper documents.

During the early 1990s, rather strong support grew inside MCC for the assumption that ICT offered possibilities to support the work in much more extensive ways, and to enable a more efficient and competitive way of organizing the company's operations. This included a worldwide ICT infrastructure linking together all offices, and the enabling of a wide range of shared applications and services, including e-mail, spreadsheets, text processing, and other desktop applications. Further shared applications would be developed on top of this infrastructure. Key among these was one supporting virtually all activities involved in the inspection and classification of ships. This application would be built around a shared database containing all information about a ship, from its design as envisioned in a shipyard until it is scrapped. The application would further contain modules supporting all work tasks, with the aim of making the classification work much more efficient. It was assumed that such an application would make MCC more competitive.

Normally, the inspection of ships takes place in harbours. This is very expensive for the ship owners if a vessel is required to wait in a harbour for the inspection to be completed. Therefore, a critical factor in making a classification company more competitive would be to shorten the time the ships need to stay in a harbour. One way to achieve this is to do inspection work in a harbour while the ship is there to be loaded, and then continue to do the inspection in the next harbour when the ship arrives there. Such a splitting of inspection jobs and collaboration between inspectors could be enabled through a shared database and the design of shared, standardized practices. The usefulness of such a system could be further enhanced by adding more modules, for instance for generating statistics supporting the identification and characteristics of 'best practices' and other ways of improving and managing the classification work.

Extensive ship classification work also takes place at shipyards, during both the design and construction process and during maintenance and repair work. This requires close collaboration between ship-classification companies and the shipyards. Such collaboration may be significantly more efficient and smooth if the information systems supporting the work processes inside these organizations are integrated and data are exchanged between those involved. Further, the competitiveness could also be enhanced by improving the communication and collaboration with the ship's owners, for instance by giving them access to information about their vessels. The information in the database would also be interesting and useful for harbour authorities in their surveillance of the ships sailing in their areas of responsibility, for example in making judgement about the danger of accidents related to individual ships.

When the idea of its new global ICT system was widely accepted within the company, MCC began working on its realization in 1994. This required redesigning the company's work processes according to the new ideas, and to develop an IS that would support all functions mentioned above. The system was planned to be designed and implemented in a big-bang process, expected to take about two years.

Moving to new paradigms

The changes planned or envisioned for MCC can certainly be seen as radical—and therefore as a paradigm change. More precisely, it covered a change of three different but interrelated paradigms: one for ship classification work (global work processes), one for software applications (global information systems), and one for the organization and management of software processes (global software development and support work). These three paradigms are not independent. There are many links and interdependencies between them. Therefore, even though each of them has the characteristic of a paradigm in itself, they have to be co-developed and adopted in parallel—a fact that makes the change process even more complex and challenging.

Ship classification work at MCC was planned to be changed from a loosely-coupled distributed organization into a tightly-coupled global network. This tightly-coupled network includes not only the different MCC offices, but also ship owners, shipyards, and port authorities. MCC's different offices did not previously have any significant interaction. Splitting inspections between them in a more or less arbitrary fashion would mean their interaction and interdependencies would become rather stronger.

In the previous model, the offices were also quite independent from the corporate headquarters: they had much freedom in deciding how to carry out a ship inspection and how to organize work locally, including paying attention to local traditions and working-life cultures. In the new model planned for MCC, work at the local offices would be much more strongly regulated from the headquarters. This would happen because of: the standardization of the work and communication pattern enforced by the new ICT system and its shared database; the standardization required to enable the splitting of inspections; and management's intention to use the system to identify and enforce best practices.

The aim of using the ICT system to improve communication with external elements would imply that these externals were linked to the tightly-coupled network of units belonging to the MCC. It is easy to imagine that this idea of tightly-coupled global networks might open up more efficient ways of organizing. If MCC were to obtain the potential benefits of the system, in particular in terms of learning and knowledge generation, standards needed to be defined in a variety of areas, such as computer components, data structures, communication patterns, and work and management practices. However, it is a major task to specify exactly how these different 'standards' define the network, although the standards are needed to support and enable appropriate ways of working—and they should be sufficiently universal to facilitate gains from the economies of scale made possible by the new global system. At the same time, these standards should be open enough to allow adaptation to local contexts, and flexible enough to permit changes as new needs emerge.

The ICT solution developed to achieve these objectives also has some flavour of a new paradigm. It includes a global infrastructure consisting of a network; personal computers; operating systems; and e-mail, text processing, and other general applications. Further, the MCC's system included a document archive: a shared database containing all information gathered about a ship through its entire life cycle. It also has a large number of more specialized applications for various user groups. This included a rather large and powerful package consisting of various modules to do numerical calculations, for example relating to the strength of a particular construction of a ship's hull. Each of these applications is integrated with numbers of others in various ways.

A globally-distributed 'package' of technologies like this was certainly rather different from conventional information systems in the mid-1990s, when the MCC systems were being designed and developed. This package has a lot in common with the Enterprise Resource Planning (ERP) systems that started to be more widely introduced in global organizations in the late 1990s, after evolving over a period of about twenty years. Despite this, our knowledge about what is an appropriate design for a package like the one used by MCC is still limited. Critical aspects where better knowledge needs to be generated include issues such as how the different standards interact, their requirements for change in the future, and how they can be designed in a way that makes such future changes as smooth and simple as possible.

Finally, this organizational change process presupposed the establishment of a new paradigm for software development. The traditional paradigm can be described briefly as: develop systems from scratch, then maintain them until they are scrapped. Software for an information system was developed by a group of people organized as a project, who start by making a requirement specification and then design a system to satisfy it.

However, the new MCC package described above goes beyond what can be captured by the traditional paradigm. It contains too many different components whose relationships are largely unknown. The whole system also takes a very long time to develop, many of its components are outside the control of MCC developers—and they keep changing quite rapidly. These components in the infrastructure include: PCs; Windows and other Microsoft products; the Internet and other networking technology; and the software used in external organizations with which MCC wishes to integrate its system. Further, much of the software-related activities need to reflect the organization it is going to support, so has to be distributed globally in the same way as the classification work. Some effort is required to capture the specific needs and requirements of the different offices, and to install and test the software and hardware around the world. When a new version is developed, people from the development organization also need to be involved in its testing and evaluation at different sites. And last, but not least, there is a need for local support.

The successful development of this kind of package depends on building a great deal of new knowledge on the challenges involved, such as the appropriate strategies, how to organize and manage the required network-oriented organization to do the work, and what kind of planning and management tools and techniques might be appropriate.

How lock-in can inhibit successful innovation

The objectives and strategies behind the re-engineering project to achieve the aims of the MCC's new global ICT system were based on a traditional approach: design and implement the new organization and the IS from scratch, in a big-bang process. However, the tasks involved at MCC turned out to be far more challenging than anticipated. The data model and the IS were becoming all too complex. For instance, lots of interdependencies were discovered between the infrastructure and the IS, as well as with other applications.

During the re-engineering, the new IS fragmented because it was decided to manage the process by splitting the system into a number of more independent units, some of which were bought as off-the-shelf products. The systems had been initially conceived as a well-structured whole consisting of two layers—infrastructure and IS—with clean and simple interfaces between the layers, as well as within each of them. But this evolved into a huge collection of various components with lots of links and interdependencies between them, both direct and indirect. Six years after the project started, and four years after it should have been completed according to the original plan, a fragmented solution came into use. Its functionality was limited compared to the original specification, as were the organizational changes—and the benefits gained.

In my opinion, the lack of success compared to the initial objectives can be attributed to the fact that the project was firmly located within the traditional paradigm, which was inappropriate for this purpose. Success demands a new paradigm; unfortunately, the design and implementation of such a paradigm may be close to impossible because MCC is trapped by being locked-in to the old one. This lock-in is based on a number of key factors:

(*a*) The old and the new paradigms are to a large extent incompatible. This is the case for the most basic assumptions and values of the existing paradigm, including those of consistency, completeness, and non-redundancy. Further, planning and control are central elements in the existing paradigm. These ideals, values, and assumptions work only in a closed world of limited complexity. In the new economy's open world (in the sense of involving in principle an indefinite number of actors and systems), those assumptions are no longer valid. A project manager or other actors cannot control a project if it involves numerous actors in other organizations which act in line with the interests of their own companies, not those of MCC.

(*b*) The knowledge constituting the new paradigm can be generated only as lessons learned from ongoing work. That means that we will not learn how to work in a way different from what we are already doing. This can be argued by applying the communities-of-practice view on knowledge, which sees our knowledge as our practice—and implies that we have knowledge only about our existing practice. We can get knowledge about a new way of working only by practising it, but we are trapped because we cannot practice a way of working unless we have knowledge about it.[5] We can change our practices, but only by maintaining a strong (but not total) degree of backward compatibility—just as in the case of standards and infrastructures.

(*c*) The assumptions upon which a paradigm is built are extremely hard to discover for anybody belonging to that paradigm, partly because we are all basically unaware and unconscious of the assumptions our habits and practices are built upon. This also makes it very hard to see the consequences of adopting another set of assumptions.

(*d*) Our assumptions heavily influence how we interpret the practical reality. If we believe in planning and control, we would usually interpret the cause of projects being in a state of chaos as a lack of appropriate planning and control tools and methods (Ciborra and Lanzara 1994).

(*e*) Management is to a large extent about control. This is an ideology having a very strong hegemony. Accordingly, changing to a new paradigm where control is partly given up will be in conflict with the larger institutional context.

(*f*) The above explanations of why we are not able to learn how to work according to another paradigm are all at the individual level. However, even if we were able to overcome these problems, we would most likely be trapped because of what could be called 'coordination problems'. These come about because the generation of the new knowledge constituting a new paradigm, and the adoption of that paradigm, must be coordinated among all those involved.

The kind of projects I am talking about here are being run across many organizations around the world. The coordination problems pose an important and complex challenge, but are critical to the successful move to a new paradigm. One individual programmer cannot generate all the knowledge alone, as this needs to be a collaborative project in which all members of the project work in a coordinated manner. Accordingly, they need to collaborate about the generation of the new knowledge, share it, agree on the 'design' of the new paradigm, and adopt it together—or at least in a coordinated manner.

However, the learning and change processes involved encompass far more actors than can be managed and coordinated in an explicit manner. Projects can therefore be successful only if the actors behave in a coordinated manner without anybody managing or coordinating the process. Such coordinated processes occur occasionally, but not often. It is exactly like the adoption of a new standard. This has high value for its users if 'everybody' adopts it—otherwise it is valueless. Accordingly, potential adopters often wait for the others, and so nothing happens—as is also the case with the establishment of a new paradigm. Even if the new paradigm would be a great advantage for everybody and was a adopted by everybody, it would be most rational for everybody to stick to the old paradigm.

It should also be emphasized that knowledge is not linked only to practices, but also tools and instruments, organizational and physical structures, etc. To change paradigm, we also need to change these other structures in a coordinated manner.

Conclusions

In this chapter, I have focused on the systemic character of knowledge and argued that the complexity of knowledge structures plays important roles in many learning and change processes in, and across, organizations. In many cases where an organization tries to change itself, the complexity of knowledge structures involved may inhibit the learning which the change processes requires. This is particularly true in processes where ICT solutions seem to enable radical change and where organizations are aiming at what we might call paradigm changes, such as changes associated with globalization and transformation from the 'old' to the 'new' economy. In such processes, knowledge may—due to its inherent complexity—appear as an agent that has obtained significant autonomy (Winner 1978). I have further argued that examples of complexity theory, like network economics, may be helpful in understanding the systemic features of knowledge, how knowledge may seem to become a mere autonomous agent, and some of its behavioural features as such an agent.

References

Akerlof, R. L. (1970). The Market for Lemons: Quality Uncertainty and the Market Mechanism. *Quarterly Journal of Economics*, 89: 488–500.

Arthur, W. B. (1994). 'Increasing Returns and Path Dependence in the Economy'. *Scientific American*, 80: 92–9.

Blackler, F. (1995). 'Knowledge, Knowledge Work and Organizations: An Overview and Interpretation'. *Organization Studies*, 16/6: 1021–46.

Castells, M. (1985). *High Technology, Space and Society*. Beverly Hills, CA: Sage Publications.

Ciborra, C. U., and Lanzara, G. F. (1994). 'Formative Contexts and Information Technology: Understanding the Dynamics of Innovation in Organizations', *Accounting, Management and Information Technologies*, 4/2: 61–86.

Fujimura, J. H. (1996). *Crafting Science: A Socio-History of the Quest for the Genetics of Cancer*. Cambridge, MA: Harvard University Press.

Hanseth, O. (2000). 'The Economics of Standards', in C. U. Ciborra (ed.), *From Control to Drift: The Dynamics of Corporate Information Infrastructures*. Oxford: Oxford University Press, 56–70.

Hutchins, E. (1995). *Cognition in the Wild*. Cambridge, MA: MIT Press.

Knorr-Cetina, K. (1999). *Epistemic Cultures. How the Sciences Make Knowledge*. Cambridge, MA: Harvard University Press.

Kuhn, T. S. (1970 [1962]). *The Structure of Scientific Revolutions*. Chicago: Chicago University Press.

Latour, B. (1988). *The Pasteurization of France*. Cambridge, MA: Harvard University Press.

—— (1999). *Pandora's Hope. Essays on the Reality of Science Studies*. Cambridge, MA: Harvard University Press.

Lave, J., and Wenger, E. (1991). *Situated Learning. Legitimate Peripheral Participation*. Cambridge: Cambridge University Press.

Monteiro, E. (1998). 'Scaling Information Infrastructure: The Case of the Next Generation IP in Internet'. *The Information Society*, 14/3: 229–45. Special issue on the history of Internet.

Piore, M. H., and Sabel, C. F. (1986). *The Second Industrial Divide: Possibilities for Prosperity*. New York: Basic Books.

Porter, M. (1998). *The Competitive Advantage: Creating and Sustaining Superior Performance*. New York: Simon & Schuster.

Rolland, K. H. (2000). 'Challenging the Installed Base: Deploying a Large-scale IS in a Global Organization', in H. R. Hansen, M. Bichler, and H. Mahrer (eds.), *Proceedings of the 8th European Conference of Information Systems*, 583–90.

—— (2002*a*). 'The Business of Risk: Global Information Infrastructures and Shifting Identity'. Unpublished draft manuscript.

—— (2002*b*). 'The Boomerang Effect in Design and Use of Information Infrastructures—Evidence from a Global Company'. Unpublished draft manuscript.

—— and Monteiro, E. (2002). 'Balancing the Local and the Global in Infrastructural Information Systems'. *The Information Society*, 18/2: 87–100.

Shapiro, C., and Varian, H. R. (1999). *Information Rules: A Strategic Guide to the Network Economy*. Boston: Harvard Business School Press.

Star, S. L., and Ruhleder, K. (1996). 'Steps Toward an Ecology of Infrastructure: Design and Access for Large Information Spaces'. *Information Systems Research*, 7/1: 111–34.

Steinmueller, W. E. (1996). 'Technological Infrastructure in Information Technology Industries', in M. Teubal, D. Foray, M. Justman, and E. Zuscovitch (eds.), *Technological Infrastructure Policy*. Dordrecht, Netherlands: Kluwer Academic Publishers, 117–39.

Wenger, E. (2000). 'Communities of Practice and Social Learning Systems'. *Organization*, 7/2: 225–46.

Winner, L. (1978). *Autonomous Technology. Technologies-out-of-control as a Theme in Political Thought*. Cambridge, MA: MIT Press.

Notes

1. The various perspectives on knowledge discussed in this paragraph are examined in, for instance, Lave and Wenger (1991), Blackler (1995), Hutchins (1995), and Wenger (2000).
2. The development of the new version of the IP Internet protocol is a striking example of this process of technological updating (Monteiro 1998).
3. Latour (1999) includes an extensive discussion on whether Pasteur could be said to have 'constructed' what is otherwise called a 'discovery'.
4. This example is based on Knut Rolland's research (Rolland 2000; Rolland 2002*a*, 2002*b*; Rolland and Monteiro 2002).
5. Fortunately, this fact does not imply that we cannot learn. It only implies that new knowledge must be closely related to what we already know. We might change our practice, but only in ways where the new practices fit well together with at least the great majority of existing ones.

6

An ecology of distributed knowledge work

RICHARD J. BOLAND, JR.

Introduction

This chapter explores distributed knowledge work as an ecology of knowledge workers, mediations, knowledge objects, documents, and data repositories.[1] These elements of the ecology are linked together through the dynamics of representation, design, and implementation. Those three aspects of the experience of engaging in distributed practice are ubiquitous, pervasive, and intertwining (Bowker and Turner 2000). In the ecology of distributed knowledge work, representations are continuously being written and read, design is being enacted and evaluated, and implementations are being produced and reproduced.

Representing, designing, and implementing are, in a sense, the 'stuff' of which distributed knowledge work is made. Because representations are being created and interpreted by actors at many levels of a distributed system, it is a multi-layered hermeneutic process. Because designs of the system are always being enacted and evaluated, it is a path-creating, innovative process. And because implementations of the system are continuously being produced and reproduced, it is a process of responsible human agency. At any moment, any of the actors could interpret their situation differently, could resist the prescribed path, or could intervene to change accepted practices.

A central feature of this ecology of distributed knowledge work is the tension that individuals experience, both as knowledge workers and as knowledge objects, between the expectations of a local logic and those of a global logic. These two logics create a symbolic space within which the individual must navigate—a space that is marked by an irreducible contradiction between competing contexts for responsible action. In this chapter, I first present an ecological model of distributed knowledge work, showing its multiple tensions between local and global logics. I then discuss some recent research on distributed knowledge work that shows these tensions in action, and explore the existential experience of humans as knowledge workers.

An ecology of distributed practice

A multi-level model of knowledge work

Figure 6.1 presents a simplified image of how knowledge workers, mediations, knowledge objects, documents, and data repositories form an ecology of distributed

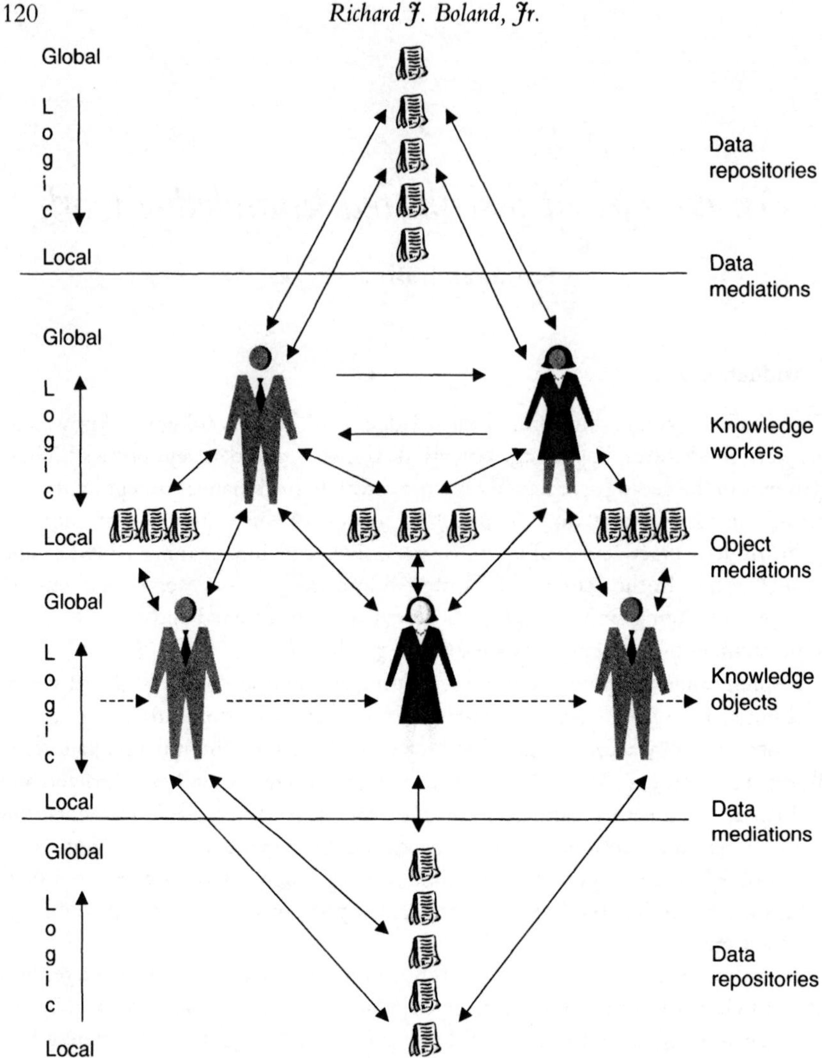

FIG. 6.1. An ecology of distributed practice

practice. It is a fragile, recursive network in which humans are both the principle knowledge workers and also the principle knowledge objects. Non-human actants (Latour 1987; Callon 1992; Law 1992) have a role to play, to be sure, but what interests me is the experience of the human being as knowledge worker and knowledge object. Judging the value of a distributed system will always involve an assessment of how it is experienced by the professionals operating within it and the human clients who are affected by it. In that sense, evaluating a distributed system will always rest on how a person is affected by the system, internally or externally, and that is where I want to focus our attention. In addition to non-human actants, there are many other important elements that are not in this simple diagram, such as culture, history, and status. My intention is to refer to them, as needed, through the human beings who are involved as its workers and objects.

The knowledge workers and knowledge objects in Figure 6.1 are engaged in reading and writing documents, thus constructing representations that mediate the interaction between those workers and objects, as well as between the documents that feed into and out of data repositories. The knowledge workers and objects, together with other supporting actors associated with larger organizational and societal information systems who are not depicted, are also engaged in the design of those very documents and interaction processes. And finally, as they work in these systems—resisting sometimes, performing sometimes, and struggling to improve them at other times—they are implementing the distributed system of knowledge work. Representing, designing, and implementing, then, are the way that the distributed practices of knowledge work are carried out.

The data repositories shown at the top and bottom of the figure are meant to be the same ones, and are therefore at the same level. These repositories therefore provide a connection between the knowledge workers and knowledge objects, forming a circle, or more accurately a Möbius strip, so that each time the circle is traversed, the surface on which one is travelling seems to be new. The image of a Möbius strip captures in some degree the (often false) sense of eternal progress we associate with technology in general, and the advanced forms of distributed practice in particular.

For each of the three layers of the Möbius strip figure, I show on the left the different levels of logic that operate within it. The levels of logic range from the most local and concrete to the most global and abstract. I see these ranges of logic, and the two extreme anchor points of the local and the global, to be important defining characteristics of the experience of action in any system, especially distributed knowledge work. Local logics, at the extreme, are the logics of face-to-face interaction in historically-bound moments of knowledge work. Culture, traditions, symbols, honour, history, and sense of future are all present and operating in determining the appropriate thing to do. Global logics, on the other hand, are experienced when a person engages in abstract analysis with logical relations among concepts, economic calculations, and attention to institutional or professional rules, so that an accountable, defensible reason can be given for a choice that is made.

Tensions between local and global logics

In Figure 6.1, knowledge workers are shown as being in a constant tension between the two logics of the local and the global. They engage in abstract, global logics when they write to, or read from, data repositories that are created and accessed by a wide range of professionals working across multiple disciplines. In dealing with a data repository, the logics they employ are not of the concrete moment, but of how the concrete moment 'fits' the closest approximation of the categories, measurements, and policies operating at the global logic level. At the same time, the documents they create and read while working with knowledge objects involve a local logic of their practice.

Take, for example, the physician who is treating a patient in a large teaching hospital. Her vocabulary will be unique to her own specialty, and even to the local categories of patients, treatments, and relations with other medical and technical staff. It is thus a local logic for the physician. However, for the 'knowledge object' (the patient), these documents represent a global form of logic, not a local one. The patient's direct experience is translated into the higher-order constructs of a medical specialty, and the local logic that unites a patient's understanding of his disease or treatment within the medical system is of little or no consequence. As the individual moves from specialist to specialist within the hospital network, his immediate experience of local logic is carried through from location to location, but it is not a part of the global logic of the documentation that is being made of them by the knowledge-worker physicians.

At the data-repository level, something strange seems to happen. The most global logics of the physicians, insurance companies, government policy makers, or social scientists are turned into very local issues of measurement, standardization, detailed definitions of particularities, and exacting techniques for ensuring common and understandable formats for data entry and retrieval. At the same time, the most abstract, summarized syntheses of those categories and measurements become the existential ground for the local logics of the everyday person as a knowledge object (patient, citizen, etc.). They become the potent categories by which we are known as individuals, to all but our most intimate friends. They become the terrain that must be navigated by ordinary people day-to-day in order to get things done in institutionalized worlds such as health care, education, or government services.

The tension of the local and global logics as they cross each other from level to level in this simple depiction of distributed practice is the tension of reason versus experience, which has been with us in Western thought since at least Plato. I believe that the emerging field of distributed mediated practice is an important step in bringing these age-old tensions to our conscious awareness. We should expect that the more technically advanced and distributed our knowledge-work practices become, the more we will encounter fundamental questions of human existence in our studies of such practices. As computing and communication technologies become universal and pervasive in our lives, the questions that first stirred humans on a reflective, philosophical path return to us in a socio-technical presence. The computing and communication technologies that mediate distributed practice and make virtual teams possible also serve to raise philosophical issues about the nature of the modern world which can remain unnoticed and unexamined in their absence. Fundamental tensions between the local and the global, which are present even in the most basic forms of human interaction, are made visible and problematic when we introduce advanced technologies to mediate organizational practice.

Examples of distributed practice

The simplified representation of Figure 6.1 could be criticized because the distinction between local and global logics can be somewhat difficult to make, with the same documents representing local or global logics depending on the writer or reader. So, I intend this conceptualization as an analytic distinction, useful for dissecting the overall structure and dynamics of distributed knowledge work and not intended as an exhaustive and exclusive taxonomy. Below, I use the analytic distinction of local and global to discuss some recent research on distributed practice that displays the multiple, overlapping logics and the Möbius-strip quality of their interrelation.

Interactions between local and global logics

Michael Barrett has studied auditors and their documenting practices in large-scale corporate financial audits (Barrett, Cooper, and Jamal 2001). He argues that as their documentation practices changed, their identities as auditors changed. I read this as saying that the documentation practices at the global level of their audit work led to a change in their local sense of identity. An earlier study by Barrett shows a similar relation of documentation practices and professional identity in the insurance industry (Barrett and Walsham 1999). As documentation practices changed, the identity of insurance brokers changed along with them.

Another example of global and local interactions is provided by Carsten Østerlund's (2002) study of a large teaching hospital in which it was hard to comprehend the intensity of writing and rewriting of patient records by physicians, with its repetitious documentation of the same patient's symptoms and history. Why, the medical informatics community asked, couldn't the patient history be collected once and be electronically transmitted across all the specialists as the patient moved among them? Why, in other words, can't a global logic be used to structure the data collection? Øesterlund's answer was found in the multiple roles that documentation played in that distributed system.

First, each specialist has unique categories that are not duplicated by others. But, more importantly, each specialist is not just treating the patient but is also reflecting on the patient, and on medical knowledge generally, in a written dialogue with themselves. Further, each physician is also communicating with specialists on their own team across shifts, as well as with other teams of specialists. Each of these modes of communication raises unique requirements of local logic and, as professionals, they will resist any attempt to homogenize that local logic into a less immediate global one through a common medical information system. At the patient level, in contrast, the physician's local logic of writing places the patient as an object in a global vocabulary of disease categories, syndromes of complications, and standardized test results. The patient's actual experience is subordinated to the local and global tensions of the physician.

Language dynamics in software development processes

Our research team at Case Western Reserve University has undertaken several projects studying distributed teams which are of relevance to the ecology of practices discussed in this chapter. Here, I will focus on two such cases which illuminate the dynamics of local and global logics through: a study of the experience of space and time by distributed team members; and the process of learning in a distributed system.

First, Alex Citurs (2002) has explored the dynamics of language use in distributed teams of software developers. He proposed that software development teams periodically experience moments of crisis during a project and that these moments are instances of learning. He also suggested that such moments of crisis are characterized by:

(*a*) a shift from the use of implicit language to the use of more explicit language;
(*b*) the emergence of new vocabulary elements; and
(*c*) an increase in the level of integrative complexity of their language practice.[2]

In 2001, Citrus conducted a six-month study of three software development teams located in the USA, India, and the UK who were working on a distributed project. He found they did indeed experience periodic moments of crisis in their work. He also found that the periods of crisis were associated with a shift from implicit to explicit language, the emergence of new vocabulary items, and an increase in integrative complexity. After the crisis passed, language patterns returned to their pre-crisis state of implicit language use, fewer new vocabulary items, and reduced integrative complexity. But what was even more interesting to me was the finding that this type of oscillation in language practice continued in wave patterns throughout the project. Moments of crisis were correlated with peaks in these oscillations, but the oscillations were ongoing, even when the team members did not identify a crisis as occurring.

I take the oscillation in language use as an element in the tension between local and global logics that is experienced by knowledge workers in a distributed practice. The tension and the resulting oscillation between the local and the global are continuous. Language is always breaking down in the face of this tension, and is always being repaired—but in ways that lead to further breakdown. As language becomes more implicit, relying on non-verbal understanding, and as integrative complexity is reduced, the knowledge worker is coming closer to the local experience and relying more intimately on the local logic of practice. As this trend progresses, the global logic is threatened. Things happen in the local practice for which no global term applies. Communication between teams suffers. Their language then begins to shift in the face of this breakdown, becoming more explicit, inventing new vocabulary items to capture the local experience that is escaping the global logic, and building more integratively complex arguments to bring the diversity of local experience within a global framework.

This trend to strengthen the global logic leads to the emergence of new problems, in which the local practice is burdened by the attention to global logics and vocabularies. It is inefficient for the team to spend time being explicit, using integrative complex reasoning, and inventing new vocabulary items. Their language practice then begins a reverse shift toward a greater use of implicit vocabulary, less integrative complexity, and reliance on the existing lexicon.

I am really excited by the image of language as always breaking down and always being repaired in ways that keeps reversing itself. I see this as an aspect of our existence which is not apparent to us in our day-to-day experience, but is highlighted for us in the study of distributed practice. That this language dynamic generally remains hidden from us is, I believe, related to the way we privilege space over time in our thought. Temporality is almost impossible for us to express in language, or to think about in its own terms. The very idea of temporality has to be translated into a spatial representation in order to be conceptualized in language. Time is shown as a line in Cartesian space. Temporal duration is discussed as movement along the line. An instant of experience is a limitlessly small point on the line. In this way, our ability to conceptualize the dynamic nature of language use and learning is made within a spatial framework that includes built-in inhibitions against a thorough-going sense of temporality. Learning is therefore associated with the concepts learned, which is a spatialized representation of learning, not the temporal process of breakdown and repair in language. But that is a topic which needs to be explored in more depth elsewhere.

Another particularly interesting finding in Citurs' studies of distributed software development teams is the phenomenon of 'unblackboxing'. He discovered that teams writing software sometimes encountered an element of code that had been created years before and had been used as a modularized 'blackbox', but now no longer seemed to work as intended. The team was then forced to engage in 'unblackboxing' in order to unearth the logics employed in a module's creation. This 'unblackboxing' required that they uncover the organizational processes and social context that were assumed in its original construction. In essence, they were confronted by an element of their global logic (the blackbox module) that no longer applied to their local practice, which made them 'reverse engineer' an understanding of the lost local logics that had gone into its construction. Once again, we see language breaking down and being repaired as part of the continuous process of learning in the team, stemming from the tension experienced between local and global logics. To paraphrase Latour (1987), we like to think of language as being ready-made, but instead it is always being made. And, I would add, being made in the tension of local and global logics.

Space and time experienced by outsourced computer administrators

My work with Ulrike Schultze revealed the importance of two ways (known as 'space' and 'place') in which space and time are experienced, through an ethnographic study of the experience of outsourced computer administrators (Schultze and Boland 2000). Place is the experiencing of space and time by having a sense of being in place, having a place, and understanding how things work because of the place we are in. Place is local logic, tradition, accepted practice, and a shared sense of 'how we do things here'. Space, on the other hand, is the experience of space and time by having a sense of an abstract, homogeneous, limitless world in which laws of nature hold universally, and in which local practices are quaint anachronisms of little significance.

The computer system administrators that Schultze studied were caught between a desire to be aligned with a sense of space, in order to strengthen their claim to professional expertise and high consulting rates, and a desire to belong in the organization—to have a sense of place that could make their day-to-day work more humanly satisfying. Schultze's study of these teams showed how their attempts to break out of the tension they experienced only served to reproduce the very conditions that made the tensions possible. In order to have professional satisfaction in their work, the administrators sought to define themselves as 'third level' computer system workers. This meant that they dealt increasingly with the most arcane and deeply embedded system-level aspects of the computer installation. Yet, this attempt to gain professional satisfaction and to enable them to be more marketable as consultants on future assignments meant that they were less able to experience a sense of place in the organization. Further, they were increasingly seen as 'commodity workers' who had so little organizational knowledge that they could easily be replaced. This, in turn, caused them to seek to strengthen their professional competence and value by moving even further into the sense of space found in the most esoteric levels of the computing system software.

In short, the tension of local and global logics as manifest in the experience of space and place resulted in behaviours that served to reinforce the field of tensions initially giving rise to their discomfort. As in the language breakdowns that were seen to drive learning in Citurs' work, the tensions of local and global logics encountered by the computer administrators in Schultze's study had created a cycle of breakdown and repair, and served to reinforce the very tensions they tried to escape.

Conclusions

I have proposed in this chapter that distributed practices can be productively viewed as a multi-level ecology in which each level is characterized by a tension between local and global logics, as they are experienced by human actors. I have given some research examples that are relevant to understanding the dynamics of these local–global tensions in distributed knowledge work settings. The research suggests that there is a recursive, self-propelling process of language use in which breakdown and repair create and recreate the very conditions that give rise to tensions in the field.

A larger issue raised by the ecological view of knowledge work presented here is the shift it suggests in the emphasis of our research on knowledge work and virtual teams. The new direction that an ecological view suggests is one that emphasizes the existential experience of individuals as they engage with the tension of local and global logics. This existential direction approaches the study of organizational life as processual, temporal, and dynamic, with a focus on individuals and their subjectivity. It also emphasizes the importance of a structuralist analysis to the study of organizational life.

By structuralist analysis, I mean an approach to studying knowledge work that unpacks the oppositions associated with the tensions experienced in organizational life, such as local versus global, and how those oppositions create a space for playing out knowledge work. An existential study of knowledge work will reveal how actors partition their ongoing flow of experience with oppositional language that symbolizes differences between self and other, core and periphery, freedom and control, and other binary oppositions forming their field of action. Hopefully, this existential approach will strengthen a more thoroughgoing sense of the temporal flow of organizational life and an appreciation of the unavoidably ethical basis for our engagement with that flow, as we produce and reproduce through language the conditions we experience. The processes used by organizational actors for representing, designing, and implementing both make and navigate a space for engaging with the tensions that characterize organizational life. An ecological approach to the study of knowledge work as a distributed practice gives us a way to begin studying the existential experience of making, and moving within, that space.

References

Barrett, M., and Walsham, W. (1999). 'Electronic Trading and Work Transformation in the London Insurance Market'. *Information Systems Research*, 10/1: 1–22.

Barrett, M., Cooper, D. J., and Jamal, K. (2001). 'Managing Across the "Friction of Space" in Auditing'. Paper presented at Rethinking Globalization and Standardization, an International Conference on Spacing and Timing, Palermo, Italy 1–3 November.

Boland, R. J., and Tenkasi, R. V. (1995). 'Perspective Making and Perspective Taking in Communities of Knowing'. *Organization Science*, 6/4, 350–72.

—— —— and Te'eni, D. (1994). 'Designing Information Technology to Support Distributed Cognition'. *Organization Science*, 5/3: 456–75.

Bowker, G., and Turner, W. (2000). 'Summary of Conclusions'. Unpublished presentation at Distributed Collective Practices 2000: Open Conference on Collective Cognition and Memory Practices, Paris, September 19–20. www.limsi.fr/Individu/turner/DCP

Callon, M. (1992). 'The Dynamics of Techno-economic Networks', in R. Coombs, P. Savotti, and V. Walsh (eds.), *Technological Change and Company Strategies: Economic and Sociological Perspectives*. San Diego, CA: Harcourt Brace Jovanovich.

Citurs, A. (2002). *Changes in Team Communication Patterns: Learning during Project Problem/Crisis Resolution Phases—An Interpretative Perspective*. Unpublished Ph.D. thesis. Cleveland, OH: Weatherhead School of Management, Case Western Reserve University.

Latour, B. (1987). *Science in Action*. Cambridge, MA: Harvard University Press.

Law, J. (1992). 'Notes on the Theory of the Actor-network: Ordering, Strategy, and Heterogeneity'. *System Practice*, 5/4: 379–93.

Østerlund, C. S. (2002). 'Documenting Dreams: Patient-centered Records versus Practice-centered Records'. Unpublished Ph.D. thesis. Cambridge, MA: Sloan School of Management, Massachusetts Institute of Technology.

Schultze, U., and Boland, R. J. (2000). 'Place, Space and Knowledge Work: A Study of Outsourced Computer Systems Administrators'. *Accounting, Management and Information Technologies*, 10/3: 187–219.

Notes

1. For background on key issues relating to distributed knowledge work, see Boland, Tenkasi, and Te'eni (1994) and Boland and Tenkasi (1995).
2. Integrative complexity is measured as the number of distinct concepts being dealt with as an integrated whole, and the number of relations understood to exist among those concepts.

7

Actor network theory and cultural aspects of interpretative studies

ERIC MONTEIRO

Introduction

It has become something of a truism to say that the development and use of information systems are socio-technical processes. From its conception as a purely technical discipline, Information Systems research has grown to encompass a broad and rich field. It is easily forgotten how cumbersome the legitimization of social aspects of IS research has been historically. What we today take as obvious are, in fact, insights gained over many years. Examples of valuable implications of understanding this socio-technical nature include how IS implementation projects get caught up with—and subsequently feed into—the unfolding dynamics that get abbreviated in corporate strategies, hence undermining simplistic notions of project management. Specification and design imply the privileging of some users and leaving others marginalized, thus underscoring the intrinsically political nature of organizations. The use and uptake of information systems hinge on training, on perception, and on structures of incentives. There is improvisation, drift, and unintended consequences. This much seems to be a relatively uncontroversial account, at least within the field of interpretative studies of information systems (Walsham 1993; Ciborra 2000).

Where does this crude summary of the current state of affairs leave us? In order to get where we are, a truly vast number of analytic frameworks have been proposed, suggested, and employed since the 1980s. Undoubtedly, this has advanced our understanding of socio-technical aspects of information systems. We have accumulated a significant body of analysed experiences and concepts. Given this, what are the challenges for this type of empirically-underpinned, analytically-framed interpretive IS research? The aim of this chapter is to argue that a key challenge is to cultivate analytical notions in *tandem* or *co-evolution* with a nuanced grasp of the constantly changing nature of IS. This argument is based on two types of observations.

First, to a disturbing extent, we—the loose network of scholars of interpretative studies of information systems—have imported theories from the social sciences

Parts of this chapter were published in: E. Monteiro (1999), 'Living with Technology', *Scandinavian Journal of Information Systems*, 11/1–2. Reprinted with permission.

and the humanities unidirectionally, and then relatively straightforwardly applied them (Orlikowski and Barley 2001). And then we, myself included, have accordingly labelled without too much effort instances of 'social constructions', 'activity systems', 'inscriptions', 'bricolage', and 'organizational politics'. However, merely decorating empirical instances with analytical notions is neither very illuminating nor intellectually stimulating. To make an analysis interesting, elements of an empirical case need to be moulded with the analytical notions that go well beyond labelling exercises.

Secondly, there is a strong empirical argument. The technology itself keeps changing, resulting in new patterns of use. Use and development of a package for statistical analysis is different from a system for accounting in a small firm, which is different from a business-to-business (B2B) e-business solution, which is different from a personal digital assistant (PDA) prototype at a hospital ward allowing physicians to write and order prescribed medications for patients, and so on. The very phenomenon—covered by the catch-all phrase of 'information system'—is anything but stable (Hanseth, Monteiro, and Hatling 1996).

The two tendencies together—a (too) mechanical application of (imported) theory and failure to break down the monolithic notion of an 'information system'—add up to the research programme aimed at developing an analytically-embedded understanding that pays justice to the empirical transformations of IS. The remainder of this chapter illustrates such a broad research programme through one effort involving a critical discussion of actor network theory (ANT). It thus assumes a certain familiarity with the basics of ANT, and will not attempt a detailed introduction (e.g. Star 1991; Walsham 1997; Latour 1999*b*; Law 1999; Monteiro 2000). Based on a critique of one aspect of ANT, we point out how this creates problems in capturing interesting, empirical changes in and around IS today.

ANT has attracted considerable interest among interpretative IS scholars in recent years. In part, this has been specifically for its alleged appropriateness in accounting for the network or infrastructure aspects of the technology (Walsham 1997). Here, however, we critically discuss the suitability of ANT to a set of relevant, empirical settings for IS research. The aim is to inquire whether ANT, as an interpretative framework, privileges certain aspects and is blind to others, a blindness that IS researchers need to reflect upon critically in studies we are already involved in, as well as those we see coming. Our concern, then, is about the seemingly (implicit) assumption that ANT can be used for everything—urban transportation, microbes, and hotel keys alike. The motivation is accordingly to contribute constructively to an appropriation of ANT within the community of practising IS researchers.

More specifically, the chapter critically explores the baggage that ANT often brings with it in interpretative studies, in particular tendencies in ANT to (over?) emphasize relatively goal-directed actions. This is not an argument that ANT *intrinsically* 'has' these tendencies, but that (too) often ANT studies *in practice* display them. It is particularly problematic, we argue, for those types of studies of IS—which might be increasing in number—that have significant elements of what crudely could be dubbed 'cultural' aspects: the process over time whereby information systems are symbolically bestowed with meaning that go beyond the hopelessly crude notion of 'use' of technology; the way cultural aspects of local use of global IS operate; or how identity formation goes hand-in-hand with 'use' of teenagers' mobiles.

The next section presents a critique of the (often overly) goal-oriented aspect of ANT. It is followed by a section focusing on everyday practices of use. This argues that everyday life lacks much of the goal-directedness that ANT presupposes, and presents the alternative notion of 'domestication' to underscore the blind zones of ANT. The subsequent section offers a few empirical illustrations of areas where domestication fits better than mechanical applications of ANT. The chapter concludes by enquiring into the scope of ANT in IS.

A critique of actor network theory

Without the ambition of being comprehensive, it is helpful to map out some of the traits of ANT in order to position the subsequent critique relative to this. Basically, ANT is an effort to develop a vocabulary to trace out how plans, goals, and intentions are translated and inscribed into alternative expressions that may or may not be material. Manoeuvring takes place in the dynamic processes with competing goals (and their translations) and unrealized intentions. A corollary of ANT, but also the basis for ANT's attention in IS, is the observation that, more often than not, inscriptions tend to be materialized in (or delegated to) technology to be effective. Another marked characteristics of ANT is its hostility to granting given categories—such as 'class', 'knowledge intensive firms', 'the state', or 'culture'—much explanatory power. Instead, ANT promotes the stance that such categories need to demonstrate their role and relevance. Our argument about 'cultural' aspects of IS should accordingly be read as a call to demonstrate the relevance of what, perhaps afterwards, might get dubbed 'cultural'.

During its relatively short-lived career, ANT has succeeded in stimulating widespread debates (e.g. Pickering 1992; Walsham 1997; Law 1999). Several types of criticisms have been launched. Among the ones fiercely debated are the provoking principle of symmetry (technical = non-technical), ANT's apparent lack of political relevance, and its vagueness with respect to status (is ANT a theory, a methodology, a perspective, a vocabulary, or an ontology?). The interpretation of ANT has no doubt been made cumbersome within a US–UK dominated IS research community by the distinctly French intellectual tradition that Latour adheres to: where arguments are implicit, references—if present at all—hardly count as relevant, and format is playful or ironic. It is well beyond the scope of this chapter to attempt a comprehensive discussion of ANT within IS. Instead, we will more narrowly pursue one thread of critique of ANT, which we take to be the source of the most substantive critique of ANT relevant to IS.

As indicated above, ANT traces out how aims and goals of projects, through various steps of translations, are sought to be realized. There is a danger in ANT of emphasizing this goal-directedness too much. This will not come as a big shock to those well-versed in ANT, as indicated by Latour's (1999*a*: 16) comment that the 'managerial, engineering, Machiavellian, demiurgic character of ANT has been criticized many times'. In this sense, ANT has been said to produce actants that are 'flat'; they are without much blood and tears—or 'hairy gorilla-like' as Latour (1999*a*: 16) expresses it.

Still, for the IS research community, it is informative to elaborate on the sources and manifestations of this tendency to over-emphasize the strategic/rational/goal-oriented aspect of ANT. This is valuable in order to minimize the danger of an instrumental employment of ANT that already 'has done harm as well as good' (Law 1999: 8).

The principal reason why ANT is biased towards strategic behaviour is fairly obvious. Given the ambition within science and technology studies (STS) of empirically examining how facts and technology get invented, this is not altogether surprising. Projects—turning beliefs into scientific facts, conquering the world from a laboratory, or manoeuvring an artefact from the sketch board into use—are in a straightforward sense about realizing goals, accomplishments. Hence, it is no coincidence that this may result in accounts that are 'excessively strategic' (ibid.: 6).

Two well-known manifestations of this bias are the privileging of the centre and privileging of one perspective. The privileging of the centre stems, in practical, methodological terms, from the way ANT starts out by identifying key actors, interests, and scenarios before tracing these over time. This centeredness became further exaggerated by the early studies of laboratories that act as natural centres. Mol and Law (1994), for instance, challenge the centeredness of many ANTish accounts by generalizing the notion of 'social space'. Space is not, they argue, intrinsically linked to our traditional, physical sense of space, which may privilege centred accounts; rather, space behaves more like a fluid (ibid.: 643), thereby emphasizing the leaking and deleting of physical boundaries. A similar modification of the centeredness of ANT is formulated by Coussins (1996) through the notion of choreography.

Part of the reason ANT could be perceived to be too Machiavellian is due to its reliance on pedagogic examples of the 'hotel-key and door-opener' type. Still, the problem of privileging one perspective has been criticized most succinctly by the work of Star (1991), where ANT is supplemented with insights from feminist theory, symbolic interactionism, and activity theory. Her suggestive example is from the McDonalds fast-food chain. In ANT terms, the practices of producing and serving McDonalds' hamburgers is a well-aligned, heterogeneous network of technology (such as the configuration of frying pans, trays, and counters) and routines (such as queuing, scripted orders, and replies) which, in total, produce the hamburger. Star's point is to think systematically about what it would take to request something that falls outside the robust actor-network, for example an order for no onions on a hamburger for someone with an allergy. In ANT terms, this would require an alternative, competing network to be viable. More generally, this reminds us not to focus on the existing networks only (e.g. McDonalds) but also to make visible the conditions for alternative networks. The gist of her critique is that, despite ANT's capacity to account for heterogeneity and multivocality, 'this is only one kind of multiplicity, and one kind of power, and one kind of network' (Star 1991: 28).

The critique about an excessively strategic bias in ANT is explicitly addressed in Latour (1999*b*) by emphasizing the 'slight surprise of action'. Action, Latour reassures us, is never really goal-oriented/strategic as 'we are not in command, we are slightly *overtaken* by the action' (ibid.: 281), 'their consequences are unforeseen' (ibid.: 288), and subject to translations characterized by 'displacement, drift, invention, mediation' (ibid.: 179).

Hence, there can be no doubt that in Latour's (1999*b*) version—the most comprehensive and systematic outline of this theory in many years—at the espoused level the strategic/goal-oriented bias of ANT is greatly exaggerated as, in principle, there can always be 'a drift, a slippage, a displacement' (ibid.: 88). The more recent versions of ANT definitely underscore explicitly that we must not exaggerate the goal-directedness of ANT. Yet, is there not, in practice, still a bias? Does not the assurance about the 'slight surprise' go only skin deep, without ever upsetting the basics of ANT?

The relevance: symbolic and cultural aspects of IS

In practice, though not on an espoused level, ANT could be seen to promote a goal-directedness that easily encourages overly purposeful action. This stems from the *de facto* tradition within STS of studying the 'biography' of projects, the pursuit of agendas, manoeuvring, and enrolment. To the extent that this critique carries any weight, what is the relevance for IS? What are the implications in practical and analytical terms for IS? In a nutshell, the implications for IS are linked to a certain class of studies, namely those which draw heavily on the more symbolic, 'cultural' aspects of human action. In what follows, we will explore situations relevant to IS that seem to me to be particularly sensitive to the critique of ANT outlined above.

Home v. work: domestication and everyday life

The critique of ANT we have outlined is reinforced within IS in an unfortunate way, as the analytical bias towards a privileging of purposeful action is reinforced by the empirical bias in IS of studies of work-related cases (Ehn 1989; Greenbaum and Kyng 1991; Kyng and Mathiassen 1997). This is crudely characterized as being:

- purposeful, work-oriented;
- performed by workers;
- delineated from other work tasks; and
- restricted to working hours.

Work and work-tasks are, at least on a certain level, purposeful and goal-directed. Our point, then, is that the analytical bias in ANT is supported by an empirical bias in IS. Of course, this need not be a problem. It is a problem only when we want to capture the trend where IS is employed *outside* of these traditional work-related settings characterized by more or less purposeful action. Does the baggage or bias also spill over to these other settings?

Our present and future use of IS will clearly not be restricted along the same dimensions as those listed above. IS has moved beyond the workplace and into leisure, entertainment, games, and homes. In short, it has moved into our everyday life. My retired father and oldest daughter use it. Use is no longer delineated but seeps into, and gets interwoven with, a number of my activities: while paying for my groceries, planning this summer's vacation, and so forth.

The aim here is not to plunge deeper into speculation about what the future will bring. The point is simply to underline the fairly obvious observation that our use of IT is changing. Changing, we would hold, so much that our traditional notion of 'use' of IT should be re-examined. It carries too much of a tool connotation: the image of a competent User mastering her Tool—even after modifying the image through an appeal to a more phenomenological understanding of a tool (Suchman 1987; Ehn 1989; Greenbaum and Kyng 1991). Instead of 'using' technology, we should look at ways to explore how we live with technology.

To avoid this, a viable—but largely uncharted—strategy is to focus on, and learn from, studies of mundane 'everyday technology', such as washing machines, phone-answering machines, and television, because they emphasize the new role for IT—where we mingle around it without much ado. The motivation for looking at how we operate, perceive of, and mingle with everyday technology is two-fold.

First and most immediately, an interesting aspect of our new 'use' of IT is the way it rapidly becomes natural, mundane, or trivialized (Lie and Sørensen 1996). Secondly, the way everyday technology is embedded underscores the need to 'see the social and symbolic as well as material objects' (Silverstone, Hirsch, and Morley 1992: 2). A focus on the everyday world encourages a richer and fuller grasp of human action—underscoring symbols, rituals, identity construction, and values—rather than traditional, more or less functional explanations of purposeful, work-oriented tasks (Silverstone, Hirsch, and Morley 1992; Lie and Sørensen 1996).

There are several proposed metaphors for conceptualizing this, including: 'appropriation', 'domestication', 'consumption', and 'taming' of technology (Silverstone, Hirsch and Morley 1992; Mackay 1997; Lie and Sørensen 1996). To focus on one: domestication is intended to emphasize the naturalization process, the way we cultivate and discipline artefacts when weaving them into the domestic sphere. It simultaneously underlines how 'the wider world of work, leisure, and shopping are defined . . . [and] are expressed in the specific and various cosmologies and rituals that define, or fail to define, the household's integrity as a social and cultural unit' (Silverstone, Hirsch and Morley 1992: 18). More precisely, the process of domestication may be seen to comprise (ibid.: 21):

- appropriation: where the access to the artefact is defined;
- objectification: through which various classificatory principles identify perceptions and claims for the status of the artefact;
- incorporation: the routinization or embedding into social patterns; and

- conversion: the presentation of the constructed artefact to the world outside the domestic sphere—the household's contribution to the currencies of meanings.

Consider an example. An bank's automatic teller machine (ATM), the automatic check-in teller at airports, and electronic commerce on the Internet are all basically the 'same', in the sense that they are artefacts that allow a service (cash, travel, or purchase) to be carried out. Despite the similarity, the 'use' of this technology varies a lot, arguably more than is reasonable to ascribe to the differences in the artefacts themselves. We all use ATMs, some use check-in tellers, and a few purchase on the Internet. The key reason, we suggest, why these differences prevail is that they are in quite different stages of being interwoven into the social fabric of life. How should this process be conceived of, when we do not simply 'use' these artefacts? How does it take place? What drives it? How important is your perception of what others do? How does trust get established? How many fellow passengers need to queue up in front of you before you insert your own ticket? These kinds of questions indicate the shortcomings of the traditional notion of 'using' technology.

This has immediate implications for design. Being trapped in the traditional vocabulary of the 'use' of technology is not merely a luxury. It is not the case that we, IS designers, can go on talking about the 'use' of technology and leave a more fine-grained analysis to, say, social scientists with their particular disposition for details. On the contrary, we believe it is essential and high time to develop further our conceptualization of the 'use' of IT in order to do better design.

Implementing an information system

The term 'implementation' is widely used by practitioners and scholars alike when describing (the status of) an information system. It is also, for exactly this reason, one of the most problematic: it is a catch-all phrase that disguises more than it reveals (see also Walsham 1993). The statement that 'we have implemented the system in our organization' may, when unpacked into its more concrete contents, signify any of a number of painfully distinct empirical phenomena; as Brunsson and Olsen (1993: 6) remind us: 'Implementation is particularly difficult if a reform requires the active participation of the reformees . . . It can sometimes be difficult to change organizational symbols, but it is much harder to change behaviour. And even if reformers succeed in changing behaviour, it may not necessarily be (only) what they intended'.

Given the insights embedded in the notion of domestication, what is really involved as an information system is gradually 'implemented' (i.e. seeps into the fabric of everyday practice)? In other words, what is involved in the process over time as an IS moves along the many, distinct meanings of 'implementation'? This focuses on the difference between: 'implementation' qua coded and prototype use; widespread installation with a few pockets of use; and a situation where the IS is domesticated by everyone, having meshed with ongoing practices and acquired a place in the symbolic universe of people's lifeworlds.

Thus, critical IS researchers need to recognize the treacherous nature of key terms like 'implementation' and 'use' of an IS. The traditional separation into phases—development, introduction, use, and diffusion—has serious shortcomings that reduce its relevance. Empirically, these terms are dramatically underspecified and, analytically, they hide precisely what a proper analysis is supposed to explain: exactly *how* is a system used and implemented?

A key challenge is to develop an understanding (empirically underpinned, conceptually framed) of how these processes unfold, based on a firm grasp of human action that includes symbolic aspects—thus moving beyond too goal-oriented accounts. The challenge is to explain terms like 'implementation' and 'use', not to employ them. In other words, to demonstrate how 'implementation' and 'use' involve processes of domestication.

History of domestication

In a similar manner to having a focus on rituals of everyday life, we believe it is useful to learn from historical accounts of domestication. The driving force behind looking at historical processes of the domestication of technology is to develop a firmer grasp of the unfolding dynamics. How is it that new, flashy technology gets transformed into mundane, invisible technology? A telling example is Nye's (1990) description of how electricity seeped into the social fabric of a typical small town in the United States at the turn of the century. Again, electricity certainly did not get 'introduced' and 'used' in any straightforward sense. It constantly transformed itself—and social order—as it gradually seeped into the pores of modern urban life.

To illustrate, the domestication of electricity in Nye's description extends well beyond simplistic dynamics of the development/introduction/use kind. Nye describes an evolving, dynamic, and reflexive process whereby electricity gets shaped by, but subsequently shapes, urban life. This unfolds through stumbling and improvised, yet comprehensive, exploration of the new technology. That process is not merely one of fitting the potential with the needs. Electricity feeds into the construction of modern, urban life. In doing so, it taps into deep sentiments and desires. For instance, the early 'use' of electricity was lighting. This involved mainly the lighting not of homes, but of spectacles: highly profiled, public buildings with iconic or mythical status, for instance the New York Stock Exchange, La Scala opera house, and Niagara falls. This implied that their 'social meaning resided not only in skilful electrical engineering but in the public perception of the new technologies as spectacle' (Nye 1990: 58), which ultimately meant that electrification contributed to the construction of national identity as it 'was becoming an essential part of experiencing both natural and national symbols' (ibid.: 61). Hence, it contributed to the 'technological sublime' by simultaneously endowing awe and reverence (ibid.: 59).

Such a perspective is illuminating when considering, for instance, the ongoing proliferation of mobile phones and mobile technology. The rapid diffusion of mobile phones makes the traditional notion in IS research of the 'use' of technology extremely crude. Mobile phones are not merely 'used'; they are domesticated and tamed as part of our ongoing, everyday life. They are cultivated as part of our self-identity. The boundary between the professional, work-oriented sphere and the private, everyday one is eroding. The 'use' of mobile phones hinges significantly on the (ongoing) manner in which they are ascribed symbolic meaning (Nye 1990; Silverstone, Hirsch, and Morely 1992; Lie and Sørensen 1996).

There does not seem to exist much detailed, empirical work on how mobile phones get appropriated into everyday life. Still, it seems reasonable to focus studies on the rich set of ways in which phones are domesticated to everyday life. This would imply starting to unpack not from the designers' inscriptions, but rather from the variety of patterns of use. How are phones used symbolically to gesture membership to certain communities? In what ways do phones express and present externalized images of identity? What does the compact and abbreviated language that teenagers use for messages signify?

Identity formation

All the areas above have, in different ways, pointed out how those of us involved in future IS studies will have to incorporate a richer, more symbolic, or 'cultural' account of human action—simply because the phenomenon of IS is expanding into new terrain. In my rather impressionistic indication of areas where this is particularly pressing, let me end by pointing out the ways that the communicative use of IS are rapidly outgrowing the traditional, smallish, semi-collocated Computer Supported Cooperative Working (CSCW) situations.

Information systems are used to communicate globally in interesting ways. These systems necessarily have to be domesticated locally. This involves, in part, the formation, presentation, and shaping of identity through IS-mediated communication. Some aspects of the 'use' of the Internet may be recognized more as expressions of identity and self-presentation than any form of strategic/goal-oriented behaviour. The Internet is but one of many manifestations whereby we present ourselves. Its 'use' carves out territories and arenas that, over time, acquire the status of communities.

To illustrate, we draw on a study reported elsewhere (Hertzberg and Monteiro 2002), which discusses how a globally-operating manufacturer of high-quality ship equipment—and their servicing requirements—organizes the communication between the key production facilities, located largely in Scandinavia, and a dispersed network of local service and sales sites worldwide. The manufacturer's management regard a local presence, with its closeness to customers, to be essential. As one manager pointed out (ibid.: 19):

> Local presence is the company's greatest differentiator in the market. The fact that we have the largest ground force in the market is one of our key strategic advantages . . . a key objective is to . . . build local customer relationships

When visiting local sites, however, there was evidently a sense of being second-order citizens of the organization, halfway between 'home' (Scandinavia) and their customers. Their electronic communication with 'home' through a variety of information systems is influenced by this. The contents, type, and frequency of electronic communication feed into their ongoing anxiety regarding their identity. Or, as one of the study's informants, a local representative, phrased it, whether their identity was that of a step-child (ibid.: 33):

> I got the feeling that they feel like, and I hate to say this, that we would misuse it. And I therefore feel like, you know, I hate to say, *feel like a stepchild*, but in that sense basically I have been here for 10 years, and I don't think I have misused too much of any information.

Thus, the company's employees' communicative 'use' of their IS is as much about reassuring identity and membership as it is about purposeful delivery of products and services.

Conclusions: A Grand Theory of Everything?

Given the elaborate and nuanced concepts in ANT which have been so convincingly demonstrated in the booming collection of cases that have employed an ANTish framework, it goes without saying that the tendency towards goal-directedness is slight. The coining of our critique as 'goal-directedness' should accordingly not be mistaken to be the ridiculous claim that ANT is comparable to simplistic accounts of goal-oriented behaviour. Of course it is not. Rather, the critique makes the more modest claim that there are some cases, including those illustrated above, that pose this question.

There are surely none who seriously hold that ANT is without boundaries, that ANT somehow is the last dinosaur of a Grand Theory. There is a fairly widespread consensus among the well-versed that ANT cannot be employed for everything. The problem for less-versed IS researchers, however, is that little or nothing of this is made explicit. The hard-core ANTers keep themselves busy refining, extending, adding nuances as 'the most important theoretical and practical questions which we confront: how to deal with and fend off the simplifications' (Law 1999: 10), or curb the 'ridiculous poverty of the ANT vocabulary' (Latour, 1999*a*: 20).

The impressionistic illustrations we have provided will, hopefully, stimulate a continual re-appraisal of analytical frameworks (like ANT) when applied to cases in IS: When do they 'fit' and how do aspects of the empirical case feed requirements for the reworking of the analytical apparatus?

References

Brunsson, N., and Olsen, J. P. (1993). *The Reforming Organization*. Bergen, Norway: Fagbokforlaget.

Ciborra, C. U. (2000) (ed.). *From Control to Drift: The Dynamics of Corporate Information Infrastructures*. Oxford: Oxford University Press.

Coussins, C. (1996). 'Ontological Choreography: Agency for Women Patients in an Infertility Clinic', in M. Berg and A. Mol (eds.), *Differences in Medicine*. Durham, NC: Duke University Press, 166–201.

Ehn, P. (1989). *Work-oriented Design of Computer Artifacts.* Stockholm: Arbeitslivscentrum.

Greenbaum, J., and Kyng, M. (1991) (eds.). *Design at Work.* Hillsdale, NJ: Lawrence Erlbaum Ltd.

Hanseth, O., Monteiro, E., and Hatling, M. (1996). 'Developing Information Infrastructure Standards: The Tension Between Standardisation and Flexibility'. *Science, Technology & Human Values*, 21/4: 407–26.

Hertzberg, D., and Monteiro, E. (2002). ' "Global but Local": Dilemmas of Global Organisations'. Paper presented at the European Group on Organisational Studies (EGOS) 2002 conference, Barcelona, July 2002.

Kyng, M., and Mathiassen, L. 1997 (eds.). *Computers and Design in Context.* Cambridge, MA: MIT Press.

Latour, B. (1999*a*). 'On Recalling ANT', in J. Law and J. Hassard (eds.), *Actor Network Theory and After.* Oxford: Blackwell, 15–25.

—— (1999*b*). *Pandora's Hope.* Cambridge, MA: Harvard University Press.

Law, J. (1999). 'After ANT: Complexity, Naming, Topology', in J. Law and J. Hassard (eds.), *Actor Network Theory and After.* Oxford: Blackwell, 1–14.

Lie, M., and Sørensen, K. H. (1996) (eds.). *Making Technology Our Own?* Oslo: Scandinavian University Press.

Mackay, H. (1997) (ed.). *Consumption and Everyday Life.* Milton Keynes, UK: Open University Press.

Mol, A., and Law, J. (1994). 'Regions, Networks and Fluids: Anaemia and Social Topology'. *Social Studies of Science*, 24: 641–71.

Monteiro, E. (2000). 'Actor-network Theory', in C. U. Ciborra (ed.), *From Control to Drift: The Dynamics of Corporate Information Infrastructures.* Oxford: Oxford University Press, 71–83.

Nye, D. (1990). *Electrifying America.* Cambridge, MA: MIT Press.

Orlikowski, W., and Barley, S. (2001). 'Technology and Institutions: What Can Research on Information Technology and Research on Organizations Learn from Each Other?'. *MIS Quarterly*, 25/2: 145–65.

Pickering, A. (1992) (ed.). *Science as Practise and Culture.* Chicago: University of Chicago Press.

Silverstone, R., Hirsch, E., and Morley, D. (1992). 'Information and Communication Technologies and the Moral Economy of the Household', in R. Silverstone and E. Hirsch (eds.), *Consuming Technologies*, London: Routledge, 1–15.

Star, S. L. (1991). 'Power, Technologies and the Phenomenology of Convention: On Being Allergic to Onions', in J. Law, *A Sociology of Monsters*, London: Routledge, 26–56.

Suchman, L. (1987). *Plans and Situated Action.* Cambridge: Cambridge University Press.

Walsham, G. (1993). *Interpreting Information Systems in Organizations.* Chichester, UK: John Wiley.

—— (1997). 'Actor-network Theory and IS Research: Current Status and Future Prospects', in A. S. Lee, J. Liebenau and J. DeGross (eds.), *Information Systems and Qualitative Research*, London: Chapman & Hall, 466–80.

8

Farewell to constructivism: technology and context-embedded action

Jannis Kallinikos

Introduction

If you are asked whether technology influences human behaviour, do not hasten to answer in the affirmative. You may be accused of being determinist, and you could attract a swarm of other, rather derogatory, adjectives. If you express the belief that technology is a distinct realm of social life marked by its own specific rules, practices, and conventions, then you may be looked upon with suspicion and declared as an unforgivable essentialist (e.g. Grint and Woolgar 1997). Easy labelling of this sort is, of course, an old rhetorical strategy that serves purposes other than analytic. Yet the issues raised by these labels are deeply serious. The view that technology is but social, and technological artefacts interpretable and reshapable to the demands of situated agents, has tended to acquire the status of an unquestionable belief, at least as far as the disciplines of information systems and organization studies are concerned. While being originally a sound counter-reaction to a simplified conception of technology as descending from engineering, such a belief is misleading, unless qualified in elaborate ways.

Brought to its extreme, the view of technology as locally negotiable ends up bypassing both the issues raised by the highly sophisticated strategies of objectification which contemporary technologies embody, and the forms by which such objectified inventions have cumulated over time. Such an understanding of technology becomes inevitably user-centric and tends to reduce the impact of technology to that slim area upon which users, or humans in general, encounter technology.[1] Technological systems, though, involve a large array of technical and organizational factors that may not be apparent at the level at which humans operate or come into contact with technology (Borgman 1984). For instance, the behaviour of passengers in airports can often be understood more adequately by reference to the overall structure of control procedures imposed for air safety and the mass handling of people, rather than by relating it to the dispositions of particular people with regard to these procedures (Misa 2003). Similarly, apparently simple software applications like word processing are supported by a complex array of software and

The author thanks Hans Hasselbladh for reading and commenting extensively on various aspects associated with the development of the major argument of this chapter. He also extends his appreciation to Chrisanthi Avgerou, Frank Land, and Carsten Sorensen for their comments on an early version of the chapter. The author retains, of course, full responsibility for the arguments put forth in the chapter.

hardware components, the history of writing techniques, and other institutions, only a small portion of which is instantiated at the user interface (Kallinikos 2002).

Legitimate and highly desirable as it may be, the user-centric understanding of technology avoids confronting the distinctive standardization brought about by technology. It re-echoes an old-fashioned humanism that re-enhances the widespread and ultimately rationalistic (despite assertions to the contrary) conception of humans perceived as being fully in command of their affairs (Heidegger 1977; Kallinikos 1996). After all, technology is a servant of human agents, and if it fails to stand up to this expectation it can be ignored, resisted, or reshaped to achieve the goals that are usually perceived as being tied to its implementation in particular settings.

In this chapter, I would like to debate—and ultimately question—that belief. Technology is certainly socially constructed (what else could it really be?). However, the historic conditions under which particular technologies emerge and develop, and the forms by which they have become institutionally and socially embedded, often coalesce in ways that can make technology a recalcitrant ally. Human inventions solidify over time, as layers of technical, organizational, and social developments get superimposed one upon another to create complex systems that impose their ways of operating. This observation is crucial to the study of technology and an appreciation of the impact it may have on human affairs. For, as suggested above, the activities that take place at the human–technology interface cover but a limited area of a wider system of instrumental relations, sustained by a huge network of technical, organizational, and social arrangements that render the functionality and contextual enactment of a specific technology possible. Some salient examples include railroads, electricity, or gas networks and large information packages (Hughes 1987; Misa 2003). Systems of this sort choreograph, as Misa (2003) formulates it, the human effort in intricate routines that evade contextual adaptation or manipulation.

The above observations foreshadow the argument to be put forth in this chapter. The study of technology and its social impact cannot be exhausted at the very interface upon which humans encounter technology. Essential strips of reality are not observable or even describable at the level of contextual encounters (Searle 1995). Situated accounts of technology must be supplemented by wider reflection that captures the complex web of dependencies, interoperabilities, and institutional relations that sustain the embeddedness of technology in local contexts. Such a complex web of relations surpasses the horizon of the present, and involves the mediation of history and culture in ways that cannot be reduced to the study of actor networks (e.g. Latour 1986; Bijker and Law 1992). History and culture reach down into mundane technological operations in a variety of ways, but a very forceful one is what Hanseth (Chapter 5 this volume) calls backward compatibilities: the need to comply with the constraints (solutions, standards, organizational arrangements, etc.) that the march of technology has produced through time.

The chapter is structured as follows. I initially reconsider in broad terms the involvement of technology in human affairs. In so doing, I revisit the constructivist understanding of technology in order to place its contribution into a wider context of technology studies, while at the same time discussing some of the claims of constructivism that I find problematic. I then outline a view of technology as a complex interdependent system of technical, social, and organizational elements, along the lines briefly mentioned above. The major claim presented is that these elements enter into multiple relations of compatibility and mutual accommodation; these constrain the very inputs the system admits, and invite human participation along highly selective paths. Local involvement of technology comes with a huge package of constraints that enable human choice through precisely the exclusion of alternative paths of action. Some of these claims are then exemplified by considering the developmental trajectory of those configurable software packages known as Enterprise Resource Planning (ERP) systems.

Constructivism revisited

The significance currently attributed to the processes through which human agents develop, encounter, and shape technology in local contexts is closely associated with the diffusion of social constructivism and the study of technology's involvement in human affairs by recourse to constructivist methods (e.g. Bijker, Hughes, and Pinch 1987; Bijker and Law 1992). Social constructivist approaches to technology have somehow managed to become identified with the term 'social studies of technology', but they are by no means the only ones that stress the social character and involvement of technology. The social study of technology has a long history (e.g. Mumford 1934; Winner 1977). It would not be unfair, I believe, to claim that the major contribution of constructivist studies of technology is the primary importance it attributes to contextual dynamics, rather than the awareness which constructivism may have created of the social involvement of technology. Whether investigating the development of new technological innovations (Bijker, Hughes, and Pinch 1987; Suchman 1987, 1996) or the forms by which existing technologies are brought to bear on local contexts (Suchman 1996; Heath, Knoblauch, and Luff 2000; Orlikowski 2000), social constructivism gives a key role to the very processes by which technology is locally negotiated.

From the constructivist view, the development or use of technological systems involves a heterogeneous ensemble of social, technical, or scientific elements that agents draw upon in order to frame, instrument, and act upon the issues they confront. Such elements often support, implicate, deny, or undo one another, in a political or micro-political game that involves the pursuit of divergent and often incompatible objectives (Bijker 2002). Nevertheless, technological systems produced this way tend to become gradually closed and consolidated, exerting significant influence upon posterior choices (Hughes 1987). However, the unquestionable impact of technology on society and organizations cannot be adequately understood; constructivism maintains, apart from the objectives, capabilities, and biases that criss-cross those contexts into which each particular technology finds itself embedded (e.g. Bijker, Hughes, and Pinch 1987; Suchman 1987; Grint and Woolgar 1997).

The complex socio-political character of the processes of technological change and innovation, and the inescapably context-embedded forms of its redeployment and remaking, constitute the basic tenet of constructivist research. Seen as a reaction to the oversimplified assumptions of a rigid and largely a-social world outlook of economists and technologists, social constructivism appears to have made a valuable contribution to the contemporary understanding of the process of technological development or use. Such a contribution underscores the haphazard, ambiguous, and local character of this process, and the social dynamics out of which technological systems and artefacts emerge and become embedded or used in particular settings.

Despite this, the contribution of constructivism remains somewhat suspended when it is placed within the wider field of the social studies of technology. Indeed, stripped of its details, social constructivism would seem to deliver a rather trivial message. As already indicated, the assumption concerning the social character of technology does not appear particularly innovative. The reciprocal and complex relationship of technology to society is one of the more persistent themes in the intellectual history of techniques. In *Technics and Civilization*, his monumental work written in the 1930s, Lewis Mumford (1934) devotes his introductory chapter of nearly sixty pages to what he called the 'cultural preparation' coinciding with the technological take-off of western modernism. With the rigour and imagination characteristic of his style, Mumford describes how a diverse system made of values (e.g. progress, materialism, equality), institutions (e.g. money, market capitalism, science), interests (e.g. classes and social groups), and diverse techniques (e.g. central perspective in painting, cartography, printing, measurement, and calculation) was essential to the technological orientation and material development of modern times. All these forces combined to form what social constructivists would today call a heterogeneous ensemble of techniques and forces out of which the development and consolidation of various technologies emerged. Other important studies on technology have also pointed out and documented the complex organizational, social, and political processes underlying the invention, development, consolidation, and deployment of technological artefacts and systems (e.g. Ellul 1964; Winner 1977; Castoriadis 1984; Noble 1984; Beniger 1986).

Studies of the sort exemplified by Mumford (1934, 1952) or Winner (1977, 1986) exhibit a macro-sociological or historical orientation that social constructivists claim is prone to overlook the specific circumstances under which any given technology is locally understood and managed. From the constructivist's view, studies focusing on macro-sociological or historical patterns of technological evolution tend to leave obscure the diversity of the forms by which any given technology is locally developed or used. It is precisely this void, created by the bird's-eye view intrinsic to the historical, macro-sociological studies of technology, that constructivism claims to be addressing. By focusing and documenting context-specific processes, constructivism claims to be better placed to disclose the irrevocable contextual dynamics by which technology is negotiated, interpreted, remade, or even misused in local contexts—and its possibilities reinvented, redeveloped, manipulated, or destroyed. Human intentionality, as revealed in the active complicity of social actors in constructing and interpreting technological objects, becomes an essential and indispensable component of the overall processes by which various technologies, as opposed to technology in the abstract, are developed and used (Suchman 1987, 1996; Leigh-Star 1992).

Beyond local contexts

The constructivist claims outlined above take us to the heart of the issues that concern us in this chapter. The disclosure of the context-embedded processes through which technology develops or becomes involved in local affairs is indeed essential to the social study of technology. Yet, the same holds true for all those wider social constituents that reflect the developmental trajectories of particular technologies and the institutional/organizational forms by which they have become involved in social life. It is woefully regrettable that constructivism has turned its back, if not in principle at least in practice, to these wider processes (Misa, Brey, and Feenberg 2003). There are various limitations following from such a choice.

First, technological characteristics that embody these wider historical and institutional processes and which partake in the constitution of local contexts are ignored.[2] Social agents are not disembodied spirits; instead, they are complex ensembles of skills, proclivities, and roles, some of which are brought into being by technology itself (Hasselbladh and Kallinikos 2000). Technology is not just an exterior force that encroaches upon local, technologically 'unspoiled' contexts, though it may be used that way; most of the time, technology partakes in the constitution of local contexts and agents. The issue of 'who' interprets and locally negotiates the meaning and functionality of technology therefore needs to be addressed.

Secondly, there is the issue of technological complexity due to backward compatibilities, functional interoperabilities, and standards—and all those technical and institutional constraints about which locally-embedded agents do not have much discretion (Hanseth 2000). When placed against the backdrop of this technical and institutional tangle, the constructivist notion of interpretive flexibility (Bijker, Hughes, and Pinch 1987) can be seen in its right proportions. Commenting on W. I. Thomas' dictum stating 'if men define situations as real, they are real in their consequences', an acknowledged interactionist such as Erving Goffman (1974: 1) provides the following instructive remark:

> [T]he statement is true as it reads but false as it is taken. Defining situations as real certainly has consequences but these may contribute very marginally to the events in progress; in some cases only a slight embarrassment flits across the scene in mild concern for those who tried to define the situation wrongly. All the world is not a stage—certainly the theatre isn't entirely.

Thirdly, there may be major technological characteristics whose adequate understanding demands to be studied in ways that transcend local contexts. For instance, the growing involvement of information systems in organizations could been seen as tending to cognitivize work and make it an extensive encounter with symbol tokens of various kinds (e.g. Zuboff 1988; Sotto 1990; Kallinikos 1999). While the study of computer-based information systems in context helps to illuminate effects of this sort, the adequate appreciation of these systems necessitates placing their study in a wider historical context that takes into account the evolution of writing techniques, the development of paper-based information systems, and the role oral forms of interaction and communication have assumed in this process (e.g. Beniger 1986; Bolter 1991). Historicity becomes crucial for understanding how technology surpasses its embeddedness in the 'here and now', and influences humans by framing and parsing reality in surreptitious ways that reflect a long process of evolution (Cassirer 1955).

Fourthly, epistemological/methodological concerns informed by the aforementioned remarks suggest that the abstract characteristics of technology that transcend time and space are not observable in the way that usually applies to human behaviour in local contexts. Ethnographic accounts of the processes by which technology is assimilated in local contexts may thus be unable to document these abstract attributes of technology. Local actors may reflect on one or other aspect of technology, but they are unlikely to give accounts that touch upon wider social or historical processes. In principle, nothing prohibits the empirical investigation of local contexts from assembling the wider picture of the historical and institutional embeddedness of particular technologies. In practice, however, the study of local processes usually ends with a number of interviews, some observations, and—in the best case—with the study of documents that are associated with the local involvement of a particular technology (e.g. archives, decision minutes, protocols).

There is often a sort of what I would call 'ethnographic positivism', where empirical data that are basically fabricated by the researcher's orientations, perceptive biases, and methods are taken as the ultimate yardstick of assessing reality—at the same time as important, but not immediately observable, aspects of that reality are ignored. Although social constructivism cannot be made responsible for this sort of revenge which empiricism takes on theoretical imagination, it is how constructivist studies of technology have virtually ended up (e.g. Orlikowski 1992, 2000; Heath, Knoblauch, and Luff 2000). The spirit of the situation is captured well by Richard Penniman's poignant formulation which Winner (1993: 371) brings to bear on the present context in claiming that, in the end, social constructivists 'got what they wanted but they lost what they had'.

Implications of the growth of cross-contextual systems

The discussion above acquires particular importance as contemporary forms of life are increasingly characterized by the rising diffusion of all sorts of artefacts and systems that cross the boundaries of local contexts, mixing remote traditions, techniques, and orientations (Giddens 1984, 1990; Beck 1992, 2000). Indeed, what we call local contexts are very different today, not only from the relatively isolated world of medieval communities but also from the corresponding contexts of early modernity. They are pervaded by systems, artefacts, and life styles that originate from other, often radically different traditions and life patterns. The issues associated with the ways social and administrative traditions and forms that are locally embedded may implicate cross-contextual systems in their reproduction therefore emerge as crucial, at the crossroads of the current age. All sorts of technologies and systems now travel rapidly across the entire globe, reconfiguring important premises of local contexts and thereby contributing to the formation of new expectations—and not infrequently to disappointment or frustration.

A few tricky questions are related to the increasing diffusion of technologies, methods, and artefacts of global reach. Local practices by definition reflect a long process of trial-and-error learning. They are the outcome of experience (adaptation in the broad sense of the word) sedimented in culture and social structure. However, the speed and comprehensiveness of change taking place today defy the languid, time-consuming forms by which situated practices develop. We must join Luhmann (1993) and admit that there is no time for locally-induced adaptation today, or perhaps that time is becoming an increasingly intractable issue for situated forms of action. The problem is solved by importing time in the form of elsewhere-constructed systems and technologies.

The increasing presence of cross-contextual systems/technologies in organizations makes evident that they cannot any longer be treated as an exterior and largely disturbing form that violates the cannons of local contexts and situated action. Cross-contextual systems/technologies participate in various ways in the making of local life. Their relationship to local contexts cannot be gauged simply in terms of exteriority. The question is not simply of interpreting, reconstructing, appropriating, or even enacting them, as Orlikowski (2000) suggests. The time is ripe to chart how these systems participate in the making of local contexts and situated forms of actorhood (subjectivities), by defining the domains of relevance and providing the means and procedures for acting upon these domains, forms of communication and information exchange, and work monitoring. The widespread assumption that people define their situation in connection with their choice to act in particular ways (the interactionist legacy), and that they therefore approach technological objects and devices through a process of interpretation and sense-making, must be attributed its right proportions.

Technology and human action, cross-contextual systems, and context-embedded forms of life cannot adequately be understood on the assumption that the one is exogenous to the other. Interpretation is not a posterior act of resistance to technology or a sinister appropriation of it. These options cannot, of course, be ruled out. There are plenty of cases to indicate a collision of values and practices that make the model of 'resistance' an appropriate tool for studying the forms by which cross-contextual systems are brought to bear on context-embedded practices. However, interpretation is not limited to resistance. It constitutes an intrinsic part of the social-life process, by means of which humans seek to enact actively the possibilities inherent in technologies, and cross-contextual systems in general, in order to construct particular forms of actorhood in local settings (Hassellbladh and Kallinikos 2000).

On the other hand, while interpretation and sense-making are indispensable components of the process of embedding any given technology in a local context, they take place from a much narrower horizon of meanings than that constructivists tend to assume. Contemporary humans do not encounter a given technology with wide-open possibilities, nor does technology itself admit similar patterns of human participation. The human–technology interaction is subject to premises that are tight, yet elusive and easily overlooked. The use of apparently simple devices such as the car, TV set, or word processing program presupposes training, development of skills, and conformity to core procedures. Most importantly, the involvement of people with technologies in formal organizations is not predicated upon the people's mobilization of their entire spectrum of interests, values, feelings, and orientations underlying their personality. Rather, in situations of this sort, contemporary humans invoke selected aspects of themselves that usually have an instrumental orientation, while suspending the invocation of all other characteristics of their personality that do not bear upon the situation (Luhmann 1995; Gellner 1996; Kallinikos 2003).These claims demand a detailed exposition.

The system of technology

Functional interdependencies and temporal patterns

The standardized and recurrent forms by which technology invites human participation can be appreciated against the background of its systemic character. I use the term 'system' here in a rather loose way to indicate the dynamic interdependence of the items, procedural steps, and processes constituting particular technologies. A given technology could thus be seen as constituting a complex arrangement of items, steps, and operations that bear upon one another in ways that create a pattern of functional interdependencies and temporal sequences. The functional interdependence and temporal pattern of the component items/processes demarcate the system in relation to its environment, and represent a crucial means upon which the self-referential functioning of the technology is predicated.

The patterns created by the functional and temporal interdependence of the various items, steps, or operations represent evidence of the systemic character of technology. Perhaps modern technology could be envisaged as a system by exploring the insights of a linguistic analogy (Kallinikos 1995, 1996). A given technological system could thus be analyzed in synchronic (metaphoric) terms that capture the population and the functional interdependence of its elements, and on diachronic (metonymic) terms that define the temporal sequence of the elements and the formation of processes. It is of the utmost significance to realize that, to a large extent, the design/functionality of particular items and operations responds to the demands for functional complementarity, compatibility, and internal accommodation with the other items and component processes that constitute a given technological system (Borgman 1984).[3] This is how technological systems are constituted and reproduced through processes that exhibit strong elements of self-referentiality ('blackboxing').

Drawing on similar observations, Luhmann (1993) suggested that the 'geist' of modern of technology is captured by the twin strategy of 'functional simplification' and 'closure'. This suggests that technology is predicated upon the selection of a function or set of functions and the layout of sequences and causal patterns which guarantee their *recurrent order* through the very exclusion of known or unknown events that may have disruptive effects on its operations (Mumford 1934, 1952; Kallinikos 1996). The predictable forms by which technology regularly operates are due precisely to the construction of causal patterns and sequences, whose recurrent unfolding is ensured through the exclusion of unforeseen factors. Functional simplification, in this sense, coincides with the controlled reduction (insulation) of causal interactions. In so doing, however, technology compresses and *intensifies* these interactions and magnifies both their power and disruptive force. Nuclear power stands as the epitome of technology's distinctive way of constructing/regulating reality, but most technologies make use of, albeit in varying degrees of intensity, the same strategy of control and regulation.

The example of nuclear technology indicates that the strategy of functional simplification does not make technology a simple accomplishment. Quite the opposite, it implies the selection of a function, or set of functions, that has to be performed in a reliable fashion, often in a magnified scale involving various types of complications. Another example is that of the technology of automobile freeways. Functional simplification here is expressed in the speed by which long-distance traffic is organized, and it is instrumented through broad, multiple, direction-separate lanes that entail functional-closure as a means of excluding incidences that may disturb traffic flows. Functional simplification and closure imply that the freeway system admits only certain types of inputs and behaviours; walking, biking, and driving at low speed are instances of inputs and behaviours that violate the premises of the technology of freeways.

The selective premises by which humans encounter any technological system, and the limited space of discretion they have in relation to that system, is due precisely to the twin strategy of functional simplification and closure—and the self-referential, recursive way by which they are called upon to join the system. Accidents and technological failures suggest that the strategies of functional simplification and closure are always incomplete, and that technological systems do differ in terms of how well these strategies are organized and implemented. It is regrettable that constructivists have paid so little attention to these cardinal differences, as they shape considerably the premises upon which humans encounter technology and define the discretion they have to modify the system locally. To a large extent, this holds true for configurable technologies, although differences do exist, as discussed in the last sections of this chapter.

The role of self-referentiality

The description of technology as a self-referential system applies equally well to both relatively simple devices, such as those exemplified by industrial appliances, and complex industrial technologies, for instance those represented by the assembly line in mass production or the refinement process in continuous-process industries. Strange as it may seem, the self-referential character of technology is better exemplified by contemporary computer and information technologies. Simpler programs or complex software packages are nothing other than arrangements of cognitive items and functions, together with a set of rules or procedures by which they relate to one another. Connectionism as a programming methodology, in particular, provides clear evidence of the recursive forms by which technology acts upon its environment. A program is able to learn by drawing on the 'experience' produced by its encounters with its environment. The environment itself is perceived through the activation of 'experiential knowledge' embodied in the learning algorithms that manipulate the strength of the connections of the various units underlying the program (Kallinikos 1998). In other words, the program projects onto the world an image of itself (its items and connections) and subsequently acts on that image.

Despite its self-referential character, no technology could avoid having some sort of relationship to the world. After all, technologies are invented and deployed to accomplish things in the world. But even such an 'allo-referential' relation is the outcome of the highly selective way by which technology comes in contact with the referential reality. For instance, automobile technology understands the physical world as an extension to be crossed as rapidly as safety would allow, and designs its processes/products with the ultimate purpose of responding to that goal. To see the earth as simply an area to be crossed involves a huge degree of abstraction, and a highly selective relation *vis-à-vis* the multidimensional, diverse, and ambiguous character of the physical world. Such a selective relation is precisely the fabrication of the automobile technology's strategy of functional simplification, which is constitutive of that particular technology. This suggests that neither the automobile industry nor any other technology can come into being unless they demarcate, in these highly selective terms, an 'object-sphere' upon which they unfold their strategy of functional simplification (Heidegger 1971, 1977).

In contrast to this slim reference which technology as a whole bears to the world, the elements and component operations of the technological system are attuned to one another in ways that may lack total reference to it. The constitution and mutual accommodation of the elements of the system (i.e. a given technology) are justified only as means that guarantee the reproduction, incorruptibility, and functional coherence of the system as a whole (Kallinikos 1995, 1996, 1998). The self-referential, inward looking character of technology, imposed by the demands of functional compatibility and temporal patterning, constitutes a major reason why separate items and component processes of technological systems cannot as a rule be torn apart from the overall framework of relations that the system imposes.

The use and reshaping of items or component processes are often meaningless outside the context defined by the technology in its entirety. This is evident in the case of the spare parts of industrial objects or technologies, but applies equally well for the items and operations by which information systems are made. The functionality, and the meaning, of component processes are always contingent on the functionality of neighbouring processes and the system as a whole. The functional compatibility of the system's elements draws technology towards its interior, as it were. It thus tends to impose a procedural standardization (an instance of recursivity) in the forms by which technology is to be encountered, as a means of complying with the patterns by which the elements of the system are attuned to one another.

The systemic and self-referential understanding of modern technology suggested here bears affinities to the description of large-scale technological systems attempted by Hughes (1987). However, our conception is both narrower and, at the same time, more abstract than his. Hughes describes large-scale technological systems in terms of functional, organizational, and institutional interdependence. Technologies like those exemplified by the automobile, chemical, or electronics industry form integrated systems. Their systemic character cannot, however, be attributed solely to the demands imposed by the functional compatibility of the processes underlying them. This character most crucially derives from the large number of heterogeneous institutions and organizations that are usually essential for the functioning of the system in its entirety. For instance, the character of the automobile industry as a system is manifested in the complex and interdependent network of roads, regulations, authorities, subcontractors, petrol stations, etc. necessary for its smooth operation. The same holds true for the electronics industry, the diversity of organizations and institutions involved in it, and the politics evolving around the crucial issue of standardization of hardware and software components. In principle, every technology implies a complex network of technical, social, and institutional elements that enhance and reproduce one another.

Such a way of viewing technology undeniably involves a highly relevant, broader perspective than the one suggested in this chapter, as it incorporates the organizations and the institutional relations that are essential to the reproduction, sustenance, and expansion of a particular technology. However, nothing precludes the expansion of the conceptual framework suggested here to take account of these highly-relevant characteristics and processes. The two perspectives are not incompatible. They just address slightly different facets of the issues occasioned by the complex character of contemporary technology. At the same time, it is important to point out that any macro-technological, historically-informed framework, like that employed by Hughes, has some difficulties in dealing with the crucial issues relating to the recursive and self-referential constitution of modern technology, other than in terms of divergent interests and power relationships.

Technology and formal organizations

The systemic character of technology and its self-referential patterns of reproduction suggest that the degree to which any given technology is amenable to contextual interpretation and reshaping is heavily conditioned by the selective premises through which humans and technologies encounter one another. The cross-contextual system that any given technology constitutes invites participation along highly specific paths of action. Such selectivity is related to the strategies of functional simplification and closure, and the specific ways in which these strategies have been implemented. The forms by which one could respond to these selective invitations may be subject to variation. However, not every course of action is admissible and variation unfolds within recognizable limits. Skill development and training in the forms by which a technology is to be deployed represent strong evidence of the selective terms on which technological systems admit human participation.

How organizations are constituted as social entities

The failure to appreciate the selective forms by which humans encounter technologies is closely associated with another prevalent confusion concerning the nature of formal organizations, perhaps the most significant arena where such encounters currently take place. It is a widespread, though often tacit, assumption that human–technology interaction within formal organizations entails the mobilization of the entire horizon of experiences underlying the lifeworld of humans. Employees or organizational members, in general, are assumed to enact their duties (and their relationship to technology) by drawing on a wide range of emotional, cognitive, and cultural resources—thoughts, feelings, values, modes of action, etc.—that define the entire spectrum of their personality. However, such a premise does not seem to withstand close scrutiny. Modern organizations are constituted as social entities on the basis of a *non-inclusive* relationship to their members. They dissociate the requirements of organizational role-taking and performance from the totality of experiences, moods, and predispositions that underlie what we commonly refer to as the personality, or the particular mode of being, of a person (Luhmann 1982, 1995, 1996; Gellner 1983, 1996).

The non-inclusive way of modulating the contributions of people in organizations implies that contemporary humans are not contained in organizations as existential totalities (Kallinikos 2003). Organizations are composed of selective action and communication patterns built by the relations of those abstract operational requirements we subsume under the label of the 'role'. Roles are enacted by the capacity of contemporary humans to 'bracket' or 'suspend' all other personal or organizational aspects that do not bear upon the role, and to undertake action along delimited and well-specified paths (as a machine operator, accountant, salesman, secretary, etc.). The living energy and the intellectual and communicative capacity of human beings are essential resources for organizations, as it is for all social life. However, despite the impressive numbers of people who everyday traverse workplaces, this should not lead one to assume that formal organizations are made of human beings qua persons (Luhmann 1995, 1996).

Modern formal organizing introduces an abstract conception of work: as a system of actions or duties that can be dissociated from the totality of the lifeworld and from the distinctive mode of being of every person. A major objective and important consequence of the innovative character by which modern formal organization modulates the individual–organization relationship is that individuals join the organization not on the basis of psychological or social characteristics in general.

Rather, individuals are brought to bear upon organizations on the basis of considerations relating to their ability or willingness to assume a role, such as on the basis of merits (or absence of merits sometimes) like education, working experience, and other indicators of adequate performance. By modulating the individual–organization relationship in non-inclusive terms, formal organizations resign from determining other aspects of a person's life that remain unrelated to organizational role performance. No matter how important, the personal characteristics that derive from education, professional specialization, and working experience cover only a part, albeit an important one, of the totality of an individual's roles and projects. An individual's interests and undertakings other than those concerned with occupation and organizational duties are kept or expressed within the context of other organizations and institutions (e.g. family, community) that are clearly and unambiguously differentiated from work organizations.[4]

Selectivity, therefore, is a constitutive relation of formal organizations that is considerably aided, structured, and shaped by the various technologies any organization employs by necessity. The systemic character of technology and the behavioural foundations of formal organization, as outlined above, provide considerable evidence that the terms by which humans encounter technologies are subject to a severe patterning. This means that technologies are deployed to structure and carry out a variety of tasks, procedures, and operations in formal organizations. Goal-oriented, highly-selective behaviour in organizations is instrumented through the means–end sequences constructed on the strategies of functional simplification and closure, while the demands of organizations for predictable, recurrent, and selective modes of action represent an inexhaustible source for technological development (Kallinikos 1998).

Again, these claims do not imply that the interaction of humans with technology can be accounted for in terms of a unidirectional, mechanical causality. In deterministic models, and not infrequently in constructivist accounts, the human–technology relationship is considered exogenous. By contrast, we strive to show throughout this chapter that this relationship cannot adequately be accounted for in such terms. However, our claims do imply that the malleability of technology in local contexts is considerably limited, most of the time. Local reshaping, and perhaps some drift away from the original goals to which a given technology was brought in to achieve (Ciborra 2000), take place along highly selective paths, and with forms other than those commonly assumed.

Dynamics of human–technology interaction

The conception of technology as a cross-contextual system has important implications for the dynamics of human–technology interaction. However, an obvious question emerges: What really happens when humans fail to encounter a particular technology in those terms implicated by the technology itself and the structural/organizational relations within which such a technology has commonly been accommodated? Experience suggests that the more likely response to this problem is the readjustment of either the 'human factor' or the technological/organizational system, or both at the same time. Betrayed expectations are often coped with through rather small incremental adjustments in technology and organizational processes, and more comprehensive changes in manpower skills and behavioural patterns. The impressive current diffusion of all sorts of educational packages, organizational restructuring associated with the implementation of new technologies, and other similar initiatives bears witness to the fact that the reshaping or appropriate formation of behavioural patterns in organizations is the most likely target of analogous endeavours.

These claims suggest that comprehensive technological change does not originate as a rule in the tension created by the abstract principles any particular technology embodies with the molten specificity of local contexts. Such a tension usually brings only small, incremental improvements in technology, even though it may have considerable negative effects for the organization. Radical, comprehensive change in any given technology usually acquires two other forms, which often combine with one another.

First, technological change and development take place against the background of considerations that relate to the lessons taught by the accumulated experience of numerous cases that cut across economic, regional, national, or geographical boundaries. Only in rare cases of high significance can a single local context itself afford to propel comprehensive technological change. Technology has no ways of understanding specificity other than those allowed for by its abstract and cross-contextual constitution (Zuboff 1988). The remaking of the premises of technological systems reflected in the specific forms by which they embody the strategies of functional simplification and closure usually demands momentous change. Most of the time, the experience of single local contexts are transformed and transcribed onto the standardized categories and procedures underlying the technological system.

The second way through which technological change takes places reflects the dynamics of technology itself, and its very constitution as a self-referential system. Forces that drive technological development in one or another direction include: the deliberate or haphazard development of new inventions, procedures, and methods; improvement of the technology induced internally or externally; scientific discoveries; and comprehensive organizational and institutional changes. A notable change in component processes or larger enclaves of a given technology opens a new space of possibilities. It sets in motion a dynamics of readjustment towards a new equilibrium, which could accommodate the functional and temporal independence of the system and the institutional relations sustaining it. The specific premises underlying local contexts reach the dynamics of technological development only as the sound of a remote and receding echo. Again, technological development can occasionally take place as an effort to respond to issues confronted by specific contexts or organizations. Yet, in similar circumstances, the local context is treated as an instance of a large group of contexts: as a representative or exemplary case to be followed and imitated by others.

Human discretion in large-scale information systems

Enterprise resource planning systems

The claims I have advanced so far can be exemplified by reference to the development of large-scale information packages known as enterprise resource planning (ERP) systems. Software packages of this sort have been widely deployed since the 1990s as a means of structuring and monitoring the activities of large and medium-sized corporations (e.g. Ciborra 2000). Such information platforms have a modular architecture and are used as computer-based administrative frameworks for planning and conducting a large array of organizational operations in ways that accommodate their intrinsic interdependencies in real time (Bancroft, Seip, and Sprengel 1996; Ptak and Schragenheim 2000). Operations that were previously conducted with the help of isolated functionally-based information systems—for example covering production planning and scheduling, accounting and finance, sales and marketing, human resource management, and logistics—can be brought to bear upon one another through the development of ERP systems. Transactions that cut across the boundaries of functions can be monitored in real time, retraced, and inspected at any moment.

The development of ERP technology responds to the needs of many organizations to obtain improved inspection and control over internal operations, external transactions, and information flows. Such inspection and visibility could previously be accomplished only afterwards, and most of the times at a considerable cost. However, the history of ERP technology shows that its development cannot be explained by reference to particular organizations and specific sets of administrative problems generated in local contexts. Rather, on the one hand, it has been the outcome of the developmental dynamics of computer and software technology, and on the other the progression over time of computer-based models for handling administrative problems in large organizations (Fleck 1994; Ptak and Schragenheim 2000). Notwithstanding the fact that ERP systems represent a significant technological innovation that has been heavily conditioned by relatively recent developments in computer hardware, the technology has clear antecedents in computer-based administrative history, such as the computer-based Material Resource Planning MRPI and MRPII systems for production planning/scheduling.

Examination of the conceptual architecture of ERP systems, and the procedures by which it is introduced and implemented in particular contexts, provide evidence in support for the claims made earlier in this chapter. The organizational architecture upon which ERP technology is based reveals an elemental functional conception of organizational structure and processes, abstracted from particular details. Organizational operations are organized in terms of functional modules and a huge number of processes, which develop either within the framework of functions or often cut across functions in ways that are supposed to recapture the intrinsic interdependencies of the various operations. Indeed, ERP systems are predicated on a rather traditional understanding of organizations as functionally-segmented units. This traditional view is complemented with newer and fashionable models of management, the so-called 're-engineering models' that understand organizations as a series of operations or sequences whose ultimate and most crucial mission is the satisfaction of their customers. The organizational architecture of ERP systems can thus be directly traced back to the development of administrative models for segmenting, grouping, and managing organizational operations.

A confluence of developments

On the basis of the above briefly-related considerations, it could be claimed that the emergence and development of large-scale information packages exemplify the dynamics of technological development as described in this chapter. ERP technology represents a developmental step in a long and perhaps crooked trajectory of technological changes, made possible by the confluence of several significant technical, managerial, organizational, and institutional developments.

ERP is clearly associated with the growth of functionally-based management information systems and other computer-based business applications. It is furthermore contingent on the expansion of the technology of processing large amounts of data through the development of relational databases. ERP technology has also been clearly enabled by the constantly-rising processing and storing capacity of computers and the development of client–server technology. These are just a few of those significant technical changes with which current software technology, of the type represented by ERP systems, is associated in an almost immediate fashion.

Taken together, these observations suggest ERP technology is the outcome of a complex interplay of hardware and software innovations and organizational and managerial developments which have taken place since the 1960s. A consideration of the complex self-referential dynamics of technological change that is the outcome of the repositioning of various elements or component processes *vis-à-vis* one another reveals that the possibilities for reshaping ERP technology in local contexts are considerably limited. Very little is understood about the effects ERP packages have on particular contexts or organizations if this larger picture is dropped, as it often is, in favour of the study of local details by which the technology is introduced in these contexts.

Consideration of the entire history of ERP technology shows it to be one of those central technological pillars around which human agency in organizations is transformed into a sort of procedure enactment that reflects an implicit understanding of organizations as procedural machines (Lackoff 1995; Kallinikos 2004). Effects of this type cannot be properly studied as instances of ERP implementation in particular settings. They demand an appreciation of the developmental trajectory of ERP technology, the confluence of various techniques and the structural/organizational relations within which these technological developments have been accommodated. These issues remain inimical to the managerialist understanding of ERP implementation that is concerned mainly with how to embed these packages successfully into organizations (e.g. Markus, Tanis, and Van Fenema 2000). But they seem also indifferent to what I earlier called ethnographic positivism, which after all descended from the social concerns that constructivism once cultivated.

Systems implementation

The procedure for introducing ERP packages in particular organizations further enhances the claims concerning the origins of ERP technology and the effects such packages seem to have on locally-embedded forms of action and communication. According to the official rhetoric, the adaptability of the package is nearly unlimited. Each organization can opt to employ those processes that better fit its operations. However, a careful consideration of the forms and processes by which adaptability is accomplished tells another story.

The implementation of ERP packages imply, first of all, the transcription of the specific realities and circumstances of the organization into the abstract language of the system. Despite a widely-diffused rhetoric that declares the system as adaptable in principle to the specific needs of any organization, such a transcription by necessity implies the reformulation of context-embedded processes to an abstract language, alien to the local context. System adaptation to local realities is indeed severely limited. What could eventually be adapted to specific organizational circumstances is the number of functional modules to be utilized, the number of processes to be monitored, and perhaps the length of the processes thus monitored.[5]

The process of implementing ERP packages constitutes a huge and costly experiment, although it is very simple in its basic conception. The organization has to be first 're-engineered'. This means basically that its routines, standard operating procedures, and communication forms must be redesigned to fit the procedural logic of the system; the processes defined in this way are then transcribed to fit the technical codification requirements of the software package. During this period, the organization becomes the object of massive re-education and retraining programmes. The aim of these efforts is to teach the organizational members those action patterns, initiatives, and behaviours they have to adopt in order to be able to comply with the requirements of the package and run the system successfully. The implementation process is completed by the acquisition or updating of the hardware infrastructure capable of supporting the system.

Large-scale information systems, like those represented by ERP systems, provide an appropriate environment in which humans and technology encounter one another in terms that exemplify the selective forms by which technology admits human participation. As briefly indicated above, comprehensive software packages of this sort define and shape the operations of local contexts in terms that reflect the logic upon which the packages are based. Such logic derives from the very organizational blueprint (functional modules and cross-functional processes) on which the package is based and the intrinsic demands that the tasks of automation and software codification impose. Unlike other software packages, however, ones like

ERP, which incorporate a conception of organizations as procedural machines and the entirety of organizational operations as an extended sequence of steps, bring into being and impose a behavioural mechanics throughout the organization. An all-encompassing computer-based application such as ERP has the potential not only of rendering visible the overwhelming part of organizational activities in real time, but also of engraving the paths along which these activities should unfold. In this respect, they invite human participation along highly-structured action patterns.

Placed in this overall context, the only local alternative that remains is to be able to manage the system as smoothly and successfully as possible. The package can certainly be resisted and undermined. But looked upon from a wider historical perspective, the battle seems already decided. Resistance is pushed to the margin. Interpretation and local reshaping of the package can take place only within those rather narrow limits allowed for by the constitution of technology as a self-referential, recursively-reproduced system. To navigate effectively in the engraved routes of large-scale information platforms entertains the hope (or illusion) of gradually acquiring control over them.

Conclusions

In this chapter, I have sought to describe technology as a system that has essential characteristics which remain beyond the reach of the negotiating and reshaping capacity of local contexts and situated agents. The geist (Luhmann 1993) of contemporary technology is reflected in the dual strategy of functional simplification and closure by which technological systems select/define a domain of the real, and instrument the standardized and recurrent cause–effect, means–end sequences underlying them. Abstract as it may be, such a conception of technology can be deployed to conduct empirical research that does pay attention to the differences by which functional simplification and closure is manifested across technologies, and to trace the effects of these differences to local contexts. In such a project, technology's involvement in human affairs could be studied in ways that allow for the abstract and the general to be tied to the concrete and specific.

Functional simplification and closure could also be used to study important differences brought about by recent technological developments. The malleability, extendibility, and configurability of software packages—together with the fact that contemporary technologies of information and communication (and perhaps genetic technology as well) break out of the secluded realms of instrumental action to the open encounters of everyday life—may be taken to imply that the strategy of functional simplification and closure that defined the technological paradigm of high modernity may currently be subject to change (e.g. Fleck 1994). These potential differences must be worked out theoretically and empirically, and with a clear awareness of the largest historical transformations involved. The deep standardization by which contemporary technologies of information and communication are made publicly available suggests, however, that the controlling strategies of functional simplification and closure are still highly viable. Indeed, the concept of 'standardization' provides an alternative lens for dealing with the issues that have concerned us here. Viewed through this lens, standardization is not just a quality of contemporary technology but one of its crucial and defining characteristics, a major convention by which functional simplification and closure are accomplished.

Finally, the understanding of technology as largely a self-referential system, or a grid that brings into being a set of relations that shape the premises by which humans are brought to bear upon the world, have important affinities to Heidegger's (1971, 1977) description of modern technology as *enframing* (Ge-stell). I have, though, deliberately left Heidegger's thinking out of consideration here, fearing that it could have complicated the much simpler argument I have been trying to deliver in this chapter. However, the meaning of enframing that constructivists find so intractable is not in the end very different from the ideas of functional simplification and closure which I have claimed, following Luhmann, are able to capture the geist of technological systems. Any given technology encounters an image of the world that the technology itself sets out to produce or organize. Although Heidegger uses the concept of enframing as a defining attribute of modern technology (which he juxtaposes to pre-modern techniques), differences in the strategies of enframing could perhaps be worked out and deployed to conduct empirical studies in context.

References

Bancroft, N. H., Seip, H., and Sprengel, A. (1996). *Implementing SAP R/3*. Greenwich, CT: Manning.

Barthes, R. (1967). *Elements of Semiology*. New York: Noonday.

Beck, U. (1992). *Risk Society: Towards a New Modernity*. London: Sage.

—— (2000). *What is Globalization?* Cambridge: Polity Press.

Beniger, J. (1986). *The Control Revolution: Technological and Economic Origins of the Information Society*. Cambridge, MA: Harvard University Press.

Bijker, B. (2002). 'Understanding Technological Culture Through a Constructivist View of Science, Technology and Culture', in S. Cutcliffe and C. Mitcham (eds.), *Visions of STS: Counterpoints in Science, Technology and Society Studies*. New York: State University of New York.

Bijker, W. E., and Law, J. (1992) (eds.). *Shaping Technology/Building Society. Studies in Sociotechnical Change*. Cambridge, MA: MIT Press.

—— Hughes, T. P., and Pinch, T. (1987). *The Social Construction of Technological Systems*. Cambridge, MA: MIT Press.

Bolter, J. D. (1991). *The Writing Space: The Computer, Hypertext and the History of Writing*. Hillsdale, NJ: LEA.

Borgman, A. (1984). *Technology and the Character of Contemporary Life*. Chicago: University of Chicago Press.

Cassirer, E. (1955). *The Philosophy of Symbolic Forms: Vol. 1 Language*. New Haven, CT: Yale University Press.

Castoriadis, C. (1984). *Crossroads in the Labyrinth*. Brighton, UK: Harvester Press.

Ciborra, C. U. (2000) (ed.). *From Control to Drift: The Dynamics of Corporate Information Infrastructures*. Oxford: Oxford University Press.

Ellul, J. (1964). *The Technological Society*. New York: Vintage Books.

Fleck, J. (1994). 'Learning by Trying: the Implementation of Configurational Technology'. *Research Policy*, 23: 637–52.

Gellner, E. (1983). *Nations and Nationalism*. Oxford: Blackwell.

——(1996). *Conditions of Liberty: Civil Society and its Rivals*. London: Penguin.

Giddens, A. (1984). *The Constitution of Society*. Cambridge: Polity Press.

——(1990). *The Consequences of Modernity*. Cambridge: Polity Press.

Goffman, E. (1974). *Frame Analysis*. New York: Harper and Row.

Grint, K., and Woolgar, S. (1997). *The Machine at Work: Technology, Work and Organization*. Cambridge: Polity Press.

Hanseth, O. (2000). 'The Economics of Standards', in C. U. Ciborra (ed.), *From Control to Drift: The Dynamics of Corporate Information Infrastructures*. Oxford: Oxford University Press, 56–70.

Hasselbladh, H., and Kallinikos, J. (2000). 'The Process of Rationalization: A Critique and Re-appraisal of Neo-institutionalism in Organization Studies'. *Organization Studies*, 21/4: 697–720.

Heath, C., Knoblauch, H., and Luff, P. (2000). 'Technology and Social Interaction: the Emergence of "Workplace Studies"'. *British Journal of Sociology*, 51/2: 299–320.

Heidegger, M. (1971). *On the Way to Language*. New York: Harper and Row.

——(1977). *The Question Concerning Technology and Other Essays*. New York: Harper and Row.

Hughes, T. P. (1987). 'The Evolution of Large Technological Systems', in W. E. Bijker, T. P. Hughes, and T. Pinch, *The Social Construction of Technological Systems*, Cambridge, MA: MIT Press, 51–82.

Kallinikos, J. (1995). 'The Archi-tecture of the Invisible: Technology is Representation'. *Organization*, 2/1: 117–40.

——(1996). *Technology and Society: Interdisciplinary Studies in Formal Organization*. Munich: Accedo.

——(1998). 'Organized Complexity: Posthumanist Remarks on the Technologizing of Intelligence'. *Organization*, 5/3: 371–96.

——(1999). 'Computer-based Technology and the Constitution of Work: A Study on the Cognitive Foundations of Work'. *Accounting, Management and Information Technologies*, 9/4: 261–91.

——(2002). 'Re-opening the Black Box of Technology: Artifacts and Human Agency', in L. Applegate, R. Galliers, and J. DeGross, *Proceedings of Twenty-third International Conference on Information Systems, Barcelona 14–16 December 2002*, 287–94. aisel.isworld.org/Proceedings/ICIS/2002/home.asp.

——(2003). 'Work, Employment and Human Agency: An Anatomy of Fragmentation', *Organization Studies*, 24/4: 595–618.

——(2004). 'Deconstructing ERP Packages: Organizational and Behavioural Implications of Large-scale Information Systems'. *Information Technology and People*, 9/17/1.

Lackoff, G. (1995). 'Body, Brain and Communication', in J. Brook and I. A. Boal (eds.), *Resisting the Virtual Life*. San Francisco: City Lights.

Lamb, R., and Kling, R. (2003). 'Reconceptualizing Users as Social Actors in Information Systems Research', *MIS Quarterly*, 27/2: 197–235.

Latour, B. (1986). *Science in Action*. Milton Keynes: Open University Press.

Leigh-Star, S. (1992). 'The Trojan Door: Organizations, Work, and the "Open Black Box"'. *Systems Practice*, 5/4: 395–410.

Luhmann, N. (1982). *The Differentiation of Society*. New York: Columbia University Press.

Luhmann, N. (1993). *The Sociology of Risk*. Berlin: de Gruyter.
——(1995). *Social Systems*. Stanford, CA: Stanford University Press.
——(1996). 'Complexity, Structural Contingency and Value Conflicts', in P. Heelas, S. Lash, and P. Morris (eds.), *Detraditionalization*. Oxford: Blackwell.
Markus, L. M., Tanis, C., and Van Fenema, P. (2000). 'Multisite ERP Implementations'. *Communications of the ACM*, 43/4: 42–6.
Misa, T. J. (2003). 'The Compelling Tangle of Modernity and Technology', in T. J. Misa, P. Brey, and A. Feenberg (eds.), *Modernity and Technology*. Cambridge, MA: MIT Press, 1–30.
Mumford, L. (1934). *Technics and Civilization*. London: Harvest/HBJ.
——(1952). *Arts and Technics*. New York: Columbia University Press.
Noble, D. (1984). *Forces of Production: A Social History of Industrial Automation*. New York: Alfred A. Knopf.
Orlikowski, W. J. (1992). 'The Duality of Technology: Rethinking the Concept of Technology in Organizations'. *Organization Science*, 3/3: 398–427.
——(2000). 'Using Technology and Constituting Structures: A Practice Lens for Studying Technology in Organizations'. *Organization Science*, 11/4: 404–28.
Ptak, C. A., and Schragenheim, E. (2000). *ERP: Tools, Techniques and Applications for Integrating the Supply Chain*. London: St. Lucie Press/APICS Series on Resource Management.
Ricoeur, P. (1977). *The Rule of Metaphor: Multidisciplinary Studies in the Creation of Meaning in Language*. Toronto: Toronto University Press.
Searle, J. (1995). *The Construction of Social Reality*. London: Penguin.
Sotto, R. (1990). 'Man Without Knowledge: Actors and Spectators in Organizations'. Unpublished Ph.D. thesis. Stockholm: School of Business, Stockholm University.
Suchman, L. (1987). *Plans and Situated Actions*. Cambridge: Cambridge University Press.
——(1996). 'Supporting Articulation Work', in R. Kling (ed.), *Computerization and Controversy: Value Conflicts and Social Choices*. San Diego, CA: Morgan Kaufman, 407–23.
Winner, L. (1977). *Autonomous Technology: Technics-out-of-control as a Theme of Political Action*. Cambridge, MA: MIT Press.
——(1986). *The Whale and the Reactor: A Search of Limits in the Age of High Technology*. Chicago: University of Chicago Press.
——(1993). 'Upon Opening the Black Box and Finding it Empty: Social Constructivism and the Philosophy of Technology'. *Science, Technology and Social Values*, 18: 362–78.
Zuboff, S. (1988). *In the Age of the Smart Machine: The Future of Work and Power*. New York: Basic Books.

Notes

1. For a recent summary and critique of the user-centric approach, see Lamb and Kling (2003).
2. One should perhaps distinguish between temporality and historicity. The terms are often used interchangeably, but they seem to refer to different processes. Temporality captures the dynamic character of technology and the fact that technological artefacts develop and change over time. On the other hand, historicity refers to how technological traditions reach down to the present, influencing the forms by which locally-embedded agents deploy the technology.

3. An analogy could be made here with the linguistic notion of the 'sign value' (e.g. Barthes 1967; Ricouer 1977).
4. See Kallinikos (2003) for a detailed exposition of the foundations of formal organizations in terms other than inclusive (i.e. essentialist), and the prospects, problems, and dilemmas associated with the organization forms thus constituted.
5. For instance, in some organizations the monitoring of, say, sales may include fewer processes (a smaller number of transactions) and fewer steps (a smaller number of external or internal actors getting involved from the initiation until the termination of the process).

9

Framing IS studies: understanding the social context of IS innovation

Chrisanthi Avgerou and Shirin Madon

Introduction

Information systems research and practice is to a large extent concerned with analysing the properties and behaviour of the organizations within which IS innovation takes place. Yet, the analytical capacity of information systems research to consider the social context of IS innovation remains limited. Overall, IS studies have not developed theories and methodologies to account for the environment within which information systems are shaped and interact with human actors.

Our interest in the analysis of the social context has been triggered by our efforts to study ICT innovation in the so-called developing countries, where many of the problems of steering IS implementation towards desirable ends appear to be exacerbated (Heeks 2002). Several authors in the research sub-field that studies information systems in developing countries have considered it necessary to broaden the focus of study beyond the immediate events and actions that comprise the development, implementation, or use of ICT artefacts. This wider scope encompasses tracing the roots of the behaviours encountered in the economic, cultural, and political domains of the social context in which innovation with IS takes places (e.g. Madon 1993; Jarvenpaa and Leidner 1998; Avgerou and Walsham 2000; Sahay 2000; Walsham 2001). The need to trace the logic of actions observed in IS innovation by expanding the focus of study in scope and time often arises in advanced industrialized settings too, particularly when the researcher is confronted with paradoxical or apparently problematic innovation cases (Ciborra and Lanzara 1994; Mitev 1996). Indeed, a number of authors have suggested the need to consider social and institutional aspects of the external context of organizations in order to understand IS innovation (e.g. Knights, Noble, and Willmott 1997).

In this chapter, we propose an approach to studying context that is capable of accounting for the diverse social settings of IS innovation. Before doing this, we need to clarify the distinction between 'ICT innovation' and 'IS innovation'. ICT innovation is primarily concerned with the design and implementation of a new technology; IS innovation goes beyond this to also address issues of information and knowledge, and it has an interest in IS-related changes of the organization and content of work tasks. Therefore, we are concerned here with IS innovation as a process intertwined with social change, that is, as a socio-technical process in the way it is understood by social constructionist theory (Callon and Law 1989). We do not set out to study the social context of IS innovation as the container of technical features and activities, but are interested in the study of intertwined technical and social processes.

The first section of this chapter critically examines the limitations of IS theory in accounting for the context of IS innovation. We juxtapose and discuss two perspectives pursued in IS studies: universalistic and situated. Universalistic studies assume that IS innovation is driven by particular values of universal relevance and validity; this makes it possible to explain any IS phenomenon in terms of a set of general concepts, and to discover general logical principles as the basis for an effective pattern of professional action and organizational behaviour. Situated studies tend to be preoccupied with the 'here and now' actions and interactions comprising IS innovation. They produce a wealth of details of contingent behaviour but do not usually seek to reveal the roots of cognitive or emotional bearings of the behaviour it uncovers. Consequently, we identify the need for situated approaches capable of accounting for the motivation and the logic of the actions comprising episodes of IS innovation.

The second section discusses contextual situated analysis as a case of framing, that is, as a matter of demarcating a social network relevant to a study and tracing its institutional underpinnings. We argue that such framing is needed to justify observed behaviour in IS innovation processes in terms of their particular historical and social setting.

In the third section, we suggest that the relevant frame for the study of IS innovation can be traced by following the network of actors involved and, consequently, by examining the institutional fields that have sustained their meanings of, and their attitudes towards, the innovation under study. To that end, we examine the way institutions are implicated in the enactment of professional practices, and in nurturing or retarding competencies for sense-making and improvisation when confronting uncertainty and novelty. We also discuss the multiplicity of institutional influences on agency in IS innovation.

Limitations of IS theory in accounting for the context of IS innovation

Differences between universalistic and situated perspectives

A distinction can be made between universalistic and situated treatments of context in information systems studies.[1] From a universalistic perspective, the potential value of technology and information, and the processes through which such value is achieved, are considered independently from the circumstances of the social actors involved in the IS implementation. ICT is seen as a means for processing information that, endowed with speed, accuracy, and reliability, is potentially capable of contributing to organizational efficiency (Hammer 1990), rational decisions (Keen and Scott Morton 1978; Sprague 1980; De Sanctis and Gallupe 1987), and productivity (Landauer 1996). The value of these technical attributes and the desirability of efficiency and rationalization effects associated with them are generally unquestioned. The process of innovation is perceived as a pattern of technical and organizational activities whose merits are judged in terms of an overall trajectory of organizational structures best suited to exploit the potential of new technologies (e.g. Scott Morton 1991; Fulk and DeSanctis 1999).

Universalistic assumptions of knowledge and conceptions of value are evident in analyses of IS phenomena in terms of intentional action, such as decisions for accepting or resisting innovation, tasks that lead to successful innovation, and models of utility maximization. Often such studies acknowledge the significance of contingencies addressed by the innovating organization (McFarlan and McKenney 1983; Porter and Millar 1984). For example, emphasis on competitive advantage highlights the differentiating characteristics of a business firm *vis-à-vis* its competitors. The literature has suggested concepts and techniques to assist managers in analysing the position of their business in its market, in order to plan for effective new information systems (Earl 1987) or pursue innovation and change that is difficult to imitate (Ciborra and Andreu 1998). However, such contingency studies still make universalistic assumptions about the logic that drives IS innovation in organizations, the goals pursued by their members, and the existence of an environment that is—or ought to be—populated by similarly minded organizations. Implicitly, if not openly, behaviour deviating from the assumed rational action is seen as problematic.

A universalistic approach has also been adopted in most critical analyses of the limitations or the negative social consequences of the narrow managerial concerns that have dominated the IS research agenda, as well as IS professional practice. For example, in the 1980s socio-technical design proponents such as Mumford (Mumford and Weir 1979; Mumford 2000) and Scandinavian scholars and practitioners (e.g. Kyng and Mathiassen 1982; Bødker *et al.* 1987; Ehn 1989) put forward the 'rights' of workers—which were often seen as safeguarded by trade unions—and workplace democracy as universal ideals to be pursued by business organizations in the context of a government-regulated free market. Such studies juxtaposed business productivity and job satisfaction as two distinct motives for IS innovation, and highlighted their reconciliation as a legitimate concern. Nevertheless, as Mumford (2000) notes with a discernible sadness, such a concern lost its perceived significance in the 1990s, even within Scandinavian societies.

A great deal of other literature in the same vein has taken a critical stance to ICT innovation on the basis of universal social ideals, such as emancipation (Lyytinen and Lehtinen 1984; Lyytinen 1986; Hirschheim, Klein, and Lyytinen 1996) or social fairness (Braverman 1974). The institutional structures and social aspirations that give rise to such concerns, counter-arguments, and suggested counter-actions—such as the fear of totalitarianism and the class system—are cast as universals in terms of the unfolding of history of society at large.

In contrast, situated approaches involve the analysis of the specific circumstances of innovation, seeking to identify the meanings and interests involved in IS innovation, the processes through which it takes place or is held back, and its consequences.

Attention within a situated perspective focuses on actors embedded in a particular social and organizational setting. Innovation unfolds as actors work out its meaning, legitimacy, and value. More specifically, the situated view emphasizes the significance of context-specific unformalizable action, such as improvisation, tinkering, and actors' negotiations. Rather than pursuing general rational patterns of behaviour, the actors involved in the innovation process construct technology artefacts and information resources while they work out organizational changes. They do this by responding to the challenges they are faced with 'here and now', according to their prior experience, knowledge, emotions, and perception of interests.

Situated perspectives have enriched IS research with insightful phenomenological studies, offering new understanding of the complex phenomena of IS innovation. Nevertheless, most situated studies of information systems limit analysis to the immediate environment of an innovation process: an organization, a group, or an office. More often, the details of the action taking place at the location of innovation are understood in relation to general assumptions about the social context of both the organization and the world beyond it. For instance, IS research typically assumes that the innovating organization is part of a competitive economy, within which it may be subject to national and international regulation and the influence of the dynamics of the ICT industry. But such assumptions remain implicit, receiving brief or no mention in the empirical narratives and little attention in the analysis of the situated action. Moreover, the authors of such studies expect the reader to be familiar with the wider context, the norms of action in organizations, and the broader life circumstances of the actors in the IS innovation process. Situated accounts tend to omit information on the institutional structures within which an organization under focus is embedded, and the ways employees live their lives beyond the organization in which the IS innovation process is enacted.

Examples of situated studies with a broad focus

A number of studies have expanded the situated perspective to look beyond the immediate unfolding of a course of action and have sought to justify it in relation to features of the organization attempting the innovation and its broader social context. The following examples are indicative of the way studies using situated perspectives have linked the observed unfolding of IS innovation processes with contextual bearings.

Orlikowski's (2000) elaboration of her 'structurational' account of the shaping of IT and organizational change through practice pointed to some links between differences of agency and consequent outcomes in the practice of the same technology tools and structural features in different organizational settings. She associated the differences in the take-up of software intended to support and facilitate collaborative work in the corporations she studied with the extent to which they fostered an adversarial organizational culture. Such associations between the emergent shaping of technology-mediated work practices and the particular organizational context within which shaping takes place were missing in her earlier analysis of situated IT and organizational change (Orlikowski 1996), in which the 'metamorphoses' traced were presented as self-justified.

Ciborra and Lanzara (1994) used Unger's (1987) concept of 'formative context' to capture the uniqueness of the dynamics of innovation in the setting of specific organizations. The notion of formative context highlights how people's perceptions, interpretations, and actions in an organization are embedded in the practical routines, transactions, and meanings of the organization's institutional setting. The enactment of innovation therefore relates to deeply-ingrained organizational culture and social relations; it draws from and at the same time challenges them. Formative contexts 'designate what binds, in a loosely connected texture, an individual or a collective (group, organization) to an established world of objects and relations, and to the associated cognitive imageries, presuppositions and meanings of which that world is the embodied vehicle' (Ciborra and Lanzara 1994: 72). Seeking to alter an existing information system that has been moulded by an organization's formative context challenges its stability, hidden background components of skilled performance, and institutional arrangements. However, formative contexts are subject to fluctuations, perpetual tinkering, and experimentation in the everyday micro-practices they comprise. Ciborra and Lanzara argue that these, together with the organization's capability to learn, to question, and to drift, are essential aspects in the enactment of innovation.

An example of a study that adopted a scope of context beyond the boundaries of an organization is Knights, Murrey, and Willmott's (1997) analysis of the development of technology-mediated inter-organizational information infrastructure in the financial services industry. This investigation not only focused on an inter-organizational case of innovation, but more importantly it questioned preconceptions about the value of 'knowledge work' and the emergence of network organizational structures in contemporary socio-economic settings. It sought to analyse these aspects in terms of the power relations of the industrial context of the case.

Further links with the societal context of the actors involved in IS innovation are analysed in the studies of the introduction of a geographical information systems (GIS) in district-level administration organizations in India, conducted by Sahay and Walsham over a five-years period (Sahay and Walsham 1997; Barrett, Sahay, and Walsham 2001; Walsham 2001). These studies examined the way forms of appropriation and rejection manifested in the effort of GIS innovation in India were 'linked to deep seated issues of cultural and professional identity' (Walsham 2001: 85). Similarly, in a longitudinal study of a series of attempts to introduce distributed IS processing and new organizational procedures in a social security organization in Greece, Avgerou (2002) sought to justify the rationality of actors' behaviour by analysing the historical formation of the role of that organization within the context of the modern Greek state. She argued that the persistently problematic behaviour of the actors involved in successive IS innovation projects, which appeared to be due to incompetent behaviour and lack of resources, could be seen in a different light by taking into consideration the political and economic history of the country.

These examples demonstrate the need to expand IS research in terms of the context it considers when trying to understand what is going on in an IS innovation episode. Narrowly situated accounts—ones that do not delve into historical, socio-structural, and cultural aspects of a case's context—rely on assumptions about commonly-held knowledge regarding the social context of the organization, its environment, and the social and cultural conditions of actors' lives beyond their work roles. If these are not familiar to the reader, or if they diverge from those assumed in common current understanding of the setting of IS knowledge, the process of innovation appears paradoxical, or just problematic. Depending on the extent to which the history of the institutional setting that affected actors' accounted behaviour is atypical for the tacitly-assumed social setting in IS theory and of intended readers, the researcher will need to trace and expose it in order to make sense of the course of actions and the observed or perceived outcomes of an IS innovation process.

The framing of contextual situated analysis of IS innovation

On framing

The question of how to delineate a study in order to account adequately for its constituting actions is not unique to the study of IS innovation. It concerns the demarcation of the time-span and the boundary of the social network of relationships under study. Callon (1998*a*, *b*) discusses the significance of such framing for the completeness of economic analysis.[2] He is concerned with the relationship between economic and social perspectives, and discusses framing in relation to the economic concept of 'externalities' in order to demonstrate the importance of demarcating the extent of the calculative behaviour of market-driven actors. In this sense, the externalities identified by an economic analysis are seen as 'overflowing' from the framing of the economic study. Considering a broader frame in terms of a more extensive network of actors can change the overflowing non-accountable externalities to explainable and calculable processes.

It should be noted at this point that Goffman's (1986) work of uncovering the constitutive rules of everyday behaviour has played a major role in highlighting the notion of a frame for actors' sense-making and calculated behaviour. He sought to understand what is happening in the interactions of small numbers of individuals through a close-up focus on the structure of experience of individuals in specific moments of their social lives. He acknowledged the significance of social structures for the experiences of individuals in co-presence, but his own research and notion of frame was narrowly situational as it referred to a framework of rules and meanings implicated within well delineated micro-interactions.

Other analysts, though, have emphasized the significance of linking localized and time-specific interactions with the institutions involved in their broader social context. In Giddens' (1984) structurational perspective, the meaningful content of interaction is constituted in relation to its social setting. For instance, he asserts that, '(c)ontext . . . connects the most intimate and detailed components of interaction to much broader properties of the institutionalization of social life' (ibid.: 119). Thus, the frame of meaning is taken by analysts of organizations to be a way of seeing things with reference to organizational context (Czarniawska-Joerges 1992).

The convincingness of a researcher's account depends on the extent to which the actors' frames of meaning are either commonly assumed, and therefore implicit in the understanding of a case under study, or made explicit by drawing the research boundaries so that they are revealed. In her discussion of the study by Spybey of change in a large manufacturer of wool textiles in Britain, Czarniawska-Joerges (1992) makes an observation which clarifies the difference between the exposure of the actors' frames of meaning and the contexts that have created these frames. Spybey had studied the change of the organization in terms of the victory of the trained professional managers over the traditional frame of meaning, according to which the company was organized. According to Czarniawska-Joerges (ibid.: 156): 'In terms of theoretical contribution, Spybey showed how frames of reference structured action and also interpreted action and inaction . . . (H)e placed all of his analysis in the context of an advanced capitalism . . . ' In other words, the plausibility and validity of Spybey's justification of that particular case of organizational change depend on the extent to which the socio-economic and cultural aspects of the alluded context are familiar to the reader and can be taken as a convincing source of the frames of reference of the two categories of action.

IS research is typically framed as a set of interrelations of pre-specified actors and roles, such as analysts, designers, project managers, users, vendors, computers, methodologies, etc. The behaviour of each of these is comprehended as rational action within existing professional norms. Gradually, however, the framing of IS projects as engineering efforts had to be extended to account for the ramifications of costs and benefits accrued to business organizations through technologically rational action. Information systems research (e.g. Bloomfield *et al.* 1997*a*) has been able to produce more convincing accounts of IS innovation processes by abolishing the distinction between technical content (which is subject to design and calculable actor behaviour) and organizational context (understood as a source of requirements rather than as subject to change itself), that is, by drawing the frame around the intertwined technical and organizational change.

Drawing an appropriate research frame to capture the contingencies of organizational context and the situatedness of action in IS innovation is particularly important when the innovation being studied overflows with controversy, unexplainable actor behaviour, obstinacy in failing to 'learn' and perform the expected technical roles, and an inability to improvise successfully. Callon (1998*a*) calls such cases 'hot'. They are situations for which conventional framing is inadequate to capture their dynamics because identifying the relevant actors, their interests, and the logic of their negotiations cannot be considered common knowledge. In such cases, the framing of the research requires tracing the actors' frames of reference, that is, their historical social settings within which their meanings and emotional attitudes are justified.

This indicates that in order to be able to account for the reason actors acted in one particular way rather than another, the challenge for contextualist research in IS is the drawing of a frame of reference and meaning: Who are the actors? What understanding and emotional attitudes drive their actions? Where do these stem from? We are interested in framing as a process of study in Rorty's (1980) terms: as an investigation that allows for justifying an observed process rather than simply explaining it. Here, 'explaining' refers to the proposing and empirical describing of cause-and-effect processes, such as how a business firm maintains a competitive position in its market. 'Justifying', on the other hand, requires the placing of such descriptions in the logical space of reasons, such as why actors in one business firm manage to mobilize IT and achieve a competitive position and in another they fail to do so.

A re-framing case study

An example to help illuminate the question of appropriate framing is drawn from one of the author's empirical research in India, which examined the role of information technology implementation for development planning (Madon 1993). Since the 1980s, administrative reforms in a number of developing countries have been directed at achieving decentralization of planning through the use of information systems. A central goal of such initiatives was to improve the efficiency and effectiveness of rural development planning. In India, one such initiative was the Computerized Rural Information Systems Project (CRISP), launched throughout the country in 1987. Its objective was to improve the efficiency and effectiveness of poverty programmes targeted at poor rural households. The system was designed and developed by a team of programmers at the central government's National Informatics Centre.

Within its first year, CRISP was deemed a success and the project received a lot of publicity in the popular press, which painted a very glowing picture of its impact. The central government measured success in terms of indicators such as whether the hardware and software had been installed, the system had been initialized with base data, and data entry of rural poor households had commenced. Data on individual districts recorded that these activities had been successfully achieved. Yet, there was no significant improvement in the efficiency or effectiveness of poverty programmes. Conventional study of the unsatisfactory impact of CRISP looked at the problem in terms of technical/rational action in the organizational setting of India's public administration. This suggested that the difficulties were due to a variety of initial teething problems, such as inadequate levels of funding, project planning, systems design quality, and human resource development. However, such an approach was not able to provide an adequate explanation of why the system continued to have poor impact over a period of five years, despite increases in budget, human resource deployment, and systems design.

Madon's research attempted to understand the reasons why CRISP did not produce the expected outcome of improving the effectiveness of the information infrastructure for the Indian government's rural development policy. She traced the perspectives of the key actors of central government, state government, district administration, and local rural households. Each of these groups was driven by a certain logic or 'rationality'. Understanding this required a rapport to be built up between the researcher and the respondents. Madon feared that the required openness of discussion would not be achieved without this. The research therefore required long spells of field work, which involved regular contacts with respondents over a period of months. This was needed to elicit information relating to the historical social settings within which the meanings and emotional attitudes of different groups of actors prevailed.

Broadening the study beyond the processes of developing, implementing, and operating CRISP to help understand the organizational setting within which the system was embedded revealed a variety of factors related to hierarchy, status, and the local culture of the key actors. One example of this concerns the value attached historically by administrators to status and hierarchy. According to central government guidelines, the computer was to be located in a room adjacent to that of the district director. In most cases, however, the popular choice of location was actually the director's room itself. This decision was influenced by the importance attached to status among senior administrators in the bureaucracy. District directors saw the computer as portraying power and authority. In many cases, the computer caused a distraction to both the director and the user: the director himself was too busy to use the machine and the user felt intimidated about using it in his presence.

A second example of the insights gained by a broader framing of the research relates to Madon's attempts to penetrate the local culture of administrators. The closed and non-participative style of bureaucracy that was part of the administrative culture had resulted in a kind of secrecy built around data and knowledge. With the manual system, it had been possible for directors to manipulate figures on the hard copy. However, it was feared that the computerized system would make information more visible to many, resulting in the elasticity of data being reduced. Faced with the prospect of a possible realignment of power, some directors actively discouraged efforts to computerize by preventing staff from attending training courses, or by insisting on the computer being kept in their room.

Such 'factors' are not new in the IS literature. These are the 'hot situations' of behaviour that deviate from the rational accounts in the literature of how IS innovation enables organizational change. In confronting such situations, IS research tends to end the empirical study with the conclusion that the problems observed are a consequence of the 'local culture'. This is understood either as historically taken-for-granted behaviour whose logic would be futile to attempt to discover, or as dysfuctionalities and irrationalities that should be eliminated in order to allow for technology to deliver its potential.

We find neither of these two attitudes satisfactory in confronting actors' obstinate refusal to behave in preconceived 'rational' ways in relation to technology innovation. Justification is needed for the way traditional IS approaches mistrust the 'rationality' inscribed in the new technology actor, the manifested contradictory meanings, and the reasons why, for instance, Indian administrators resisted a realignment of power. The rationality of the apparently irrational behaviour needs to be uncovered. And this is where re-framing is required, as it can redraw the boundaries of the study to look at various aspects of the context of the IS project. Such analysis can show whether the IS intervention was misconceived and inappropriate, or whether additional interventions are required for the IS project to achieve its objectives.

Madon's case study traced some of the aspects of actors' behaviour in the Indian government administration, the life conditions of the rural poor, and the caste system. For example, she found that complexities in the Indian administration have created confusing lines of control and reporting structures, which affect communication between agents involved in the management of rural development programmes. This overlapping of contradictory formal and informal behaviour patterns is characteristic of Indian development administration. Singhi's (1974) study, for instance, revealed that the surface behaviour of Indian bureaucrats display hierarchy, division of labour, a system of rules, and impersonality. However Singhi and others identified two important components of Indian culture that in reality influence the value system and behaviour of Indian administrators.

First, there are the close and intimate relationships experienced from an early age in the joint family system. The belief system of Indian administrators is based on this image of a nurturing, caring, dependable, sacrificing—yet demanding and authoritative—father figure which individuals learn to value and look for in working life. In return, the individual develops respect and reverence for superiors and a willingness to accept authority (Gupta 1991).

A second important component of the culture of Indian administrators is the caste system that tends to permeate domestic and working life. Superior–subordinate relationships within the administration have a clear allocation of rights and duties across the boundaries that determine the social hierarchy (Srinivas 1989). This social hierarchy within the administration is constantly stimulated and sustained by the 'casteism' that is still the predominant feature of the Indian social order. A study of the caste system is extremely complex and clearly beyond the scope of Madon's study. Nevertheless, since it is a social and political phenomenon of enormous importance in Indian life, no analysis of the context of development administration in the country can afford to ignore it.

For instance, Singhi's (1974) study of the value orientation of Indian administrators found that some 70 per cent of administrators agreed with the view that consciousness of caste and status influence the relationship between officers of different ranks. The study also found that the prevailing values and code of conduct among the rural poor have sometimes prevented them from taking advantage of government-designed rural development programmes, because they imposed an individualistic approach of identifying beneficiary households for assistance that is totally alien to certain primitive tribes residing in many parts of the country. In one area, the district agency had to abandon rabbit and goat rearing schemes because, instead of generating income from their new assets, tribal beneficiaries consumed the animals at the time of religious festivals. As a semi-monetized economy prevails in many of these tribal communities, the concept of credit appears alien.

By framing her study to include these perspectives, Madon revealed associations among the problems faced by CRISP with some fundamental aspects of the value systems of administrators and the rural poor. We do not go into details of these wider contextual factors in this chapter because our main objective here is to signal the importance of going beyond the focal organization, in this case the district agency. Viewed in such a context, the experience of the 'failure' of CRISP takes a totally different meaning and, consequently, informs further action quite differently. It is neither a matter of technical fixes nor an inevitable consequence of an entrenched and fixed local culture; and, of course, it is also not a matter of bending the obstacles of problematic behaviour by brute force. Rather, Madon's analysis points out that the unfolding of the IS intervention in CRISP is inexorably linked with the unfolding changes within several institutions of Indian society: the administration, rural poor, and caste system.

The CRISP innovation failed to produce its intended rationalization effects, but it acquired a different significance within the larger picture of the continuous efforts being made towards social change and administrative reform, as indicated by the 'left-overs' of CRISP that Madon found in later visits. Since 1990, Madon's findings have revealed patterns of change from the existing order of things in terms of subtle incremental shifts in the balance of administrative power between central, state, and district levels. End-users are acquiring the confidence to develop useful applications as they begin to perceive that they have some control over their use.

Determining the relevant frame for the study of IS innovation

Formation of institutions

Without a pre-set, commonly assumed frame of IS action, the starting point of an IS study is necessarily the observation, or the narrative tracing, of situated episodes of action. Such episodes may be clustered as IS development projects with a relatively short duration of events and actions. Or they may be clustered in longer histories of IS innovation efforts in a particular social setting, for example efforts to form an IS infrastructure in an organization over a period of many years. From such observations and narratives, the framing task is one of tracing the institutional underpinnings that justify the actions involved. The actions that comprise an IS innovation process are either enactments of institutionalized patterns of thinking and behaving or instances of spontaneous sense-making. Creating an appropriate frame for understanding such processes requires the identification of the institutional bearings of both these types of action.

Hasselbladh and Kallinikos (2000) elaborate on the link between action and institutionalization using the concept of 'actorhood'. In their analysis of the way the institutionalization of organizations occurs, actorhood is the patterning of action in recognizable and recurrent social and organizational roles, which is part of the means of constituting long-lasting social structures: 'Institutionalization marks the passage from intuitive and haphazard modes of understanding and acting to the construction of durable social artefacts. Rationalized patterns of this sort specify the content and the means of organizational action and thus shape, indirectly but decisively, particular forms of actorhood' (ibid.: 713).

In this sense, the development of forms of actorhood is a process of shaping an individual's way of interpreting events in their context and acting. Hasselbladh and Kallinikos argue that such a process accompanies the embedding of ideas and relations in an organization's formal structures, techniques, and procedures. This makes them reproducible, durable, and transferable across organizations, as suggested by Berger and Luckmann's (1967) theory of institutions. In organizations, particular forms of actorhood result from the diffusion and perpetuation of instrumental ideas, codified as methods and techniques, amidst discourses that articulate and maintain their validity and value.

However, situationist authors have convincingly argued that actions in IS innovation may be improvisations, part of sense-making rather than enactments of institutionalized roles. This indicates that IS innovation includes the behaviour of actors confronted by the surprising and uncertain unfolding of events (Weick 1995; Ciborra 1999). Such actions also draw from the lived experiences of the actors concerned and the ideologies that have shaped their beliefs, outcome preferences, and expectations of appropriate behaviours.

Lived experiences take place in social contexts, thus being subject to the formative influences of institutions. Ideologies are nurtured in the discourses that constitute the culture of broader social collectivities, such as communities, nations, and industries (Trice and Beyer 1993). The tracing of institutional influences on actors is necessary in order to answer questions such as: Why are actors in certain situations capable of finding cues that help them make sense of the possibilities that the course of an innovation process presents them with, and to pursue competent improvisations? Why, in other situations, do they follow apparently tortuous, idiosyncratic routes of action which reach results that seem 'odd' to an observer familiar with the general body of IS knowledge? Or why, in yet other situations, are actors unable to act competently and follow through the innovation process to some kind of positive results?

Constructing appropriate frames for the study of IS innovation

The analytical process of constructing appropriate frames for the understanding of episodes of IS innovation is the reverse of the task of tracing the formation processes of institutions, as undertaken by Hasselbladh and Kallinikos (2000). Of interest in the framing process is the revealing of the already institutionalized ideas, claims of truth, and codified practices enacted by actors taking part in an IS innovation. The framing effort is intended to shift the focus of attention from the observed specific actions of an IS episode to the roles or the mental and emotional states these actions enacted, the discourses responsible for the claims of their validity, and the mechanisms that reproduce them. Such a shift helps to reveal those forces which compel, inspire, provide cues for, or—simply but vaguely—enable the observed actions.

This is a complex task because the institutional influences on the actors in an innovation episode originate in multiple and often uncharted sources. The most obvious are: the organization, or set of organizations, within which the IS innovation takes place; the professional communities engaged with the various technical tasks of the construction of ICT artefacts and the working out of relevant organizational change; and the societal background within which actors have been living their lives. But none of these three institutional sources of actorhood and competencies is amenable to *a priori* analytical categories.

There is a substantial literature on the institutional nature and features of organizations in advanced western economies (Powell and DiMaggio 1991; Scott and Meyer 1994; Brunsoon and Olsen 1998) and several analyses of the institutional character of professional practice, in particular in IS innovation (Bloomfield and Danieli 1995; Swanson and Ramiller 1997; Avgerou 2003). Nevertheless, a categorization of institutions that are likely to bear on the behaviour of IS innovation actors is inhibited by the ongoing processes of de-institutionalization of established organizational forms, and the continuous remaking and recombination of professional norms within a public discourse that is preoccupied with the necessity of perpetual change. Business organizations in western economies are experimenting with formal organizational structures and techniques that depart from those of the hierarchical bureaucratic organizational model, but no new model has been institutionalized as the dominant new form to replace it (Applegate 1994, 1999).

Several researchers have used Pettigrew's contextual methodology for the study of organizational change (Pettigrew 1985, 1998; Pettigrew, Ferlie, and McKee 1992) in order to explain how IS innovation is subject to extra-organizational cultural, social, economic, and regulatory processes (Avgerou 1989; Madon 1993; Walsham 1993). However, Pettigrew's concept of layered contexts—the inner context being the organization, the outer one the national environment—does not necessarily capture the multiple influences that shape an innovation process in the contemporary condition of globalization.

Information systems innovation is often shaped through the involvement of actors with diverse organizational allegiances, including: the organization hosting the innovation, whose formative context is directly relevant to— affecting and affected by—the innovation; IS consultants from either the local or international IS services industry; and sponsoring institutions, sometimes regional or national, and very often international. While Pettigrew's analyses have convincingly made the point that what happens in an organization cannot be adequately understood without consideration of the 'outer' context, questions must be raised about the validity of the way he, and those who adopted his method, pursued the delimiting of context in terms of national institutional forces, such as government policy or social and economic structures.

Indeed, the burgeoning literature of globalization suggests changes are taking place in the established significance of national institutions in shaping the cognitive imagery of people, the symbolic basis of legitimization of action, and the socio-economic conditions within which organizations operate (Robertson 1992; Appadurai 1996; Bauman 1998; Beck 2000). What happens in a particular locality is also influenced by regional or 'global' institutions with supranational missions, rhetoric, and reach, such as the ideas and actions circulated and supported by the various organizations of the United Nations. There are also 'disembedded' institutions that are not rooted in a particular national society, although they may serve the interests of particular nation states more than others; these include geographically-dispersed financial and scientific institutions (Avgerou 2002).

Moreover organizations vary in terms of the way they are governed even in the same local social context, which implies different forms of actorhood. Small enterprises run by a proprietor, state-run companies, enterprises of collective ownership, and public services are some of the most common types of organizations in which there are most likely to be roles, workplace identities, and meanings of IS innovation other than those in the professionally-managed organizations that are assumed to be the setting of innovation in much of the IS literature. Professional identities, gender, official ranking, and political status may constitute important sources of difference of meaning for the significance of an IS change.

For example, hospitals in the British National Health System (NHS) have been implementing IS infrastructures since the early 1990s, at the same time as a complex set of changes have been undertaken in professional roles within the NHS, including those of managers, administrators, doctors, pharmacists, nurses, and technicians. Such roles are typically found in the hospitals of most countries, and important insights on the relationship between technology and social context have been gained by situated studies focusing on the interactions of particular technologies and professional roles without much reference to their institutional setting (Barley 1986). Nevertheless, the justification of actors' perceptions and responses within IS innovation episodes in hospitals may require the exposure of the hospital's institutional setting, which in the case of the British NHS in the early twenty-first century comprises the discourse of crisis of British health care systems, the deep-rooted values about health care as a universal right, and the experiments in NHS governance transformation attempted as a national solution to the contradictory expectations of maintaining high-quality universal health care services while spending modest public financial resources (Bloomfield *et al.* 1997*b*).

The enactments and competencies of IS professionals have also varied significantly. Since the 1970s, generic IS professional roles of analysts, designers, programmers, etc. have been created and equipped with standard methods and techniques (Avgerou and Cornford 1993). In the 1990s, these roles began to be reconsidered in order to accommodate new perceptions of good practice as 'knowledge workers' (Scarbrough 1999), or to fit the industrial settings within which most ICT applications are now developed in the form of market products (Cusumano and Selby 1995). If the actors involved in IS innovation episodes do not share a common extra-organizational background, or if the researchers and the readers of studies of such episodes are not familiar with the actors' background, the research frame needs to encompass the social structures whose enactment or capabilities are implicated in the observed actions—the institutional circumstances in which actors have been living their lives.

In the literature on information systems in developing countries, there is a tendency to address differences of meaning and significance attached to an IS change, differences of actorhood, and differences of competencies in improvising in terms of distinctions between local and foreign, mainly western, actors. But the institutional influences that stem from the actors' life contexts are likely to be much more complex. Local discourses are rarely homogeneous or consensual. Even if a dominant way of appraising the value of an innovation is manifested, the subjugated rationalities—whether loud and clearly noticeable or silent and covert—often have the power to frustrate the innovation translation and render it ineffective. Local differences regarding the meaning and significance of an IS innovation may stem from the stratification of the local society, ethnic divisions, or education experiences (Shoib 2002).

The multiple institutional fields that have a bearing on the meanings, the emotional dispositions, and the actions of IS innovation episodes defy a general categorization and *a priori* delineation. The framing of IS studies therefore requires an analytical approach that allows for the emergence of the boundary of an enquiry by its own unfolding. The interest attracted by Actor Network Theory (ANT) since the mid-1990s in IS research provides an opportune methodological orientation to that end.[3] From our perspective, an important feature of the sociology of translation in the form of ANT is its agnosticism regarding the identities of the actors involved in the processes of innovation. While this has been criticized as unacceptable relativism (Russell 1986; Fujimura 1991), we see it as an important element of freedom from preconceived boundaries and institutional structures which allows for the tracing of encountered frames of meaning in the contemporary diversity of IS innovation settings.

Conclusions

We argue in this chapter for a broadening of the perspective of situated research in IS in order to make explicit the contextual origins of the meanings, emotional dispositions, and competencies actors bring to bear on the interactions that constitute the processes of IS innovation. Such broadening is required when the institutional background of an innovation process differs from taken-for-granted assumptions historically relied on in the IS field regarding actors' life experiences, and consequently the meanings these actors give to the attempted innovation and their sense of what is desirable and feasible. These assumptions are often made explicit in the universalistic literature of the field, which seeks to determine what factors and/or behaviours drive towards generally desirable socio-economic goals, such as various types of economic value, industrial democracy, and individuals' emancipation or 'empowerment'.

Such assumptions are often implicit in situated studies, but they are important in making sense of the behaviour of agency-endowed actors and the interactions among them. The degree to which findings of such studies are convincing depends on the extent to which the implicit assumptions regarding the contextual setting within which the situation under study unfolds—and from which actors' cognitive and emotional resources are drawn—are understood and are relevant to that context. We suggest that positioning studies within an appropriate frame of historical and institutional background may elucidate cases in which actors' behaviour appears persistently problematic or paradoxical, according to common expectations about IS innovation and current theoretical understanding.

We see the framing process as an analysis of the institutional allegiances from which actors draw the meaning and emotional attitudes that support their professional role, their sense-making in confronting the uncertainties and surprises of the innovation processes, and their improvising competencies. The framing of an IS innovation study is contingent and emergent. A study can begin only by tracing the network of actors involved in episodes of the innovation process under investigation. But the initial boundaries of a study should be redrawn as the study progresses to probe further in history and institutional domains, with the focus shifting to look more closely within specific institutions or to understand the relationships among institutional actors.

A number of studies in the IS literature show how this framing task may be done (Madon 1993; Walsham 2001; Avgerou 2002), but more research is clearly needed to flesh out the contextualist approach we suggest here. Overcoming the micro–macro sociology distinction is a thorny problem in social sciences, particularly in debates over appropriate theoretical perspectives and trustworthy methodological practices. The IS research field, with its focus on organizations and its long and tortuous conflict between positivistic and interpretivistic epistemologies, is particularly demanding in terms of the effort required to legitimate the shifting of the boundaries of its studies and to develop methodological confidence.

Empirical studies are needed to explore how appropriate framing can be pursued, and to demonstrate the insights gained from tracing the institutional contexts that have a bearing on IS innovation. Nevertheless, one implication of the suggested approach on existing conventions in IS research is that it needs to accommodate existing knowledge on the domain of specific tasks and institutional background of innovation, such as the tasks and institutions of health sectors in cases of IS innovation in health care. Such knowledge almost certainly exists in all situations of innovation, whether it is documented by specific disciplines—for example in public administration, education, and health in every country, and the financial services industry in most countries—or is scattered in a more diverse social, political, and economic literature. Our view is that IS theory should develop its concepts and theoretical propositions with an explicit awareness of their relevance to diverse contextual circumstances. This would enhance IS studies by ensuring they are better informed by specialist domain knowledge, and in many cases engage in a dialogue with it.

References

Appadurai, A. (1996). *Modernity at Large: Cultural Dimensions of Globalization*. Minneapolis, MN: University of Minnesota Press.

Applegate, L. M. (1994). 'Managing in an Information Age: Transforming the Organization for the 1990s', in R. Baskerville, S. Smithson, O. Ngwenyama, and J. I. DeGross (eds.), *Transforming Organizations with Information Technology*. Amsterdam: North-Holland, 15–94.

—— (1999). 'In Search of a New Organizational Model: Lessons from the Field', in G. DeSanctis, and J. Fulk (eds.), *Shaping Organization Form*. Thousand Oaks, CA: Sage.

Avgerou, C. (1989). 'Information Systems in Social Administration: Factors Affecting Their Success'. Unpublished Ph.D. thesis. London: London School of Economics.

—— (2002). *Information Systems and Global Diversity*. Oxford: Oxford University Press.

—— (2003). 'IT as an Institutional Actor in Developing Countries', in S. Krishna and S. Madon (eds.), *The Digital Challenge: Information Technology in the Development Context*. Aldershot, UK: Ashgate: 46–62.

—— and Cornford, T. (1993). 'A Review of the Methodologies Movement'. *Journal of Information Technology*, 5: 277–86.

—— and Walsham, G. (2000) (eds.). *Information Technology in Context: Studies from the Perspective of Developing Countries*. London: Ashgate.

Barley, S. R. (1986). 'Technology as an Occasion for Structuring: Evidence from Observations of CT Scanners and the Social Order of Radiology Departments'. *Administrative + Science Quarterly*, 31/1: 78–108.

Barrett, M., Sahay, S., and Walsham, G. (forthcoming). 'Information Technology and Social Transformation: GIS for Forestry Management in India'. *The Information Society*, 17/1: 5–20.

Bauman, Z. (1998). *Globalization: The Human Consequences*. Cambridge: Polity Press.

Beck, U. (2000). *What Is Globalization?* Cambridge: Polity Press.

Berger, P. L., and Luckmann, T. (1967). *The Social Construction of Reality*. London: Penguin.

Bloomfield, B. P., and Danieli, A. (1995). 'The Role of Management Consultants in the Development of Information Technology: The Indissoluble Nature of Socio-political and Technical Skills'. *Journal of Management Studies*, 32/1: 23–46.

—— Coombs, R., Knights, D., and Littler, D. (1997*a*) (eds.). *Information Technology and Organizations: Strategies, Networks, and Integration*. Oxford: Oxford University Press.

—— —— Owen, J., and Taylor, P. (1997*b*). 'Doctors as Managers: Constructing Systems and Users in the National Health Service', in B. P. Bloomfield, R. Coombs, D. Knights, and D. Littler (eds.), *Information Technology and Organizations: Strategies, Networks, and Integration*. Oxford: Oxford University Press, 112–34.

Bødker, S., Ehn, P., Kammersgaard, J., Kyng, M., and Sundblad, Y. (1987). 'A Utopian Experience: On Design of Powerful Computer-based Tools for Skilled Graphic Workers', in G. Bjerknes, P. Ehn, and M. Kyng. (eds.), *Computers and Democracy*. Aldershot, UK: Avebury, 251–78.

Bourdieu, P. (1997). *Méditations pascaliennes*. Paris: Le Seuil.

Braverman, H. (1974). *Labor and Monopoly Capital: The Degradation of Work in the Twentieth Century*. New York: Monthly Review Press.

Brunsoon, N., and Olsen, J. (1998). 'Organizational Theory: Thirty Years of Dismantling, and Then ... ?', in N. Brunsoon, and J. Olsen (eds.), *Organizing Organizations*. Bergen-Sandiviken, Norway: Fagbokforlaget, 13–43.

Callon, M. (1998*a*). 'An Essay on Framing and Overflowing: Economic Externalities Revisited by Sociology', in M. Callon (ed.), *The Laws of the Markets*. Oxford: Blackwell, 244–69.

—— (1998*b*). 'Introduction: The Embededdness of Economic Markets in Economics', in M. Callon (ed.), *The Laws of the Markets*. Oxford: Blackwell, 1–57.

—— and Law, J. (1989). 'On the Construction of Sociotechnical Networks: Content and Context Revisited'. *Knowledge and Society*, 9: 57–83.

Ciborra, C. U. (1999). 'A Theory of Information Systems Based on Improvisation', in W. L. Currie, and B. Galliers, (eds.), *Rethinking Management Information Systems*. Oxford: Oxford University Press, 136–55.

—— and Andreu, R. (1998). 'Organizational Learning and Core Capabilities Development: The Role of IT', in R. D. Galliers, and W. R. J. Baets (eds.), *Information Technology and Organizational Transformation*. Chichester, UK: John Wiley, 87–106.

—— and Lanzara, G. F. (1994). 'Formative Contexts and Information Technology: Understanding the Dynamics of Innovation in Organizations'. *Accounting, Management and Information Technology*, 4/2: 61–86.

Cusumano, M. A., and Selby, R. W. (1995). *Microsoft Secrets: How the World's Most Powerful Software Company Creates Technology, Shapes Markets, and Manages People*, New York: Free Press.

Czarniawska-Joerges, B. (1992). *Exploring Complex Organizations*, Newbury Park, CA: Sage.

De Sanctis, G., and Gallupe, B. R. (1987). 'A Foundation for the Study of Group Decision Support Systems'. *Management Science*, 33/5: 589–609.

Earl, M. J. (1987). 'Information Systems Strategy Formulation', in R. J. Boland, and R. A. Hirschheim (eds.), *Critical Issues in Information Systems Research*. Chichester, UK: John Wiley, 157–78.

Ehn, P. (1989). *Work-oriented Design of Computer Artifacts*. Hillsdale, NJ: Lawrence Erlbaum Associates.

Fujimura, J. (1991). 'On Methods, Ontologies and Representation in the Sociology of Science: Where Do We Stand?', in D. Maines (ed.), *Social Organization and Social Processes: Essays in Honor of Anselm L. Strauss*. New York: Aldine de Gruyter.

Fulk, J., and DeSanctis, G. (1999). 'Articulation of Communication Technology and Organizational Form', in G. DeSanctis, and J. Fulk (eds.), *Shaping Organization Form: Communication, Connection, and Community*. Thousand Oaks, CA: Sage, 5–32.

Giddens, A. (1984). *The Constitution of Society. Outline of the Theory of Structuration*. Cambridge: Polity Press.

Goffman, E. (1986). *Frame Analysis*. Boston: Northeastern University Press.

Gupta, R. K. (1991). 'Employees and Organisation in India'. *Economic and Political Weekly*, 26: M68–M76.

Hammer, M. (1990). 'Reengineering Work: Don't Automate, Obliterate'. *Harvard Business Review*, 90: 104–12.

Hasselbladh, H., and Kallinikos, J. (2000). 'The Project of Rationalization: A Critique and Reappraisal of Neo-institutionalism in Organization Studies'. *Organization Studies*, 21/4: 697–720.

Heeks, R. (2002). 'Information Systems and Developing Countries: Failure, Success and Local Improvisations'. *The Information Society*, 18/2: 101–12.

Hirschheim, R., Klein, H. K., and Lyytinen, K. (1996). 'Exploring the Intellectual Structures of Information Systems Development: A Social Action Theoretical Analysis'. *Accounting, Management & Information Technology*, 6/1–2: 1–63.

Jarvenpaa, S. L., and Leidner, D. E. (1998). 'An Information Company in Mexico: Extending the Resource-based View of the Firm to a Developing Country Context'. *Information Systems Research*, 9/4: 342–61.

Keen, P. G. W., and Scott Morton M. S. (1978). *Decision Support Systems: An Organizational Perspective*. Reading, MA: Addison-Wesley.

Knights, D., Murrey, F., and Willmott, H. (1997). 'Networking as Knowledge Work: A Study of Strategic Inter-organizational Development in the Financial Services Industry', in B. P. Bloomfield, R. Coombs, D. Knights, and D. Littler (eds.), *Information Technology and Organizations*. Oxford: Oxford University Press, 138–59.

—— Noble, F., and Willmott, H. (1997). '"We Should Be Total Slaves to the Business": Aligning Information Technology Strategy—Issues and Evidence', in B. P. Bloomfield, R. Coombs, D. Knights, and D. Littler (eds.), *Information Technology and Organizations*. Oxford: Oxford University Press, 13–35.

Kyng, M., and Mathiassen, L. (1982). 'Systems Development and Trade Union Activities', in N. Bjørn-Andersen (ed.), *Information Society, for Richer, for Poorer*. Amsterdam: North-Holland, 247–60.

Landauer, T. K. (1996). *The Trouble with Computers: Usefulness, Usabiliy, and Productivity*. Cambridge, MA: MIT Press.

Lyytinen, K. (1986). 'Information Systems Development as Social Action—Framework and Critical Implications'. Unpublished Ph.D. thesis. Jyväskylä, Finland: Department of Computer Science, University of Jyväskylä.

—— and Lehtinen, E. (1984). 'On Information Modeling through Illocutionary Logic', in H. Kangassalo (ed.), *Report of the Third Scandinavian Research Seminar on Information Modelling and Data Base Management*. Tampere, Finland: University of Tampere, 35–118.

McFarlan, F. W., and McKenney, J. L. (1983). *Corporate Information Systems Management*. Homewood, IL: Dow-Jones-Irwin.

Madon, S. (1993). 'Introducing Administrative Reform through the Application of Computer-based Information Systems: A Case Study in India'. *Public Administration and Development*, 13: 37–48.

Mitev, N. (1996). 'Social, Organisational and Political Aspects of Information Systems Failure: The Computerised Reservation System at French Railways', in J. D. Coelho, T. Jelassi, W. König, H. Kremar, R. O'Callaghan, and M. Saaksjarvi (eds.), *4th European Conference on Information Systems, Lisbon, July 2–4, Volume II*, 1213–22.

Monteiro, E. (2000). 'Actor-network Theory and Information Infrastructure', in C. U. Ciborra (ed.), *From Control to Drift: The Dynamics of Corporate Information Infrastructures*. Oxford: Oxford University Press, 71–83.

Mumford, E. (2000). 'Socio-technical Design: An Unfulfilled Promise or a Future Opportunity', in R. Baskerville, J. Stage, and J. DeGross (eds.), *Organizational and Social Perspectives on Information Technology*. London: Kluwer Academic Publishers, 33–46.

—— and Weir, M. (1979). *Computer Systems in Work Design: The Ethics Method*. London: Associated Business Press.

Orlikowski, W. J. (1996). 'Improvising Organizational Transformation over Time: A Situated Change Perspective'. *Information Systems Research*, 7/1: 63–92.

—— (2000). 'Using Technology and Constituting Structures: A Practice Lens for Studying Technology in Organizations'. *Organization Science*, 11/4, 404–28.

Pettigrew, A. M. (1985). 'Contextualist Research and the Study of Organisational Change Processes', in E. Mumford, R. Hirschheim, G. Fitzgerald, and

A. T. Wood-Harper (eds.), *Research Methods in Information Systems*. Amsterdam: North-Holland, 53–78.

—— (1998). 'Success and Failure in Corporate Transformation Initiatives', in R. D. Galliers and W. R. J. Baets (eds.), *Information Technology and Organizational Transformation. Innovation for the 21st Century Organization*. Chichester, UK: John Wiley, 271–90.

—— Ferlie, E., and McKee, L. (1992). *Shaping Strategic Change: The Case of the NHS*. London: Sage.

Porter, M., and Millar, V. (1984). 'How Information Gives You Competitive Advantage'. *Harvard Business Review*, 63/4: 149–60.

Powell, W. W., and DiMaggio, P. J. (1991) (eds.). *The New Institutionalism in Organizational Analysis*. Chicago: University of Chicago Press.

Robertson, R. (1992). *Globalization: Social Theory and Global Culture*. London: Sage.

Rorty, R. (1980). *Philosophy and the Mirror of Nature*. Oxford: Blackwell.

Russell, S. (1986). 'The Social Construction of Artefacts: A Response to Pinch and Bijker'. *Social Studies of Science*, 16: 331–46.

Sahay, S. (2000) (ed.). 'Information Flows, Local Improvisations and Work Practices'. Paper presented at the IFIP WG 9.4 Conference, Cape Town, 24–6 May.

—— and Walsham, G. (1997). 'Social Structure and Managerial Agency in India'. *Organization Studies*, 18/3: 415–44.

Scarbrough, H. (1999). 'The Management of Knowledge Workers', in W. L. Currie, and B. Galliers (eds.), *Rethinking Management Information Systems*. Oxford: Oxford University Press, 474–98.

Scott Morton, M. S. (1991). *The Corporation of the 1990's, Information Technology and Organizational Transformation*. Oxford: Oxford University Press.

Scott, W. R., and Meyer, J. W. (1994) (eds.). *Institutional Environments and Organizations: Structural Complexity and Individualism*. Thousand Oaks, CA: Sage.

Shoib, G. M. (2002). 'The Politics of Rationality: An Ethnographic Study of Sales Force Automation in a Multi-national Company in Egypt'. Unpublished Ph.D. thesis. Oxford: Judge Institute of Management, Oxford University.

Singhi, N. K. (1974). *Bureaucracy: Positions and Persons—Role, Structures, Interactions and Value Orientation in Rajasthan*. New Delhi: Abhinav Publications.

Sprague, E. H. (1980). 'A Framework for the Development of Decision Support Systems'. *MIS Quarterly*, 4/4: 1–24.

Srinivas, M. N. (1989). *Social Change in Modern India*, New Delhi: Orient Longman Limited.

Swanson, E. B., and Ramiller, N. (1997). 'The Organizing Vision in Information Systems Innovation'. *Organizational Science*, September/October: 458–74.

Trice, H. M., and Beyer, J. M. (1993). *The Cultures of Work Organizations*. Englewood Cliffs, NJ: Prentice Hall.

Unger, R. M. (1987). *False Necessity*. Cambridge: Cambridge University Press.

Walsham, G. (1993). *Interpreting Information Systems in Organizations*. Chichester, UK: John Wiley.

—— (1997). 'Actor-network Theory and IS Research: Current Status and Future Prospects', in A. S. Lee, J. Liebenau, and J. DeGross (eds.), *Information Systems and Qualitative Research*. London: Chapman & Hall: 466–80.

—— (2001). *Making a World of Difference: IT in a Global Context*. Chichester, UK: John Wiley.

Weick, K. E. (1995). *Sensemaking in Organizations*. Thousand Oaks, CA: Sage.

Notes

1. In philosophy, the opposite of universalistic is relativistic. IS studies rarely raise explicitly the epistemological distinction between universalism and relativism. In the 1990s, nevertheless, the most visible challenge to the long tradition of universalistic research came from a stream of situated studies. By focusing on the innovation event in its social setting, these studies have tended to avoid general assumptions and conclusions on what social actors are striving for and what is rational behaviour. This chapter draws on prior IS research like this to juxtapose universalistic to situated approaches, rather than to the relativistic perspective.
2. Callon's analysis of framing draws on Bourdieu's (1997) analysis of the time dimension in relation to the question of whether gift economies rely on disinterestedness, and on Goffman's (1986) studies of everyday close interactions.
3. For examples drawn from ANT, see the chapters in this volume by Monteiro (Chapter 7), Kallinikos (Chapter 8), and Smithson and Tsiavos (Chapter 11). Walsham (1997), Monteiro (2000), and Avgerou (2002) provide other descriptions of ANT and discussions of its relevance in IS research.

PART III

Substantive issues and applications

10

Bridging the digital divide: new route to development or new form of dependency?

ROBERT HUNTER WADE

Introduction

The Catholic Church has the practice of appointing a 'devil's advocate' (*Advocatus Diaboli*) when considering a person for sainthood. The role of the advocate is to ferret out all the reasons why that person should *not* be made a saint and present them forcefully in the discussion among the cardinals who make the decision. The current campaign to promote the uptake of ICT in developing countries and to get aid donors to redirect their aid budgets needs devil's advocates to challenge what J. S. Mill once called 'the deep slumber of a decided opinion'.

In the first half of this chapter, I argue that ICT is being oversold as the key both to higher efficiency of corporate organization and public administration, and to stronger responsiveness of government to citizen-customers. ICT tools can help people to learn how to absorb knowledge generated elsewhere and combine it with local needs and local knowledge, and they can help to raise real economic returns to investments, but they are being touted in the development community as though they can leapfrog over the more familiar development problems. This is like saying that cheap books can cure illiteracy. Once the illiteracy problem is solved (as it was in the south Indian state of Kerala), cheap books are a great boon; but giving illiterate people cheap books does not solve illiteracy. My purpose is not to throw out the whole idea of an ICT-for-development campaign, but to signal traps that will cause the campaign to lose credibility. By engaging with these issues, ICT proponents can ensure that the ICT fad ends up with benefits greater than resource misallocations, in contrast to some earlier development fads.

In the second half of the chapter, I suggest that efforts to bridge the digital divide may have the effect of locking developing countries into a new form of dependency on the west. The technologies and 'regimes' (international standards governing ICT) are designed by developed-country entities for developed-country conditions. As the developing countries participate in ICT, they become more vulnerable to the increasing complexity of the hardware and software, and to the quasi-monopolistic power of providers of key ICT services. Worse, the western aid industry, by linking aid to good governance and good governance to programmes to digitalize the public sector ('e-governance'), may be reinforcing the overall

A version of this chapter first appeared in 2002 in *Global Governance*, 8: 365–88.

dependency of developing countries. Governments of developing countries should not take the technologies and international regimes as given. They should press for standards and pricing regimes that make it easier for entities in their countries to access the global information economy. They need more representation in the standard-setting bodies and more support in the ICT domain for the principle that 'simple is beautiful'.

Beware the assumption that ICT is a top development priority

In taking for granted that 'bridging the digital divide' is the central issue of development, literature from the World Bank, Organization for Economic Cooperation and Development (OECD), the Group of Seven (G-7) governments, and individual academics displays a high aspiration-to-evidence ratio and a disregard of trade-offs between specific ICT investments and alternative investments. The unwillingness to grapple with choices is signalled in the commonly-heard refrain 'it is not either/or', meaning that ICT investments do not compete with other investments. A kind of 'groupthink' has emerged such that, as expressed on 21 August 2001 by a well-known World Bank ICT expert who has become sceptical of the claims (and who requested anonymity):

> People react as though you are mouthing obscenities in St Peters if you suggest that there might be problems with treating the digital divide as the number one development priority. For that reason, I've been very cautious in sharing my scepticism within the Bank, although I'm trying a process of gentle acclimatization . . . The last big public forum where I heard someone arguing against the centrality of the Internet was in Seattle about a year ago, during a meeting of the IT rich and famous to discuss their role in overcoming the Global Digital Divide. He made the point that people living on a dollar a day had a lot of other needs to be met first, before Internet access, and was rounded upon by the audience for being a blinkered cynic.

The speaker was Bill Gates!

The ICT campaign rests on the assumption that the digital divide is large. Certainly the west has a much higher ICT density relative to population than developing countries. But relative to income, the divide hardly exists. At the start of the twenty-first century, according to the World Bank (2001), developing countries had a bigger share of world televisions (61 per cent), telephones (25 per cent) and Internet connections (28 per cent) than their share of world income (20 per cent, measured at current exchange rates). The digital divide may have become simply a reflection of the more familiar income divide.

The ICT campaign also rests on the assumption that ICT has some inherent quality that enables it to leapfrog institutional obstacles, and skill and resource deficiencies on the ground, provided that specific supply-side complements are also available, such as trained computer technicians (for more on ICT as a new development paradigm, see Hilbert 2001).[1] The evidence, if any, typically comes from pilot projects, where the alleged success is generalized with little attention to the 'scaling up' problems of providing teaching and maintenance over a much larger number of sites.

Consider the World Bank (2000) policy paper on ICT, called *The Networking Revolution: Opportunities and Challenges for Developing Countries*. Like much of the ICT-for-development literature, it talks about plans, intentions, and opportunities provided; and it blurs the distinction between these and verified actions on the ground. It talks about benefits and not costs. It makes no comparisons between returns to investments in various types of ICT and returns to other kinds of investments, for example in improving crop varieties in tropical Africa. It largely ignores issues of sustainability—such as the financing of recurrent costs of computer servicing and training. And it explains cases of failure, when noted, in ways that protect the assumption that ICT investment is a top priority. The following sections raise questions about some key solutions proposed in this kind of literature.

Multipurpose community telecentres?

Buried at the end of the thirty page World Bank (2000) paper on the 'networking revolution' is an account of multipurpose community telecentres (MCTs) in rural Mexico. These are facilities that provide public access to a variety of information and communication services. It turns out that of the twenty-three MCTs built recently in rural Mexico, only five were working two years later. This is a failure rate of 80 per cent.

The policy paper comments: 'Problems encountered included insufficient maintenance funding, inadequate political interest and will, and cultural constraints which hamper community interest in the projects'. The paper gives no hint why 'political interest and will' might have been inadequate, and why community interest might have been constrained by that hold-all excuse for failure, 'culture'. This fact about the performance of a Mexican MCT programme is the only piece of information in the entire World Bank paper on the relationship between plans and reality.

The paper concludes that the Mexican case, 'underscores the importance of participatory design and attention to sustainability issues in the development of such programs'. The premise is clear: the MCTs are inherently good investments and problems can be solved by paying more attention to unspecified participatory design and 'sustainability issues'.

In the Mexico case, we need to know the answers to big questions in four main areas:

(*a*) Was there a good use for the services the MCTs offered? Or, more generally, what sense did the MCTs make to local people? If their business was mainly local, they may have seen little gain from the sudden access to worldwide opportunities.

(*b*) What would it take to get twenty out of twenty-three working, instead of just five? What sort of analysis was done to develop alternative organizational models of access and to match ICT applications with existing and preferred modes of communication and information exchange within the local settings in which the MCTs were placed?

(*c*) How do benefit/cost ratios for MCTs compare against benefit/cost ratios for other developmental investments in rural Mexico?

(*d*) Can the MCT programme, even with its high failure rate, be justified on the grounds that early projects are bound to have high mortality but give experience necessary for later success?[2]

E-governance?

The World Bank (2000) paper also talks enthusiastically of the ICT programme in the Indian state of Andhra Pradesh:

> Andhra Pradesh was the first state in India to design a statewide computerization programme covering all levels of the administrative spectrum from the smallest—the Mandal Revenue Offices—to the largest... ICTs have provided opportunities for government to redefine its relationship with citizens by encouraging (1) economies resulting in efficient government, (2) personalized service, (3) equal access to government for all, (4) speed and responsiveness, and (5) responsible and measurable government.

The paper refers only to opportunities intended to be provided. According to my own informal survey of friends who live in Andhra (where I used to live), the changes on the ground are largely confined to:

(*a*) easier registration of property transactions, especially land; and

(*b*) easier communication through videoconferencing between government ministers and the government secretariat, on the one hand, and district collectors on the other, although at a very high cost per minute.

All the other programs hardly function, for example computerized one-stop Integrated Citizen Service Centres to handle services such as the payment of utility bills and property taxes, issuing of certificates, issuing permits and licenses, providing information, and facilitating common transactions like change of address and transfer of vehicle ownership. Yet, very costly infrastructure investments have been made. Government offices up and down the state are equipped with computers, but the computers do little. What is the ratio of benefits to costs, including benefits of public procurement for India's computer industry?

In another World Bank ICT-for-development paper, Kenny, Navas, and Qiang (2000: 10) report that an Andhra Pradesh 'programme to computerize the issuance of caste certificates, essential for obtaining government services and access to educational scholarships, managed to decrease the time for certificate issuance from twenty to thirty days to only ten minutes'. My contacts in the capital city of Hyderabad suggested that the real time facing any real applicant is scarcely less time than before. The authors of the World Bank report forget that substantial organizational change is also required to realize the technological potentials. The change is often not induced and the infrastructure is used in a grossly inefficient way.

The Andhra case may be only a matter of a time lag between providing the infrastructure, which can be set up relatively quickly, and writing and adapting the software for all these various tasks, training staff, and making needed organizational changes, which all may take an inordinate time. Perhaps the lesson for the Andhra case is that it is too soon to tell, perhaps the situation is no worse than expected at an early stage, perhaps the experiment has diffuse benefits in mass acceptance stemming from the simple fact that computers are sitting on the desks of offices, 'domesticating' them in the eyes of the citizenry. The World Bank papers do not begin to talk about these complications. Nor do they see the significance of the government charging for issuing certificates and providing other services that were previously free (on the grounds that these services previously required bribes but now no longer do so, making the net cost the same). In the programme of e-governance, citizens become customers.

Computerization within local government?

Aid donors to Tanzania, one of the poorest countries in the world, have championed computerization in local government as part of a reform of local government that began in 1998. As one component, they designed a computer-based program to monitor and evaluate the local governments' service delivery. This in an environment where, according to Hirschmann (2002), 'there was no money [in local government budgets] for paper, pencils, storage of any materials or to pay for transporting anything that data collectors might gather or for checking on data collection . . . Electricity is absent in many locations and unreliable in others'. One aid agency reported that one quarter of its own computers were not working at any one time. Yet Hirschmann (ibid.) commented that the computerization drive proceeded 'with little interest [on the part of the aid agencies and the Tanzanian government] in caution and in questions of how appropriate this might be for rural areas or how it might affect the government's budget . . . One donor said that such a cautious approach would have the effect of keeping Tanzania technologically backward'.

From control to drift

The ICT-for-development movement underplays the well-documented difficulties that have attended efforts to introduce integrated software infrastructures even in large, successful western organizations with no shortage of resources or technical talent. A study by Ciborra (2000) of seven multinational corporations found that, far from realizing substantial gains in control and productivity, the ICT restructuring projects tended to cause 'drift' and a loss of control and productivity. Management interventions to stem the drift tended to cause more drift. This study emphasizes that drift can be positive in encouraging a period of experimentation and search, but more commonly has net negative effects.

One reason for drift is the unanticipatable interactions between different components of the ICT infrastructure. There is no central point, no 'eye', that can decide standards for all components, thereby ensuring compatibility. Complex ICT systems have 'layers' of components—including PCs, computer hardware, telecommunications, cables, software—and decisions made about standards for one layer, in one part of a large organization, can easily interfere with decisions about standards for another layer made in another part of the organization. Compatibility can take years to achieve at a huge cost; by which time new incompatibilities may have arisen.

A second and even more decisive factor is related to the encounter between the technology and the social organization of work. Complex ICT systems imply rigid ways of working that are often incompatible with the ways of working that people actually use when performing inherently vague and fuzzy tasks. The effort to make people change their business practices to fit the technology ('business process re-engineering') often fails. The technology is absorbed into unchanged business practices, with the technology itself becoming a source of inefficiency.

In the public sector, the dangers are compounded by public sector culture and incentives. The introduction of ICT integration programmes is commonly allied with the expectation that ICT will bring private sector efficiency into the public sector. But the incentive to save money rather than spend more to increase efficiency results in contracts being awarded on the basis of price rather than value, and low civil service salaries make it hard to hire project managers with the right technical skills that could save money later on. When the result turns out to be a bureaucratic mess, demoralization sets in. Soon another ICT guru comes along promising to cure the problems, expectations are raised that this time . . ., and the cycle begins again. The *Economist* (2002: 37) describes 'a long list of [British] government IT horror stories', including the Passport Agency where a Siemens-designed system costing £40 million resulted in passport application processing going from ten days to eight weeks, and the magistrates' courts where a £178-million system designed by Fujitsu to link the courts with other organizations, such as the police, customs, and the Crown Prosecution Service, was on the point of collapse, causing remand prisoners to face justice delayed and therefore denied, and victims to face criminals who re-offend before their cases can be heard.

Then there is the World Bank, and its efforts to introduce a unified software program (called SAP) that hobbled it for several years after 1998 and cost $60 million for initial installation and $16 million for a second tranche. A staff survey in 2002, four years after SAP's introduction, showed that, in the words of a Staff Association Newsletter, 'staff suffered mightily when the promised 6-month dip after SAP's inauguration stretched far beyond that period, and many people are still finding SAP a struggle . . . SAP and related systems are still, in some areas, a drag on the Bank Group's efficiency' (*SA Newsletter* 2002).

More than this, e-governance programmes typically carry a political agenda of 'government by the market', based on the assumption that market relations in the public sector are better at meeting people's needs than conventional forms of public administration: the state is shrunk, operational functions are separated from advisory functions, employees are on job-specific contracts, pay is related to performance, and citizens are defined as paying customers, all made possible by ICT. The public administrations of developing countries will be unable to escape these effects when the ways of working required by the technology (designed with OECD business processes in mind) encounter the work practices of developing countries (see Avgerou 2002).

Supply or demand?

The ICT-for-development literature is biased towards the supply side and gives scant attention to demand. It is striking that at a time when the 'digital divide' has become a buzz phrase among researchers, we have little knowledge of the relationship over time between, for example, human development indicators and ICT indicators. Such knowledge would help to make investment decisions about ICT that do not simply assume that ICT investments must have a high priority relative to other investments. Where is the evidence on the relative benefits of investment in ICT infrastructure compared to education, health, roads and dams, and industrial parks? Instead, the literature presents a pot-pourri of anecdotes and correlations, where the criteria of inference are so elastic that correlations become causations. Area A is rich, integrated into market relationships, and has a lot of telephones; area B is poorer, less integrated into market relationships, and has fewer telephones; therefore, a telephone rollout will make B richer and more integrated. Treating ICT as a single lump carries the danger that stories about the success of one component—mobile phones—generate optimism about the success of other components—Internet access, for example.

This kind of thinking has something in common with the campaigns to promote tractors in African agriculture from the Second World War into the 1970s. The tractorization campaigns started with the notion (but not the phrase) of a 'tractor divide'. Developed countries had lots of tractors, African agriculture had hardly any; therefore the US Agency for International Development (USAID), the British government, the World Bank, and the East Germans promoted tractors as a techno-fix that could avoid or help to moderate the institutional obstacles to higher land productivity in Africa. When evidence of high failure rates became too obvious to ignore, the solution was more training and more political will (e.g. Dumont 1962).

On the other hand, the spectacular success of the Mondragon ('Dragon Mountain') cooperative business group in the Basque region of Spain illustrates an on-the-ground sequence from:

(*a*) investments in learning to learn how to absorb ideas from elsewhere, to
(*b*) adoption of latest technologies as they come along.

In only thirty years, the group went from having no modern technology beyond metal-bashing to constructing its own robots and Computer Numerical Control (CNC) machine tools, without the Internet or mobile phones. It was because they had learned how to learn that they could use the latest technologies; it was not technology that somehow gave them the capacity to learn where previously the learning capacity was lacking. In particular, the group helped establish a university (now called Mondragon University), and also technical research institutes (such as Ikerlan) specializing in scanning for innovations that could be turned into new products or product enhancements in the Mondragon firms.

The East Asian experience—by far the most successful experience in the annals of post-World War II development—illustrates the same broad sequence, but with the governments playing an important role (for more on state-private partnerships as a crucial ingredient for successful ICT industries in East Asia, see Wade 1990 and Hanna 1996). For example, the Taiwan government intervened aggressively to force the expansion of engineering education at a much faster pace than any other subject, starting in the early 1960s. With the facilities and incentives in place to produce a surging flow of people through vocational engineering courses, the Taiwan government also mounted a sector-specific industrial policy to build up the demand and supply of electronics. As early as 1965, it established the publicly-owned China Data Processing Centre to promote the use of computers throughout Taiwanese industry. Taiwan's later emergence as a major player in ICT was the result of a long gestation period in which the government steered market-based activity so as to enlarge the capacity to learn, while simultaneously providing incentives to adopt new technologies.

There is a general lesson from these and other experiences of regions and countries that have become active suppliers and participants in ICT. They have all given very high priority to education, including higher education in technical subjects. They were able to take up ICTs quickly because their supply of skilled manpower was relatively high and therefore its cost relatively low. But this was not just a matter of opening up training centres for computer technicians; the education push had to be more broadly based, not ICT specific. The implication is that the ICT campaigners working in and around aid agencies need to press for changes in the current focus on 'poverty reduction' and primary education. If ICTs are to go beyond quite specific uses (like mobile phones), then higher education has to be improved in technical subjects more broadly. This is expensive and difficult, but development practitioners can capitalize on the current enthusiasm for ICTs by placing higher education on the aid agenda. For many developing countries, the promotion of ICTs is tied to not only the education domain but also the domain of civil and political freedoms, particularly freedom of the press; and one of the collateral benefits of ICTs may be expansion of these freedoms.

Developing country disadvantages in accessing the global information economy

In this section, I will look at the impact of the global ICT industry and the international ICT regimes on less developed countries (LDCs). ICTs seem to have a large quotient of public goodness about them, and therefore high spill-over benefits.

'The Internet was created in the United States, but its cost-slashing consequences for information and communications enhance people's opportunities everywhere', in the words of the *Human Development Report* of the United Nations Development Programme (UNDP 2001: 95) This might be taken to imply that LDCs are not much disadvantaged by the structure of the global ICT regimes. Not so. In several ways, developing-country users are being tied more tightly into hardware and software escalation with ramifications difficult to anticipate (like the more abstruse parts of derivatives markets). Escape is increasingly limited, and the costs grow as the dependence of the users increases. This is a new form of international digital dependence for which we need a new version of the 1970s dependency theory.

The US/Microsoft privilege in software

Developing countries are placed at a growing disadvantage by the software–hardware arms race in the global market for savvy computer users. The software is constantly being rewritten to take advantage of unused slack in memory and speed, and then the hardware manufacturers escalate the headroom so that the new software can run as fast. The resulting complexity is attractive to the minority of younger, wealthier, and better-educated people; but it is a deterrence for others, who perceive that software is actually getting worse for many word processing and spreadsheet uses because the new features cause glitches and slowdowns. The effect of this technological arms race is to keep widening the digital divide between the prosperous democracies and the rest of the world.

Every time Microsoft comes out with a new version of its software that can run only on the latest generation of chip, LDCs either face more costly and hassle-full communication with customers and suppliers in the OECD countries, or spend scarce foreign exchange to replace their old machines and software. This is an inbuilt gravitational force against their ascent across the digital and incomes divide. It is where Microsoft is most abusing its monopoly power, checked only a little by the unorganized anger of the 'orphans' left with incompatible document formats (see Hanseth and Braa 1998).[3]

The software–hardware race and the constant need for further investment almost certainly represents a huge misallocation of resources, not only in LDCs but even in the OECD countries. The rush to upgrade software, and hence hardware, comes first from large companies, ostensibly because they 'need' the new capabilities. Large multinationals promote the idea of 24-hour global working, for which they need to have messages and designs winging their way to Tokyo as Los Angeles closes down, and vice versa. But the rush to upgrade software is probably driven as much by organizational pressures from ICT staff and from top managers. Since salaries are linked to budgets, ICT staff press for whatever keeps ICT budgets high and fully spent by year end. The top managers worry that the 'brand' of the company will be downgraded if the company does not boast the latest ICT. Smaller organizations have to follow this trend, because otherwise they have difficulties exchanging attachments and various other text- and graphics-bearing bits of software with the large companies with which they do business.

In short, software upgrading in big companies is driven in large part by ICT staff and organizational concerns about brand and reputation, though even big companies tend to use only a fraction of the software's capacities. Other companies then need to upgrade their own software so that they can communicate with the big ones, though they use an even smaller fraction of the new capacity. The software upgrading is thus diffused from big to small and from OECD companies to LDC companies that wish to deal with the former. The hardware makers then follow with bigger capacity machines.

This makes it hard for an organization with important outside electronic communication to decide that its existing infrastructure is good enough for the next five years. One line of solution is open source software development (e.g. Feller and Fitzgerald 2002). Open source is based on the idea that people using a piece of software should be able to examine the programming instructions, or source code, in order to identify problems, make improvements, and otherwise adapt it to their requirements. Such software is usually put together by loose coalitions of volunteers and given away more or less free on the Internet. Open source software (particularly the Linux operating system) would get around the Microsoft monopoly. Not surprisingly, Microsoft executives have denounced open source as a threat to intellectual property. For instance, Gallagher (2001) reports that some software companies, including Microsoft, are beginning to experiment with a software protocol that makes some parts of a given piece of software open source, in the spirit of 'greater openness behind closed doors', while other parts are closed (see also Lessig 2001).

Some governments are pushing forward with public programmes of support for the development of open source software, and the open source movement continues to flourish. But many of the key developments are being incorporated within systems being developed by the dominant private suppliers. The upshot is that LDCs are unlikely to be able to escape the significant costs associated with upgrading Microsoft and other software. Much depends on how easily Linux or other open source software can communicate with Microsoft software, and on the extent to which open source software can be used with 'low' technology computers. It is a major problem that Linux takes high-level computer skills to install.

In this very important issue of software standards and Microsoft's monopoly, there is no clear answer about what LDCs should do. China and India may be a source of hope, because they both have a large and independently incentivized and resourced pool of relevant skills that may push the development of enduring open source software.

Appropriate technology, in addition to open source software, can also help. Developing country producers could make 'Volks' computers with open source software that would give people the necessary basic capacities without getting caught up in the race to get the latest power-enhancing bells and whistles from the USA. The laptops could be cheap, stripped down, suitable for beginners, and solar powered. We might even see in laptops the equivalent of the trajectory from simple hand calculators with four functions, which cost fifty dollars with special batteries, to the current throwaway solar-powered units. Aid donors could promote the development of such computers. Indian and Brazilian producers are already starting to make $200 'simputers', going against the grain.

However, without regulation in developed countries, the opportunities for LDCs presented by open source software and simple computers are limited by the negative effects of network economies. Much of the value of a networked technology comes from its compatibility. Much of the value of a specific piece of ICT comes from its compatibility with the broad stock of ICT—which resides in the developed world. As a general rule, the value of LDC ICT is much reduced if it is not compatible with OECD ICT. If India's simputer does not allow rural Indian users to e-mail their relatives in the UK, its value is diminished. If it does not even allow rural Indian users to e-mail their relatives in the city (because they have had to upgrade in order to e-mail their relatives in the UK), its value is further diminished. But India controls too small a stock of global ICT to ensure that all computers remain 'simputer compatible'.

The only solution seems to be a US regulatory requirement that all Internet/email/word processing software should be made open source or simputer compatible. Whatever bells and whistles are on the software, it must allow open source software (or some global standard minimum-level software) to connect with it. And it would help greatly if the professional bodies of software and hardware developers—in the USA, India, and elsewhere—lent their prestige to a 'simple is beautiful' movement. At present, the surging software industry in developing countries like India and Brazil is oriented mainly towards servicing the US market, and therefore takes maximum software capacity as the objective; it neglects the servicing of domestic needs, as shown by the fact that India's own simputer is not taken seriously by India's ICT industry or by the Indian government. Redefinition of professional standards could encourage professionals to tune themselves to simplicity, robustness, and cheapness. Failing this, LDCs are trapped in the software–hardware arms race.

The US privilege in Internet access

Developing countries, like the rest of the world outside the USA, operate at a cost disadvantage in Internet access compared to the USA. To oversimplify, US Internet users—or at least Internet service providers (ISPs)—pay nothing to access a website in foreign countries, whereas foreign users (ISPs) pay when US users visit their sites and pay when they access US sites. This is a scandal which has not been studied much, and no international agreement is in sight.

ISPs in developing countries are at an especially big disadvantage. As Internet traffic charging arrangements have shifted from the early 'peer-to-peer' relationships (each party bore its own costs and no revenue was settled between them, hence they were 'peers' in a non-hierarchical network) to commercial relationships, the big ISPs—with large capacity and crash-free infrastructure—have become able to aggregate traffic to themselves and then have the clout to set prices for their traffic with smaller ISPs. ISPs in developing countries cannot invest in large capacity networks, experience more crashes and congestion, and so have more difficulty attracting traffic; users try to divert around them. The problems are similar to those of telephone-revenue settlement arrangements (see next section), with the difference that the telephone arrangements are subject to some scrutiny from the International Telecommunication Union (ITU) whereas those for the Internet are strictly company-to-company.

At the same time, some technical experts say that the Internet architecture would collapse if it was organized any other way. Whatever side one comes down on, it is clearly the case that the structure of the network peering arrangements and the hierarchy of traffic-handling agreements is a significant issue for improving developing countries' access. One line of solution is for other countries and regions to build up their own network hierarchy on a regional basis, without having to route their Internet traffic through the USA. But this runs into the 'diseconomies of small scale' problem of all networks: the global reach of these regional networks is limited by their small capacity. The costs of installing the necessary infrastructure would be large and the results (certainly for the poorer countries) might still leave users in developing countries facing more network congestion and lower-quality service.

The US privilege in telephone services

In a similar way to the Internet, the issue in telephone services is the accounting system by which international payments for voice traffic between countries are made. The ITU used to set and manage the rules of who paid whom how much, in a system largely dominated by national telecommunications monopolies. Since the late 1990s, that system has broken down under the galloping impact of new facilities (satellite networks and submarine cables), new facilities-based operators, and subsidiaries of foreign telecommunications companies which have sprung into liberalized telecommunication sectors through the liberalization promoted by the Basic Telecommunications Agreement of the World Trade Organization (WTO), effective as of February 1998.

The moves to encourage competition and market liberalization of telecommunications in developing countries have begun to restructure prices and make governments in developing countries more willing to renegotiate the terms of the accounting-rate system, especially through the opening of markets to foreign telecommunications subsidiaries that can exchange telecommunications traffic with their home base or other countries at prices unrelated to the generally higher costs facing a domestically-based company. A major upheaval is underway, driven not only by the supply-side changes already mentioned but also by unhappy consumers who want to know why, in India, they pay ten times as much to call a relative in a neighbouring country as to call a relative a similar distance away within India—and why, in Latin America, it costs two to three times as much to call Chile from

Argentina, Brazil, and Mexico as in the opposite direction (Braga, Forestier, and Stern 1999).

However, the US government and major global firms have shown little willingness to negotiate new arrangements that will allow developing-country providers to charge prices that reflect the higher real costs in providing telephone services in developing countries. Instead, they use their dominant market position to force down global costs of telephone services and thereby save US consumers (makers of international calls) billions of dollars a year; but at these prices, telephone companies in many of the poorer LDCs cannot compete or invest in upgrading.

Box 10.1 describes schematically the nature of the accounting-rate problem in international traffic. The situation is actually made messier than that depicted in the box because most LDC telecommunications operating companies do not have anything close to reliable data on the costs of service by category of service, or reliable in terms of the kinds of accounting systems now established by the dominant world companies like AT&T, BT, and Bell Canada (which themselves did not have this data before they faced competition). So, when LDCs try to argue that their costs of service are higher than in the developed countries, they cannot demonstrate it with empirical data. The ITU attempted in the late 1980s to conduct a study of several LDC companies and found no reliable data. Several subsequent attempts have floundered for the same reason.

Box 10.1. The accounting-rate problem in international traffic

Stage I, the ITU regime, pre-WTO Basic Telecommunications Agreement 1998

Monopolist telephone companies A* and B* in countries A[a] and B[b] negotiated an agreed rate for calls between A and B and an agreed (perhaps 50:50) settlement split. Typically, calls from A to B outnumbered calls from B to A. Therefore, B* (and B government) earned net foreign exchange. The deal was often very profitable. Profits could be used for upgrading the domestic network—or could be captured by central government and used for other purposes.

Stage II, post-WTO Basic Telecommunications Agreement, 1998

Countries agree to give up monopolies in long-distance telecoms, to allow foreign competition, etc. The old regime crumbles. Now, A and its companies seek to find partners in B (perhaps own subsidiaries) who will accept a lower price for international calls between A and B. Prices start to fall. But A companies can provide calls from A to B at much lower cost than B-owned companies can provide calls from B to A. The B companies therefore face higher traffic from A (because calls are cheaper), but at lower profit margins. They find it more difficult to invest in network expansion, etc. B government does not like it because, despite higher traffic, net foreign exchange payments from A fall.

Notes: [a]Country A: developed; [b]Country B: developing

Creative approaches would provide stronger incentives for inward investment in developing-country infrastructures, thereby increasing the capacity of networks to offer services at prices that are affordable by more users in developing countries. One might argue that since northern consumers benefit from new connections being made in LDCs, some sort of subsidy by northern telecommunications companies of their southern counterparts is justified. Also, LDCs could address on a regional basis some inequalities created by existing accounting-rate systems. They could cooperate regionally to revise traffic transit agreements at the same time that issues are addressed on a global basis. Whatever happens, they have to strengthen their generally archaic national accounting-rate systems for telephone services. But this means more resources allocated to data collection and to building up the relevant skills, which are in very short supply. Aid donors can help.

World Bank control of the development portal to the Internet

The Internet is often presented as a 'democratizing' force, a subverter of hierarchies of power and privilege. The claim rests on the image of the Internet as a public library with equal access for all. But the library is staffed by experts in learning who have chosen certain books because they meet its purposes, and organized them in a way that makes them easy to find. The Internet—the web—is more like a huge pile of papers than a library. Only a small proportion of the 26 million sites are catalogued or retrievable at will (most sites are not covered by search engines), and the majority are of little value for anyone with a rational economic or social purpose. The Internet becomes like a library only when some expert chooses useful bits of material—'wisdom' as well as 'fact'—and organizes them to be accessible to users from a single portal. The point applies to ICTs more generally. Except in the rather restricted way of Kerala fishermen using mobile phones to do their own arbitraging, their use for learning requires an infrastructure of people and systems.

This raises the questions of who controls the infrastructure, and what the potential is for knowledge to become more and more centralized. The big players can develop glossy and comprehensive 'mega portals' that, with heavy promotion, can hope to capture all but the most resistant users, leading them round and round a network of 'linked sites' that will exhaust the curiosity of almost everyone who enters, making them feel that they have learned all there is to learn about the subject.

This is one of the main questions being asked about what appears to be an excellent idea, the establishment in 2001 of the World Bank's multi-stakeholder, $70 million 'supersite', called the Global Development Gateway. The site is intended to be 'the premier web entry point for information about poverty and sustainable development'. It will provide easy access to: data about aid agency projects; data about organizations working on development; an online bookstore; nested country websites; and a selection of links to over one hundred policy topics. The site will be heavily marketed to officials, journalists, students, non-governmental organizations (NGOs) and others. Words like 'transparency', 'inclusivity', and 'interactivity' buzz through the promotional literature, but many criticisms of the Gateway plans have been made by a variety of officials and web experts.[4]

The World Bank is right to say that the Internet is a powerful but confusing medium for people working on international issues. Even apart from the Gateway, the World Bank gets over four million hits per month on its main website, www.worldbank.org. But critics of the Gateway say that the World Bank is drastically overestimating what can be achieved by one website; it is 'trying to kill five birds with one stone'. In particular, critics worry about editorial control of the content accessible through the portal. The Bank has recruited the editors for each topic area, called by the user-friendly term 'Topic Guides'. These editors are given the task of examining websites across the world to see what exists on their issues. They then post links to whatever reports they feel match their 'quality' standard, highlighting certain reports as priorities. But the Topic Guides will not only provide links to existing sites; they will also add value, interpretations, and judgements. Since the World Bank has an official view about development that is not shared by all, the danger is that it is using its influence to ensure that non-Bank views are weeded out, leaving the field unchallenged for bedrock assumptions of typical World Bank development thinking, such as that 'openness is good for growth and poverty reduction' (meaning developing countries should keep tariffs very low on imports of both goods and services, and allow free entry and exit of foreign firms).

Weeding out contrary views would be justified on grounds of quality and of being 'unsubstantiated opinions rather than validated research'. In this way, the World Bank—and behind it, the US state and US NGOs—could reinforce their intellectual dominance of the development field; while developing countries would continue to be enlisted as recipients of (northern) knowledge rather than co-partners in knowledge generation (Wade 2002). The Development Gateway would become the Development Gatekeeper. The problem is not only the Topic Guides' discretion to screen out dissident views, but also the lack of incentive for Topic Guides from the Bank to devote any time to it or develop it for purposes beyond publicizing World Bank publications. For instance, the Gateway had an ICT page for a time, but the Topic Guide treated it as a low priority, the external contributors contributed little, and the page was removed. When the Bank established the Development Gateway Foundation, it appointed an ex-Bank official to head it. The foundation immediately contracted the key functions back to the Bank, with no tendering. Some NGOs have been campaigning via the Internet for NGOs not to join or participate in the Gateway.

The value of increasing the voice of developing countries in organizations setting ICT standards and rules

Developing countries complain that they do not have enough meaningful representation on the standard-setting bodies, notably the ITU and the Internet governing body—to the extent there is one—the Internet Corporation for Assigned Names and Numbers (ICANN), a US-based organization made up of representatives from industry, government, and Internet users set up in 1998 to operate as a privately-run corporation.[5]

Although it is true that developing countries have little representation in organizations like the ITU and ICANN, the problem is not just formal representation. It is also that, while developing countries and groupings like the Caribbean states can—and do—argue a position in the ITU which is different from that of the USA, they find it difficult to fight for their position once they present it, because of thin backup. The G-7 countries have more staying power. Moreover, there is evidence that standards setting is becoming more dominated by private firms and consortia, and in these private entities people from developing countries are even less represented than in the more public bodies (Cohen 2002).

But does a weak voice for developing countries in standard-setting bodies harm their take-up of ICT? One line of argument says no:

(*a*) The standards-setting bodies are too weak in what they do for more LDC representation to make much difference to LDC access to the global information economy.

(*b*) Developing countries (aside from the Koreas) are consumers of ICT services, not producers, and as such they are standards-takers rather than standards-makers; their principle interest is in the 'uniformity' of standards, not in the precise content of the standards—uniformity is a public good, which once supplied to some is available to all.

(*c*) Hence, developing countries' near-exclusion from standards setting does not harm their ICT take-up.

The counter-argument is that ICANN decisions do filter down to affect such things as the architecture of the Internet and price-cost relationships in ISP networks—who pays for what. In the short term, the impacts of ICANN decisions are probably small, and so more representation by developing countries will not quickly improve their access to global networks. But over the longer run, each new generation of technology contains latitude for tweaking in a direction which looks the 'natural' way to go from a North American and European perspective, or towards one that makes access cheaper for developing countries. The accounting-rate regime that settles international payments for telephone traffic between countries is an example.

In the general case, ICT standards-setting bodies are only rarely neutral coordinating bodies, because ICT standards are only rarely purely public goods. People who participate in these bodies freely acknowledge that they are there because their company or country wishes them either to slow down the negotiations until their own approach is *de facto* dominant in the market, or to push for the approach closest to their own and least useful to their rivals, or to get advanced knowledge of how to future-proof their products. This is why firms, industry associations and states are willing to meet the substantial expenses of their representatives on the legions of standard-making bodies, who typically have to spend many days a year in a foreign country. They are willing to pay because they see privatizable gains at stake. For instance, the Internet Engineering Task Force, the main global forum for software developers, operators, and vendors that sets the protocols designed to maximize connectivity between computer systems, has seen determined efforts to capture it by dominant industry representatives (see Spar 2001 for an analysis of the same idea in broader historical perspective). Only when the market is very new, and perhaps when standards setting has high-level government backing, may cooperation in standards setting be driven mainly by the potential gains of enlarging the whole market.

All this suggests that more LDC representation in the ICT standard-making bodies could help to shrink the digital divide on terms more favourable to LDCs. But they need technical support to ensure that they can participate actively. Those in charge of standards setting (whether in private consortia or in public/private fora) could formulate a code of conduct in which they commit themselves to require representation from developing countries. 'Good citizen' corporations might commit themselves to include entrepreneurs from developing countries in the teams they send to selected standard-setting meetings, as part of a training package. Aid donors can help in the same way. And donors can also provide funds for developing countries to hire first-class people to negotiate on their behalf in standard-setting bodies and with multinational corporations.

On the face of it, less dominance of the Internet by the English language, with a higher proportion of the content in LDC languages, would help build the base of Internet users. At present, something over 60 per cent of Internet content is in English, the mother tongue of 10 per cent of the world's population.

Conclusions

I have argued that a kind of groupthink grips the burgeoning ICT-in-development field. This groupthink says the following:

- The digital divide is the site of the major unequalizing force in the world economy today.
- The divide can be bridged by supplying more ICT to developing countries.
- ICT is an inherently enabling meta-technology that can bypass or leapfrog institutional and infrastructural obstacles.
- ICT investment is exempt from normal comparisons of costs and benefits between alternative kinds of ICT and non-ICT investments—because it reflects a 'new paradigm' of thinking and because 'it is not a matter of either/or'.
- Evidence of high failure rates in ICT projects shows only the need for more training, or lifting of cultural constraints, or strengthening of political will, and does not question the priority of ICT investment.[6]

This argument assumes that the spread of computers and Internet access itself will cause efficiency gains in firms and public administrations, and lower transactions costs for all. I have suggested that ICT is more likely to be 'captured' by the existing organizational inefficiencies and itself made to function inefficiently; and that developing countries are in danger of locking themselves into a new form of e-dependency on the west as they introduce software and hardware systems that they have no capacity to maintain for themselves and that become crucial to the very functioning of their corporate and public sectors.

The case of the 'e-governance' movement is especially worrying. Western governments and the World Bank have said that aid should be targeted at countries with good policies and good governance. Good governance is shown by transparency, quick response, and accountability—achieved by creating unified ICT infrastructures through the public sector. Who can do this? The big western ICT firms. The donor governments steer the recipient governments towards their national champions (the French to Alcatel, the Germans to Siemens, etc.). The earlier-mentioned difficulties that many sophisticated western organizations have experienced as they tried to reconfigure themselves with a unified ICT infrastructure are brushed aside. In this way, developing country governments have an incentive to commit themselves to introducing ICTs for e-governance simply in order to get more aid. They then tie themselves to the standards of the ICT suppliers that those providing their aid choose and enter an open-ended dependency on these suppliers for the continued functioning of their public administration.

ICT-for-development can be seen as a fad, a focal point for enthusiastic action by many kinds of actors concerned with development. The history of development thinking since the 1950s is full of fads: not only tractors in Africa in the 1960s and early 1970s, but also 'integrated rural development' (IRD) in the 1970s, and environment-in-development in the 1980s and early 1990s. Fads are analogous to financial bubbles. They have advocates—who (by definition) have a solution looking for a problem—who skew information to favour their cause and disfavour alternatives, who build up alliances to promote their cause, who want action now, and who insulate themselves from disconfirming evidence. Fads can help development by channelling enthusiasm into action. But channelling enthusiasm into action may require poor decision making in the sense of no serious consideration being given to alternatives—because serious consideration of alternatives might make consensus around one alternative more difficult. Better to claim that the route to the future is clear. Good decisions, on the other hand, require open-minded consideration of several goals and alternative paths, and open-minded consideration of evidence for and against. However, it is then more difficult to get consensus that one of the alternatives is the best, and therefore more difficult to get concerted and enthusiastic action. The big worry about the ICT-for-development movement is that it reflects a rationality of *action* that is obstructing rational *decisions* about development investments.

This being said, a big push, perhaps with some public funding, for universal telephone access does seem worthwhile, especially for mobile phones (unless a mobile phone programme would eclipse the completion of a public wired system where conditions of topography and population density mean that the wired system would be cheaper and benefit more people). The Internet is much more problematic. It is an electronic 'Andrew Carnegie', who made free libraries available to already literate middle America, when what is needed first in much of the developing world is the Internet equivalent of 'Paulo Freire', aimed at making people literate. The 'everyone-in-telecentres' vision has to be given up.

ICT capacity, including Internet capacity, is clearly important in specific sites of an LDC economy, notably at sites of exporting and international banking. Here, targeted ICT interventions to upgrade to international standards may have big payoffs. This is also the case at the regional level: the World Bank, aid donors, and private companies can sensibly promote infrastructure, like the circum-African submarine cable, as a way of lowering the cost of communication with poor regions; and they can promote regional communication networks between scientists, such as the Association for Strengthening Agricultural Research in East and Central Africa (UNDP 2001: 113).

Indeed, I would argue, dropping my devil's advocate hat, that the World Bank and the other multilateral development banks should be doing more than they are currently doing in the ICT field. The missionary Bank literature cited earlier comes from a set of people who are marginal to the organization as a whole. The operational managers tend to know very little about ICT potentials, and are inclined to think that ICT applications are too complex and fast-changing for them to get involved. They turn away from informing themselves about the technologies to the point where they could see how to exploit specific potentials while managing the risks.

I have also argued that developing countries are disadvantaged in their access to the global economy not just by their lack of income, skills, infrastructure and the like, but also by the very standards and rules that are built in to the international systems. These standards and rules ensure that western suppliers benefit disproportionately as developing countries become more integrated into the international ICT system. The most obvious case in point is the private standards being set for software by Microsoft, and its continual upgrading to meet the wishes of savvy computer users in the north, which places developing countries at a big disadvantage. The public and semi-public standards being set in other areas—the pricing of Internet access, for example—may be having a similar disadvantaging effect. In principle, regulation in developed countries provides a way out. Software and hardware makers in the north could be required to keep their products compatible with some global minimum-standard software in order to overcome the negative effects of ICT network economics on LDCs. The politics are hardly encouraging.

In any case, this is something to which development organizations like the World Bank and UNDP should be drawing attention. They are not. Might this have something to do with the businesses behind the ICT-for-development campaign—companies like Microsoft, Hewlett Packard, Siemens and Alcatel? The tractors-for-African-agriculture campaign was not the brainchild of USAID, the UK government, and the World Bank alone. It was pushed by tractor companies facing stagnant tractor sales and seeing LDC markets as their vent for surplus; for which purpose they needed a complementary push from public bodies to create the market. The big ICT firms may be playing a similar role. Their interest is presumably to get the public bodies to create demand for their products in LDCs—and not have LDCs worry themselves about the international regimes or the dangers of dependency. Cisco Systems, for example, is training people around the world in the use of IT products through the Cisco Networking Academy Program that 'enables the Internet to bring digital opportunities to every corner of the earth' (to quote from the promotional material), in 'strategic partnership' with USAID, UNDP, and United Nations Volunteer Program (Friedman 2002).

It will be interesting to follow how the dotcom crash that started in early 2000 affects the ICT-for-development movement. The market bubble was driven by supposedly impartial experts from consulting firms and academia who spun unrealistic tales of dotcom success for investors; and many of the experts went on to reap extraordinary profits from advising clients to invest in online ventures that turned out to be unprofitable. The question is whether the ICT-for-development campaign has received added impetus from these same consultants and academics who are now spinning equally unrealistic tales of ICT success in developing countries.

References

Avgerou, C. (2002). *Information Systems and Global Diversity*. Oxford: Oxford University Press.

Braga, C., Forestier, E., and Stern, P. (1999). 'Developing Countries and Accounting Rates Reform—A Technological and Regulatory El Nino?', *Private Sector, Note 173*. Washington, DC: World Bank, January. www.worldbank.org/html/fpd/notes/173/173braga.pdf

Ciborra, C. U. (2000) (ed.). *From Control to Drift: The Dynamics of Corporate Information Infrastructures*. Oxford: Oxford University Press.

Cohen, E. S. (2002). 'Allocating Power and Wealth in the Global Economy: The Role of Private Law and Legal Agents'. Paper presented at the Globalization, Growth and (In)Equality conference. Warwick: Centre for Study of Globalization and Regionalization, University of Warwick, 15–17 March. www.warwick.ac.uk/fac/soc/CSGR/5th_Annual_Conference

Dumont, R. (1962). *False Start in Africa*. London: Earthscan Publications.

Economist (2002). 'Government and IT: Your Health Depends on It', *Economist*, 2 May.

Feller, J., and Fitzgerald, B. (2002). *Understanding Open Source Software Development*. London: Addison-Wesley.

Friedman, A. (2002). 'Using IT to Fight Poverty'. *International Herald Tribune*, 12 April.

Gallagher, D. (2001). 'Open Source Leads to New Approach: "Gated Community"'. *International Herald Tribune*, 16 November.

Hanna, N. (1996). 'The East Asia Miracle and Information Technology'. Working Paper. Washington, DC: World Bank.

Hanseth, O., and Braa, K. (1998). 'Technology as Traitor: Emergent SAP Infrastructure in a Global Organization', in R. Hirschheim, M. Newman, and J. I. DeGross (eds.) *Proceedings of the Nineteenth International Conference on Information Systems*. Helsinki: ICIS: 188–96.

Harvey, F. (2002). 'A Storm Brews for Surfers'. *Financial Times*, 18 March: 16.

Hilbert, M. (2001). 'From Industrial Economics to Digital Economics: An Introduction to the Transition'. Santiago, Chile: CEPAL.

Hirschmann, D. (2002). 'Aid Dependence, Sustainability and Technical Assistance: Designing an M&E System in Tanzania'. Working Paper. Washington, DC: International Development Program, School of International Service, American University, May.

Kenny, C., Navas, J., and Qiang, C. (2000). 'Information and Communications Technologies: A Toolkit for Designers of Poverty Reduction Strategy Papers'. Discussion Paper. Washington, DC: World Bank, 29 August. www.worldbank.org/poverty/strategies/chapters/ict/ict.htm

King, K. (2000). 'Towards Knowledge-based Aid: A New Way of Working or a New North-South Divide?'. *Journal of International Cooperation in Education*, 3/2: 23–48.

Lessig, L. (2001). 'The Internet Under Siege'. *Foreign Policy*, 127, November/December.

SA Newsletter (2002). 'SAP: A Work in Progress', *SA Newsletter*, March/April.

Spar, D. (2001). *Pirates, Prophets and Pioneers*. London: Arrow.

UNDP (2001), United Nations Development Programme. 'Global Initiatives to Create Technologies for Human Development'. *Human Development Report 2001*, New York: UNDP, Chapter 5.

Wade, R. (1990). *Governing the Market: Economic Theory and the Role of Government in East Asian Industrialization*. Princeton, NJ: Princeton University Press.

—— (2002). 'US Hegemony and the World Bank: The Fight over People and Ideas'. *Review of International Political Economy*, 9/2: 215–43.

Wilks, A. (2000). 'Development Through the Looking Glass: The Knowledge Bank in Cyberspace'. Paper presented at 6th Oxford Conference on Education and Development: Knowledge Values and Policy, September. www.brettonwoodsproject.org/briefings/index.html

World Bank (2000). *The Networking Revolution: Opportunities and Challenges for Developing Countries*. Washington, DC: Global Information and Communication Technologies Department, World Bank, June. www.infodev.org/library/NetworkingRevolution.pdf

—— (2001). *World Development Indicators*. Washington, DC: World Bank.

Notes

1. The same assumptions are made by the G-7-sponsored Digital Opportunity Task Force (or DOT Force), a public-private partnership of G-7 governments, multinational corporations, the World Bank, NGOs, and developing country governments. This is indicated by a comment from a spokesperson: 'The DOT force's work can be a new model for the battle against poverty' (Friedman 2002).
2. Even if the MCT failures yielded nothing, the economic rate of return on the whole investment could still be a possibly acceptable ten per cent if the return on the successful twenty per cent of cases is forty per cent.
3. I am indebted to Carsten Sorensen for discussion of some of the points made here.
4. See: the publicly-accessible Internet consultations at www.bellanet.org/gdgprinciples and www.globalknowledge.com; the September 2000 open letter at www.brettonwoods project.org (King 2000); and, especially, Wilks (2000).
5. The ICANN organizational structure before new byelaws were adopted on 15 December 2002 is a good example of a dynamic technology hitting institutional constraints (see

Harvey 2002). That structure of a Board of Directors with nineteen elected members, including five representatives of the Internet community, was said to generate management paralysis, resulting in long delays in decisions about new domain suffixes which would open up enormously the number of new website names. The new byelaws introduced changes to the appointment and composition of the ICANN Board (see www.icann.org).

6. An updated draft in August 2001 of the paper by Kenny, Navas, and Qiang (2000) on a Poverty Reduction Strategy Paper Toolkit shows a somewhat lower aspiration-to-evidence ratio.

11

Re-constructing information systems evaluation

Steve Smithson and Prodromos Tsiavos

Introduction

Within organizations, information systems evaluation is a key part of IS investment decisions, for example in deciding whether to go ahead with a particular project, or to modify or scrap a development. The evaluation becomes a basis for comparison with other projects, both inside and outside the organization. With increasingly large investments, these become decisions with significant financial implications. Moreover, such decisions can have significant implications for both groups and individuals within the organization. People's careers or working lives, or a group's command of resources or status, may be contingent on the evaluation of a particular information system.

Information systems evaluation has therefore been a popular and long-standing research topic for many years, but it is a slippery one: 'a wicked problem' (Farbey, Land, and Targett 1999). It has haunted the discipline since the 1970s (Land 1976), spawning countless conferences, journal articles, and books aimed at academic or practitioner audiences. The objective of this chapter is to examine the underlying role of IS evaluation through a focus on its common core, amidst the numerous accounts of what evaluation represents.

In the chapter, we clarify the differences and links between the act—or instance—of evaluation, the method of evaluation, and the outcome of evaluation. We argue that an evaluation is not the result of the efforts of a single group of stakeholders, but rather of the complex interplay of various actors, both human and non-human. We explore how evaluation is used by managers as a means for managing selected aspects of information systems at a particular time, and in a particular situation, as well as the inherent limitations of such an effort. Finally, we approach evaluation as a series of open episodes, as a process that comes from the past but looks towards the future. Evaluation may sometimes seem rational, orderly, and stabilizing, but we argue that it is never static; rather, it is a dynamic process that is always changing.

We hold that the concept of IS evaluation, as it is practised within organizations, is inherently linked to instrumentalist and functional assumptions (Farbey, Land, and Targett 1999). The act of evaluating is an act of categorizing, classifying, and reducing the world to a series of numbers—or constructed classifications—to enable this representation to be compared and managed (Cooper 1991), and thus to be able to function as the basis for rational decision making. The world is viewed as a series of resources to be mastered. In that sense, evaluation is the archetype of the technological world as described by Heidegger (1977*a*) and Mumford (1952).

Furthermore, the object of evaluation is not the world itself but a reduced version of it: the information system (Kallinikos 1993). Through the requirements specification, which takes place during the construction of the system, the world is reduced to its representation within the information system. Moreover, the evaluation itself is normally conducted through the use of information systems. By combining this realization with the fact that evaluation is mostly used to justify future decisions, we are faced with a phenomenon of reflexivity and recursiveness. The assumptions contained in the information system are projected into the future, and amplified through the evaluation process. In that sense, the evaluation operates as a 'time gateway' through which various aspects of the organization's past and its information systems are inscribed into future systems.

After revisiting the literature, this chapter presents a rather different analysis, based on treating the various aspects of evaluation in terms of their representations. We examine, from a new perspective, questions concerning who does evaluation, on what, how, and why. The theoretical work of Heidegger (1977*b*) and Mumford (1934; 1970; 1986), as well as Kallinikos (1993), Ciborra and Lanzara (1994), and Ciborra (2000), is employed to conceptualize evaluation. In order to understand the evaluation process, we also use the work of theorists from Actor Network Theory (ANT). Our analysis is underpinned by an examination of a longitudinal empirical case study.

Help in understanding IS evaluation

There is a huge literature on IS evaluation, stretching back more than thirty years. While most authors differ on the solution(s) to the problem, they mostly agree that IS evaluation is a very complex issue (e.g. Farbey, Land, and Targett 1993). By synthesizing a number of definitions (Willcocks 1992; Farbey, Land, and Targett 1999), we can summarize the view of most authors along the following lines: 'IS evaluation is an organizational process to establish by quantitative or qualitative means the worth of an IT system to the organization. In other words, it is a management judgement of the value of a particular system in a particular organizational context.' This rather vague definition provides tremendous scope for subjectivity and numerous contingent factors, which can breed complexity in both practical and academic discussions. In order to illustrate the extent of the contingencies and subjectivities surrounding IS evaluation, we present the literature review in this section using a basic 'W-word' framework, based on the questions what, why, who, which, etc. This serves to demonstrate clearly the slipperiness of the concept.

Before presenting the main review, it is worth noting that there is often ambiguity in the use of the term 'evaluation' in the literature and in everyday parlance. This relates to whether the term refers to the method used for an evaluation, a particular instance of the evaluation process of an IS, or the outcome of a particular instance. We will try to qualify and clarify our usage of the term throughout this chapter.

What is 'the system' being evaluated?

What constitutes 'the system' being evaluated may not be totally clear. Is it just the software? Or does it include the hardware, network, data, and the people operating the system? Having taken a probably fairly subjective decision regarding what is the object of the evaluation, and where to draw its boundary, there is still considerable scope for debate. Farbey, Land, and Targett (1995) argue that different types of system (they use the term 'application') require different evaluation approaches. Their eight categories, arranged as a benefits 'ladder', range from mandatory changes, on the first rung, to business transformation on the top rung.

Moving up the benefits ladder increases the return on the investments, but also increases the complexity of the evaluation, as well as the risk and uncertainty. Thus, they argue that applications based on new tax laws, which are 'mandatory', require a different evaluation approach to an investment in new networking facilities, which they categorize as an infrastructure investment. This contingency approach to evaluation implies a good (subjective) understanding of the content and context of the development and evaluation, in order to categorize the application sensibly.

Why carry out the evaluation?

A key issue within IS evaluation is to understand why it is being carried out. Farbey, Land, and Targett (1999) note the instrumental nature of evaluation: it is normally done for a purpose. This may be because of a problem or opportunity inside or outside the system, or it may be initiated as part of the normal process of IS development, or of an organizational evaluation policy that stipulates regular evaluations. However, the stated official purpose may mask hidden agendas, which tend to be more political, as well as being difficult to disentangle from the official rationale. Thus, the 'why' of an evaluation has considerable implications for the design and the expected, or permitted, outcomes of the evaluation. These outcomes may have considerable implications for both individuals and groups.

Within the literature, and in practice, issues of intentionality are difficult to disentangle. Serafeimidis and Smithson (2003) use a matrix to discuss the four 'orientations' of IS evaluation. On one axis, this is based on whether the impact of the system on the organization is more tactical or strategic. The other axis refers to the extent to which the objectives of the evaluation and the system are clear and widely accepted, as opposed to ambiguous or lacking consensus. Thus, 'organizational control' refers to the relatively simple situation where there is: limited, tactical impact; clear objectives; and no stakeholder conflict. 'Sense-making' has the same tactical impact, but there is conflict or ambiguity concerning the objectives. Where the impact is more far-reaching or strategic, the orientations are 'social learning' (clarity of objectives, no conflict) and 'exploratory' evaluation (ambiguity or conflict).

Serafeimidis and Smithson argue that these very different contexts require different approaches to evaluation, although it again requires a subjective understanding of the situation.

Roles and modes of evaluation. The political role of evaluation is important because this may involve the allocation of blame or credit and the expansion, survival, or contraction of organizational empires (Walsham 1993). A related role is in consensus achievement, where agreement is sought between stakeholders (Avgerou 1995). Another relatively high-level role is organizational exploration, which seeks to examine where the organization should go next in terms of future systems (Robson 1994). Finally, an evaluation has a role in supporting organizational learning to preserve best practice and avoid repeating past mistakes (Symons 1990).

A somewhat lower level of analysis is concerned with what may be described as modes of evaluation. These include the familiar feasibility study as a form of *ex ante* evaluation that seeks to determine whether a particular development proposal is likely to be successful in terms of meeting its objectives (Willcocks and Lester 1994). Similarly, an investment analysis attempts to evaluate whether an IS investment is likely to be worthwhile financially (Irani and Love 2002). Other modes include:

- impact analysis, looking at the effect a new system has had—or is likely to have—on a particular group or organizational function (Korac-Kakabadse, Kouzmin, and Korac-Kakabadse 2001);
- performance measurement, focusing on how well a particular system is working, typically using fairly low-level technical or financial metrics (Banker and Kemerer 1992);
- problem diagnosis, seeking to account for, and solve, a particular IS problem (Hawgood and Land 1988);
- risk assessment, examining the risks of projects failing in some way (Boehm 1988; Sauer 1993); and
- portfolio optimization, where an organization wishes to maximize its portfolio of projects, rather than judging individual projects (see the PAM case study below).

Evaluation in decision making. The recognition of instrumentality can be seen by considering the close relationship between IS evaluation and decision making, in a sense regardless of the nature of the decision. In principle, evaluation should inform decision making, as the decision should be based on the outcome of the evaluation (Jain, Tanniru, and Fazlollahi 1991). Thus, many IS evaluations have a 'what next' quality about them, where the subsequent decision may include answering questions such as:

- Shall we build the system described in this proposal?
- Shall we continue using this system, or replace it with a new one?
- Shall we continue with this development project?
- Shall we extend this pilot project?

- Shall we invest in a customer relationship management system—or a knowledge management system (or whatever the latest fad happens to be)?

In practice, however, this evaluation may not always precede decision making. Decisions may be taken intuitively, on the basis of a hunch, without a formal evaluation—or they may be followed by a formal evaluation, in order to legitimate the prior decision (Legge 1984). Alternatively, evaluations may not be followed by decision making; they may be carried out as part of a ritual. For example, where post-implementation evaluation is actually carried out, it may just be for project closure (Kumar 1990). In many cases, this does not take place at all because it is seen as unnecessary, too difficult, against the organizational culture and philosophy, and perhaps leading to adverse effects from poor results (Remenyi and Money 1993).

Who is involved in the evaluation?

The subjectivity arising from the management judgement of worth necessarily means that the people aspects (the 'Who' questions) can be especially important: Who initiates the evaluation? Who designs it? Who carries it out? And who reads the results of the evaluation?

The evaluation may be initiated by a senior manager faced with a problem or opportunity, but another lower manager probably designs the evaluation and selects the method and the evaluators, as well as the details of time and place. These issues may be part of a policy or procedure that has been stipulated and documented at some time in the past, or the manager may have considerable freedom of choice. Serafeimidis and Smithson (2003) distinguish between the evaluation strategist and the evaluator. In the context of the freedom allowed within the evaluation design, much may depend upon the evaluator's preconceptions, norms, and value systems, which are in turn likely to be strongly influenced by the evaluator's previous experience and stakeholder-group membership. An evaluation normally results in a document that will be read by particular decision makers. These could include senior or middle management, shareholders, trade unions, customers, or, conceivably, local or national government. Each of these groups has their own interests and objectives.

Which aspects are to be evaluated using which methods?

The questions of what is being evaluated, and why the evaluation is being carried out, are likely to influence strongly which aspects of the system are to be evaluated, and how the different aspects are weighted. For example, the system may be costly but also user friendly, or it may be fast but difficult to maintain. In addition, certain standards may need to be met, such as for response time, input error rate, or down time.

These 'target' aspects are likely to influence how the evaluation is done: which evaluation methodology or approach is adopted (e.g. cost benefit analysis, user satisfaction survey, technical benchmarks) and the precise way that the methodology is applied. Again, this may be highly subjective. The choice of methodology constrains the potential outcomes, as does the actual carrying out of the evaluation. These arguments are rehearsed in Mason and Swanson's (1981) work on measurement and management.

When are the evaluations to be undertaken?

When the evaluation takes place is also important, for example deciding at what stage in the development cycle it should be done. Furthermore, systems evolve in use; 'failures' become 'successes' and *vice versa.* For instance, the London Ambulance Service's computer-aided despatch system was a notorious 'failure' in 1992 but, after further development, won a British Computer Society award five years later (McGrath 2001). Larsen and Myers (1999) describe an example of success turning into failure in the financial services industry. The evaluation outcome may depend upon traffic levels at the time: a network may perform well when traffic is slack but degrade with heavy traffic loads. The time element may also be important, for example in deciding for how long the evaluation team will collect data. Techniques such as the total cost of ownership attempt to assess the total cost of procuring and using a system over a period of time (Smith David, Schuff, and St Louis 2002).

When looking at the future, evaluations tend to take a rational form, rather like rational planning, with the assumption of a stable and deterministic future (Beenhakker 1980). In the field of planning, Dror (1971) and Lindblom (1959) call for more incremental, holistic approaches, especially for 'messy' situations. Just as organizations should perhaps be continuously planning, so perhaps they should be continuously evaluating, rather than relying on an out-of-date plan or evaluation.

Where is the evaluation to be performed?

For multi-site systems, the place where the evaluation is performed may influence the outcome, as the business and political context may vary considerably between locations. For instance, an ambulance command-and-control system may work well in a rural area but not in the capital city. There are two aspects of the 'where': the location of the system that is evaluated and the location where the actual evaluation is carried out. Very different outcomes may result from carrying out the evaluation centrally or locally, or in a finance, operating, or IT department. Thus, the 'who', 'when', and 'where' of an evaluation can be important decisions that are typically made subjectively, as opposed to routine applications of standard methodologies.

According to some authors, this complexity—or 'mess'—of IS evaluation implies a need for a greater understanding. For instance, Hirschheim and Smithson (1988) classify IS evaluation approaches into three zones: efficiency, effectiveness, and understanding. They argue that the understanding zone is the most important one for making progress in this difficult area, pointing towards a more interpretivist paradigm. They suggest this can be found in the contextualist approach (Symons 1991), stakeholder analysis (Gregory and Jackson 1992), and the various cognitive approaches based on mental models or technological frames (Orlikowski and Gash 1994). However, all these approaches are difficult to operationalize in an organizational setting and have been largely ignored by practitioners. Similarly, Farbey, Land, and Targett (1999) offer five learning themes, which involves studying and understanding more about: evaluation itself; its stakeholders; how decisions are taken based on evaluation; the longitudinal nature of IS project dynamics; and how managers learn from experience and evaluations.

Theoretical frameworks: evaluation, representation, and ANT

The complexity of the evaluation phenomenon calls for a deeper understanding of its intrinsic assumptions. In this chapter, we employ a series of conceptual tools to render this analysis. In particular, we view evaluation as an act of representation, what Heidegger (1977*a*) calls '*Vorstellung*', complemented by an analysis based on actor network theory to explore the different stages and implications (Callon 1986; Latour 1988, 1999*a*; Monteiro 2000).

'Representation' here means the act of substituting something for something else. The crux of this is the selectiveness, the substitution, and the deconstruction/re-construction of the artefact to be represented. Through the representation 'process', the world is reduced to a more manageable, portable form. Thus, the 'act' of representation does not entail a substitution of equals, but rather an inevitable alteration of form and content; there is no translation without transformation (Law 1999). In the case of IS evaluation, we have a representation of an IS in, for example, financial or technical terms. When an information system is evaluated, it is 'represented-as', or it 'becomes', a series of numbers and tables that express a subset of its characteristics, in this case its economic viability or technical performance. The manager does not have to deal with the system itself but with its more manageable representation, which is a 'product' of the evaluation process.

An evaluation is normally a form of measurement or classification, a type of inscription that translates a particular entity or situation, such as an IT investment, into a quantifiable outcome against certain criteria. In that sense, any semiotic approach could be appropriate to analyse such a translation (Barthes 1967). In this chapter, we hold that ANT could provide the appropriate tools for gaining additional insight into the evaluation episode. According to ANT, networks are created and maintained through the translation of actors' interests, and this translation process can be seen in the 'four moments of translation' (Callon 1986; Latour 1999*b*). These moments comprise: problematization (the definition of the actors and their identities), *interessement* (locking the other actors into place), enrolment (motivating the actors), and mobilization (actually making the translation happen). Using this notion, we show how the interests of various stakeholders, both humans and non-humans, are inscribed, or represented, by the evaluation.

ANT and the sociology of translation deny any essential differentiation between humans and non-humans (Latour 1987) and approach artefacts as open, and therefore constantly changing, outcomes of translation processes. The term 'translation' is used here interchangeably with that of 'representation'. Both human and non-human actors are represented in the evaluation network; they have certain affordances, as discussed later, but are typically 'reductions' of the real world actors. For example, the finance director's penchant for foreign languages and his sporting prowess are unlikely to feature in the evaluation process.

The representation process clarified

As noted above, in the literature (e.g. Irani and Love 2002) we often find the term evaluation treated as an abstract entity: a 'method' with which to assess, for instance, the economic performance of a particular information system. However, evaluation is also used to describe the actual 'instance' of the use of this method, i.e. the act of evaluation (e.g. Jones and Hughes 2001); the 'product' of the evaluation instance is sometimes also referred to as the evaluation (e.g. see Legge 1984 for a discussion of the crisis of utilization). This ambiguity suggests links between the three notions of instance, method, and product, which are demonstrated here by treating evaluation as a form of representation.

The method used in an evaluation. First, consider the way in which evaluation is conducted, the particular instance of the use of an evaluation method. In using the method, the evaluator represents certain aspects of the system, such as its economic viability, rather than approaching it as a whole. This segmentation of the artefact and its reproduction in a substitute form (Derrida 1982) is done according to certain criteria, and these criteria are the crux of the act of representation.

Following the work of Heidegger (1977*b*), Baudrillard (1983; 1988) and others, as transferred in the IS field by Kallinikos (1993), we may approach these criteria as being closely related to the intentionality of the acting subject. In other words, when evaluating an information system, in those instances where we have the choice we select the characteristics of the IS and its context that fit our own objectives. Thus, the world becomes easier to manage by dealing only with selected aspects (Kallinikos 1993). The real world becomes distant from the representations selected, as the representation replaces reality. This becomes the reality we choose to deal with.

The selected criteria to perform the representation constitute a representation of the intentionality of the subject, but are not the intentionality itself (Heidegger 1977*b*). In practice, few individuals have sufficient autonomy to select the evaluation criteria solely according to personal taste. Normally, there is a history of IS evaluation embodied in methods, procedures, and norms that organizational members are expected to follow. These save 'reinventing the wheel', but also confer legitimacy on the evaluation. Thus, the scope of this intentionality must be considered, in its broadest sense, to include the objectives of current users plus all the other past and present influences encapsulated in objects and structures that still impact on the evaluation. So, the disembodiment of the world, through its representation in the IS, does not take place solely according to the intentionality of the evaluator—but in accordance to one of its representations, the allowable evaluation criteria.

Currently, the representation of an IS for an evaluation is usually made according to financial or technical criteria. These are supposed to represent the intentions of the evaluator for achieving greater efficiency or effectiveness; but they may not be the real intentions of the evaluator, who may have something totally different in mind. In that sense, the act of representation may tend to distance the representation from both the evaluator and the world itself. Thus, an IS person may intend to introduce a 'fashionable' knowledge management system, while an accountant may want to reduce total expenditure at all costs. In this way, the process of translation is reversed and the criteria being used inscribed into the intention of the subject. The IS person's goal can therefore be displaced into cost reduction, while the accountant becomes an enthusiast for IT. This goal displacement (Angell and Smithson 1991) can be seen when, for example, managers who want to improve customer service find themselves instead working hard to 'reduce average queuing time', or whatever the quantitative representation for customer service happens to be, even when this reduces the overall level of customer service.

New forms for representing the re-embodiment of the fragmented world. The re-embodiment of the fragmented world in a new form is the next stage of representation. The charts, numbers, and tables produced by the application of the evaluation method replace the IS as a tangible artefact. These new forms constitute the product and process through which the world is 'compressed' into a more convenient form. The numbers in a report can be more portable, they reduce the complexity of the real world, and can be used across contexts (Kallinikos 2002). A report summarizing average performance is less complex than a detailed minute-by-minute account of traffic loadings within the numerous, complex links of a global IS infrastructure. By documenting the possible benefits that an information system will contribute to the organization, the senior management does not really need to know the technical details of the functioning of the system. Its context is re-created through the reports.

However, in order for the subject to interpret the re-embodiment, or representation, of the new world in the evaluation report, the new form needs to be familiar with—and adhere to—the assumptions made during the disembodiment of the world. From this perspective, the subject has to accept the criteria of representation and interpret the results according to them (Weick 1985). A table of numbers is nonsense if the reader is not familiar with the assumptions behind them. Thus, the subject may itself be 'de-constructed' and re-constructed during the interpretation phase—this may be the only way for the subject to stay in the evaluation network. For example, an accountant may have to accept the logic and legitimacy of an increase in expenditure. One of the basic attributes of the representation, namely its cross-contextuality, stems precisely from the ability of the artefact to re-construct and embody a context by forcing the recipient to 're-produce' himself in accordance to it during the act of interpretation. This can be seen as a form of hostage taking, where some of those involved in the evaluation are taken hostage by the evaluation itself, such as users who find themselves in a situation where the representation is dominated by developers.

One of the often-quoted problems with formal evaluation is that it is either not done or else it is ignored by the organization (Legge 1984; Irani and Love 2001). This may be related to the process of representation entailed. By reducing the world to a series of representations that can be easily managed, the evaluator only repositions in time the problem of dealing with a complex world. If the world is compressed during the phase of evaluation, it is going to be de-compressed when the evaluation has to be translated into action. The complexity of the world is reintroduced and the manager has to find ways to deal with it.

The process of re-production is dependent on a conceptualization of time as a linear phenomenon that can also be objectified and fragmented. This becomes apparent in the case of evaluation, which may operate as a time gateway for the organization (Mumford 1934; 1986). In the case of post-implementation evaluation, the accumulated experience of the past—or that part inscribed in the evaluation method—is further inscribed in decisions that will affect the future. In that sense, the past is transferred to the future. In the case of exploratory evaluation, the construction of a possible future, again with reference to the past, is inscribed into the present decisions of the organisation (Heidegger 1977*b*).

While the concept of representation is potentially very useful for providing a general framework for understanding evaluation, it lacks the detail to describe how the evaluation network is formed and evolves. That is why we apply actor network theory as an additional layer of analysis in the following case study.

Case study of the Project Appraisal Method (PAM)

Our case study involves a large, well-established insurance company.[1] At the time, it was the biggest British provider of general and individual life insurance. The study was set within the insurance company's Home Service Division. This accounted then for around three-quarters of the firm's UK turnover, employed around 20,000 people, including 160 IS staff, and had an annual IT budget of between £70 and £95 million.

The data for the study are based on interviews with senior and middle business managers and some developers of the evaluation method in question. These were carried out in the late 1990s and supplemented by documentation, including internal reports and press announcements, that stretched back to the early 1990s.[2]

Phase 1: Constructing PAM

In the 1980s, the IS evaluation process in the insurance company comprised mainly two tools: a capacity planning tool and a budget estimation system. These were used primarily for assessing technical performance, aimed at the planning and control of IS resources. In the 1990s, the attitude towards IS evaluation changed towards assessing the degree to which business objectives were achieved. In particular, the goal became to establish a link between IT expenditure and business performance.

From the perspective of actor network theory, the evaluation method and tools can be regarded as 'actants' that incorporate particular programmes of action. Thus, in the 1980s the programme of action was the planning and control of IS resources, and in the 1990s it became the level of service for the business. The metamorphosis of these actants can be traced through the experience of human actants. For instance, after the reorientation the then Systems Strategy Manager commented:

> My job is to ensure that in spending our limited [IT] resources we deliver the best service possible for the business . . . It is always substantial to get down to bottom line results. For instance, if the business case says: 'The new sales management system will improve productivity by 20 per cent', it is important first to identify the link between the technology and the business benefit. Before constructing a business case, it is essential to establish the logical credentials and really understand where that 20 per cent saving is going to come from.

Aims and development of PAM. The new evaluation method would function as a representation of the business performance of that particular system (e.g. the new sales management system). Both technology and business benefit would be represented in the new method. We argue that the evaluation method and tools employed to represent the actants influenced the representation, as discussed below.

The new method was known as the Project Appraisal Method (PAM). The programme of action inscribed into this method was 'the delivery of high quality systems and the achievement of both hard and soft benefits desirable by the business'. The shift from viewing IT investment as general business support to seeing it as an integral component of the strategic plan was reflected in the evaluation method. This shifted from a simple cost-benefit analysis, highlighting tangible benefits, to a more sophisticated understanding of the complex returns and strategic nature of IT investments. That led to the metamorphosis of the evaluation tools from capacity planning and budget estimation to a three-tier evaluation tool. PAM was supposed to allow a better choice of the right portfolio of IT projects and effective risk analysis and benefits management.

The development of PAM began in the early 1990s, as a bottom-up process from a departmental level that would be migrated to corporate level. No external consultants were in the development team. The project was led by the manager of the Business Systems Co-ordination Group and championed by the manager of the IT Systems Steering Group. These origins, in the Information Systems Department, provide an indication of the worldviews inscribed into the PAM project.

As already indicated, the construction of the evaluation method should be viewed as a process of inscription. When presenting the structural elements of PAM, we therefore do nothing more than to identify the inscriptions of different actants into it. These inscriptions and representations are materialized in different ways: the structural elements are a direct result of what is inscribed into it, while the programme of action co-develops with the materiality of the inscription.

Tools used in PAM. PAM's theoretical foundations lay in information economics (Parker, Benson, and Trainor 1988), which was popular with the then Finance Director. However, PAM adopted various concepts and techniques that reflected the interests of other actants.

PAM consisted of three main streams: appraising the hard financial costs and benefits; the risks; and the strategic and intangible benefits. In terms of tools, a financial model of the costs and benefits, a risk management plan, and a benefits profile were employed. In particular, the financial appraisal used spreadsheets to model the cash flows, automatically calculate certain measures—such as the project payback period, the internal rate of return, and the net present value—and to encourage sensitivity analysis. The risk analysis used computer-based questionnaires to identify the main risks to the project, to estimate their probability, and to calculate their potential impact. The final module was used to evaluate intangible benefits against nine key organisational objectives in order to produce a total score and a graphical benefits profile. This was to be used to assess the initial business case and to form the basis for a benefits-delivery plan.

It is in these three modules that we can identify the inscriptions of other actants. The financial model was an inscription of the previous evaluation tool. The risk analysis was influenced by business considerations and was essentially the bridge between the first and third part of PAM. Finally, the tool to evaluate intangible IT benefits constituted an effort to translate the business interests and IT characteristics symmetrically into the evaluation method. Thus, this evaluation method was to function as an obligatory passage point that would align the business and IT strategies through the evaluation of their relationships.

The evaluation of intangible benefits, in particular, comprised successive inscriptions of different actants. The business objectives had to be translated to the IT investment requirements. The project team, largely made up of systems people, analysed the division's business goals, measures, and key performance indicators before identifying a subset of seven (out of the initial 200) as key objectives: customer base; customer service; sales effectiveness; unit cost; exploitation of business opportunities; company's brand; and product design. Two extra objectives were added later: management information and staff attitude to IT.

The criteria were selected using this top-down approach, but they were prioritized by the business stakeholders using a Delphi approach (Linstone and Turoff 1975). Although the development was initiated by the Information Systems Department of one division, the rest of the organization was made aware of the project through seminars, reports, and leaflets. Various user groups were engaged in the process through training sessions. User managers, traditionally reluctant to participate in IT decision making, seemed to be interested in this particular project.

The construction of the method was in itself an inscription of different groups' interests. ANT's four moments of translation can be traced in this particular actant. In the problematization moment, where the roles are defined, the definitions of success were set. The criteria set by the project team were the final product of negotiation with different actants: the business goals as well as the measures and key performance indicators were retranslated by a particular division and the development team to produce the seven (later nine) criteria for intangible benefits evaluation. The participation process, training sessions, seminars, leaflets, and reports functioned as *interessement* devices that would cut off the actants from alternative options. By the end of this phase, it seemed that enrolment had actually taken place since most of the actors (e.g. user groups and managers) were keen to participate and a number of them functioned as spokesmen for others. The Delphi approach used representatives of the different stakeholders in the final development, and the training sessions, leaflets, reports, and seminars represented the evaluation process 'under construction' to the users themselves. The financial appraisal represented the previous cost-benefit analysis. As the actants were becoming represented by new actants, a process of translation, transcription, and transformation was taking place.

Phase 2: Using PAM

A separation of PAM's life cycle into its construction and its actual use is purely analytical; PAM's development was never really completed. Through its use, PAM was retranslated, transformed, and finally converted into different evaluation methods within the same company. As its formal development was concluding, significant issues of cost reduction became uppermost in the corporation's thinking, rather than intangible and strategic aspects. At this point, PAM stopped representing the particular set of interests that had led to its initial balanced form. At the same time, the development team ceased to represent PAM in the organization as most of the team, including the leader, were moved elsewhere or made redundant. The dissidence phase was reflected in the team's failure to obtain full support for PAM from the new Finance Director and the Marketing Department.

It is important to understand who was considered to be PAM's main user, because a new translation-transformation took place once PAM's implementation started, and thus a new actant was created. This new hybrid actant comprised both PAM and the Impact Assessment Group (IAG), a new unit close to the Information Systems Department. The IAG and PAM seemed to function as an organic whole, with the IAG using PAM to validate project proposals and choose the optimum portfolio of developments. It thus operated as an obligatory passage point: any proposed system had to be approved by the IAG. Moreover, the majority of stakeholders regarded the use of PAM as being the only way to communicate with the systems people and the IAG. The latter was responsible for allocating systems development resources and controlling IT-related changes and their impact on the business.

During its first six months, PAM was considered to be successful and was credited with £1.8 million in savings. However, its real function at the time seemed to be a communication platform between the business and systems people. PAM was constructed as an actant because all the moments of translation had been successfully completed, but this translation process was never fully resolved. Although PAM was accepted at the time as a common platform of interests, this could not be sustained in the long run.

PAM was successful in terms of evaluating individual projects, but not in optimizing the whole portfolio. As a result, a new prioritization process was introduced to translate the 'approved' projects into a successful portfolio. This meant a new translation process came into being, and that the new prioritization process—and not PAM—would function as the common communication platform or 'spokesman' for the various stakeholders. The *interessement* devices used by PAM were its capacity to function as an effective tool for evaluating projects and as a communication tool with the decision makers. However, its shortcomings in terms of portfolio optimization meant that management did not go ahead with many projects even though they had been 'approved' by PAM.

Clearly, PAM could then no longer function as an effective representative of different stakeholders. The result was that PAM gradually dissolved within the organization. It is important to note that PAM's fate was not the result of its ineffectiveness to provide meaningful evaluations, but rather the result of it stopping to function as the interface for communication between different stakeholders.

Phase 3: PAM's metamorphoses

As the concerns and structure of the organization changed, PAM's role changed accordingly. A directive in the mid-1990s suggested that the business and systems groups should collaborate in project selection and prioritization. The role of PAM was to reveal the opinions of different stakeholders as a first step towards communication and consensus. However, the actant we identified in the previous phase had changed. Instead of PAM being used by the IAG, this function passed to PAM as used by a new organizational group: the Development Directorate, made up of senior business managers headed by the Finance Director.

The plan was to move decision making away from the Information Systems Department, although the disentanglement of IAG and PAM happened gradually. In an intermediate phase, another group, the Account Managers, was formed as a bridge between the different business areas and the IAG. Account Managers were senior to the IAG and the latter's decision-making capacity was significantly reduced. At the same time, another actant was introduced; this was an extra cross-assessment of project proposals, performed according to various business criteria. The new cross-assessment was a more representative spokesman since it could be used not just to compare similar projects but to take into account different-sized projects as well.

A phase of dissidence therefore started to emerge. The emphasis on costs, the existence of the cross-assessment tool, and the new organizational structure caused the dissolution of PAM. The intangible-benefits tool and the risk-analysis components gradually fell into disuse and the IAG team was absorbed into the Systems Resource Management unit. The decision-making process was altered accordingly. New 'delivery managers' carried out the impact assessment exercise and were accountable for project delivery. They worked with the Account Managers in forwarding proposals to the Development Directorate for approval.

The role of the Development Directorate was upgraded. It became responsible for: the approval of the IT budget for business units; maintenance of the IT project portfolio for the division; and, more generally, it shifted from IT planning to become more of an overall business planning and advisory board. PAM's three tiers could now be traced in different evaluation functions within the same organization. The evaluation process was conducted with the significant involvement of the Finance Department. The tools used were accordingly changed. The financial elements were evaluated by standard return-on-investment techniques, including net present value, internal rate of return, and payback (Radcliffe 1982). Non-financial costs and benefits were identified and attempts made to quantify them in high-level terms. The risk analysis elements were incorporated into the project management plans and covered mainly development risks. These were represented in a text format; questionnaires or forms were no longer used. A high-level benefits plan identified the activities needed to realize the benefits and allocated responsibilities to accountable individuals to deliver the benefits.

Understanding the mechanics of evaluation

In the previous section, we discussed evaluation as representation involving selection and transformation. Here, we return to the 'W-questions' used earlier in the chapter and examine the additional insight from the use of ANT and the consideration of evaluation as representation. In so doing, we try to address questions such as: How much are evaluators constrained by existing and emerging evaluation methods, or even by the potential outcomes of the evaluation process? How are past decisions depicted in future ones? And how can we speak of a single evaluation concept when it is in a state of constant flux?

Who constructs and who conducts the evaluation?

ANT is heavily influenced by the concept of representation (Cooper 1993). The translation process can be seen as a series of representations of various actants into new actants (Latour 1992). The construction of an artefact, such as an evaluation method or tool, is the result of an inscription (or representation) of other actants into a new actant.

A superficial view of 'who constructs and who conducts the evaluation' would show only the 'visible' people involved. This would include the official 'author' of the method, perhaps the person credited with its construction, together with those staff officially designated to carry out the evaluation at the time. However, based on the above argument, we should also consider all inscriptions made during the construction phase of the method. Some of these would have been inherited from inscriptions made to the previous method. The latter can be seen as a non-human actant involved in the construction of the new method. This potentially includes many (human) actors that have long since retired or changed jobs and do not feature in the currently visible set of actors. These 'old' inscriptions may or may not still reflect the actors' original intentionality; this could have been distorted over time by further inscriptions. Since its political nature emerges during the construction stage, any examination should be initiated during this phase (Orlikowski and Iacono 2001).

Furthermore, we should take account of those human and non-human actors that influence those who are officially conducting the evaluation. Evaluations may be carried out by representatives of various groups within the organization and these representatives may be strongly influenced, formally or informally, by their colleagues or related non-human actors. For example, a representative from the Human Resource Department would probably be strongly influenced by their superiors, and by the legislation and norms guiding the human resource function.

From the case we studied, a plethora of actants can be seen as being involved in the construction and operation of PAM. Its predecessor, the 1980s technical and cost estimation methodology—itself inscribed by various actants from the 1980s and 1970s—was inscribed into PAM. The PAM development team, mostly from the Information Systems Group, canvassed a wide range of organizational stakeholders for their views and an interpretation of these views was inscribed into PAM, especially into the risk-assessment and intangible-benefits modules. The then Finance Director also inscribed certain principles from information economics into PAM.

Initially, in Phase 2 as described above, the operation of PAM was the preserve of the Impact Assessment Group, who also came largely from Information Systems, following the demise of the development team. However, Phase 3 saw the collapse of the Impact Assessment Group, and the arrival of the Development Directorate (headed by the new Finance Director) and the account managers, who also came from a more finance or business background. These new groups inscribed their own interests into PAM through their use of it, and allowed the risk-analysis and intangible-benefits modules to wither away. They also brought the new cross-assessment tool into use. Thus, in a relatively short space of time, a large number of people with different backgrounds inscribed their interests into PAM through its construction and operation. However, it should be noted that, for any particular evaluation, many of these actants would not be visible.

The what, which, and how of evaluation processes

The questions lying at the crux of the evaluation-as-representation issue are: What is the object of the evaluation? Which aspects of the system are evaluated? And what evaluation process is followed?

The element of selection in the evaluation-as-representation already starts from the choice of a particular 'object' to be evaluated: the information system. Moreover, it is often the case that only certain facets are chosen to be evaluated. Finally, the evaluation is conducted by following a certain method, using certain tools, and employing certain individuals to undertake it. We have already highlighted the importance of these parameters in the selection process. However, such an approach can be misleading if it implies a form of hierarchy between who is conducting the evaluation and what or how something is evaluated. In this conceptualization, the evaluator is depicted as an active entity, the object as a passive artefact, and the method as a neutral means for accomplishing the whole task. In practice, none of these assumptions holds true (Knorr-Cetina 1981; Latour and Woolgar 1986).

All the evaluation methods and tools, as well as the objects of evaluation, are active in a certain way: they possess affordances, allowing other actants to 'do things'. They inscribe as well as being inscribed. An evaluation method is realized through certain means, such as a particular series of steps, as well as questionnaires or software tools. All these means are 'objects' that have certain qualities. The same is true for the particular information system, or its aspects, that is chosen to be evaluated. What is important is not so much what these 'objects' do, as what they *allow* you to do. 'The agency of things does not come from within but from the "inscriptions" generated by a heterogeneous network upon . . . [its] blankness' (Hetherington 1997).

The particular evaluation allows certain stakeholders to be represented better. For instance, a cost-benefit analysis tool is more likely to represent the considerations of the accountants than the long-term planners. The choice of this tool is itself a political decision that allows certain interests to be inscribed into these means. Evaluation allows inscriptions, but also predisposes the type of inscriptions permitted. It is not what an evaluation method says, but rather *what it allows to be said* that makes a difference. For instance, by choosing an evaluation method that emphasizes financial elements, other elements—such as ergonomic ones—may lack representation.

The evaluation method itself is a 're-presentation', and in that sense it is an effect or outcome of an inscription procedure. When certain facets of the world are chosen to be represented in the evaluation, it is not usually predetermined which attributes are to be selected. The world is re-constructed according to certain criteria, which are often not under the control of a single entity and it may be difficult to know in advance what it is going to look like. The series of metamorphoses to which an evaluation method, as an actor-network, is subjected can be described by the metaphor of 'rhizome' (Deleuze and Guattari 1988), which Latour (1999*a*) suggests is the most appropriate term for describing the evolution of an actor-network. A rhizome resembles a root that, as it evolves, constantly changes. New possibilities are created and old ones become redundant. As time passes, the network may alter beyond recognition. In that sense, Lynch suggested that ANT should have been named 'actant-rhizome-ontology' (Latour 1999*a*).

We have described above how PAM itself evolved and, clearly, there were also changes to the object of evaluation and the aspects of the proposed IS investment to be evaluated. Thus, in pre-PAM days, it was a technical and cost representation that was evaluated. In its initial incarnation, PAM used three representations of the investment proposal: tangible costs and benefits; risk analysis; and intangible and strategic benefits. Of course, this also later changed to a largely tangible costs-and-benefits representation of the proposed investment. Furthermore, it should be noted that this evaluation exercise was carried out on a varying representation of a proposal for an information system, not on the system itself.

Why is an evaluation process initiated?

The evaluation of an information system may be triggered by 'real world' events or phenomena, as discussed earlier in the chapter. However, it could also be triggered from within the evaluation actant, the network of people and methods linked together by the notion of evaluation. In other words, the evaluation is triggered by the concerns inscribed into an evaluation method, for example, of an evaluation committee or by a commitment to regular evaluation. The mere existence of an evaluation actant may be enough to trigger an evaluation.

The rhizome metaphor used above indicates that evaluation methods do not remain the same through time, that they make sense only through their use, and that they operate as an important feedback mechanism. We suggest that the different roles of evaluation (Serafeimidis and Smithson 2003) may reflect the evolution of the evaluation method through time, and its whole *raison d'etre* may change in the process. For instance, an evaluation method may start as a control mechanism to be later translated into an exploratory mechanism or sense-making device. The evaluation actant has to be placed in a temporal as well as spatial context. The time factor is definitive for the constitution and evolution of the actant.

In the case of PAM, the 'why' of evaluation shifted considerably, in terms of the programme of action. The pre-PAM evaluations were driven by the notion of technical and cost feasibility, while the first incarnation of PAM was largely concerned with the level of service for the business. However, this could not deal with prioritizing investments, which is a rather different programme of action, and PAM had to be modified accordingly.

Furthermore, as noted above, PAM became an obligatory passage point for IS investment proposals. It was not possible to continue with an IS investment without going through the PAM evaluation procedure. Thus, the 'why' of evaluation, in the case of PAM, changed according to various rationales, including a mandatory administrative procedure. This would not be apparent to the casual researcher but requires a good understanding of the context of the construction and operation of PAM.

When an evaluation is carried out

It follows from the above discussion that it would be a mistake to conceptualize an evaluation as being a 'true' representation of the real world at a particular point in time. Rather, we have argued that the evaluation method is not a 'fresh-faced' neutral tool, but has evolved over time and probably retains many inscriptions from times long past. Furthermore, it operates on a representation of the real world that may not be in real time. For example, it may use last month's or last year's data in parts of the representation.

We have argued that PAM should not be seen as something totally new, and neither was it something that remained immutable in time. It originated from the previous evaluation method and, at various points in time, it crystallized the interests of diverse groups. Once the creation of PAM was initiated, it remained under constant transformation. At a particular point in time it represented the consensus of more than one diverse group of actants. However, stability is never fully attained and the process of translation is never resolved. Every day is a working day.

As a time gateway, PAM allowed the resource management concerns of the 1980s to be applied to information systems proposals during the 1990s (despite the official shift in policy), some of which were not delivered until the next century. This demonstrates the importance of the temporal dimension. Furthermore, PAM's evaluations were carried out while the information system was still at the investment-proposal stage. It would be particularly interesting, although the data are not available, to compare the investment proposals as approved by PAM with the final systems that emerged at implementation, and evolved during operation.

Conclusions

The way much of the existing literature approaches IS evaluation only scratches the surface of the problem. We would argue that issues concerning measurement—'what gets measured gets managed' (Willcocks and Lester 1996) or the politics of evaluation—are symptoms of a lack of questioning of evaluation's fundamental assumptions. By choosing a different mode of analysis, we are able to gain a deeper understanding. This is important, as we question not just the outcomes or performance of a particular evaluation mechanism but its constitution and purpose. In this chapter, we have approached evaluation as a process of representation, which has been illustrated through the case study.

The basic assumption behind the evaluation-as-representation concept is the idea of selecting certain aspects of an artefact (the information system) that will be reproduced and inscribed into the final product of the evaluation process, such as the evaluation report. An evaluation method is the crystallization of a series of evaluation instances. In the choice of the criteria, and the act of the evaluation itself, evaluators are not free to transcribe their intentionality into the artefact. They are limited by the method that was chosen and the *modus operandi* of the organization itself. The way the representation takes place is limited by the physical properties of the final product of the evaluation. If that is a financial report, the aspects that are evaluated are limited to those compatible with this financial outcome.

The evaluation criteria represent the intentionality of human actors and the dynamics of inanimate objects. The criteria are themselves represented by the means and methods used for the realization of the evaluation. At this point, the thread of intentionality may be lost and the distinction between the represented and the means of representation is blurred. Are the criteria represented by technical means or are they a representation of the means available? The word 'representation' may not be optimal for describing the phenomenon. It would be more accurate to talk about 'co-development' of criteria and intentionality. The phenomenon could also be described as an 'inversion' phenomenon: it is not clear *who* is the inscriber and *what* is the object of inscription; it is not clear who produces and who is being produced. The evaluation phenomenon is a clear illustration of this inversion and many of the problems in the existing literature of measurement and methodology can be traced here.

A deconstruction of evaluation through such an analysis poses some interesting questions. Evaluation 'should be' the apotheosis of intentionality and instrumental thinking; it is supposed to re-construct the world in a way that would render it more manageable and closer to the intentionality of the subject. However, this may not always be the case. In the selection of the criteria, the intentionality is not just that of the subject; it is the outcome of a series of existing dynamics and intentionalities of different groups of stakeholders, some of whom may no longer be with the company.

During the evaluation phase itself, the intentionality of the subject becomes even further deluded: the technical means, methods, and schemes used for the evaluation are inscribing their own dynamics in the final product of the whole process. The world is represented selectively and compressed in the evaluation, to be later de-compressed in the process of action. Although the evaluation may be highly convenient as a more reduced and portable version of the world, the reality itself remains inherently *irreducible*. By examining the fundamental assumptions behind evaluation, we realize its operation as a mechanism for dealing with the complexity of the real world that provides only a temporary solution. The manager *will have to* deal with a complex world, and the simplifications of reality contained in an evaluation document can serve only the spreading of this complexity across time and space.

We emphasized the fact that evaluation is not a simple event and is very closely linked to concepts of time within the organization. Evaluation is a time gateway, which distorts time within the organization through a deconstruction of real events within a desired classification and re-allocation of those events within a different time-span. Evaluation necessarily contains numerous inscriptions from the past and thus allows the present decisions to be formed by it. The past is pushed through the evaluation into the present. When possible versions of a future system are evaluated, the time gateway operation of evaluation comes to the fore again: the projected future will be inscribed in the present decisions; a possible future pulls the present towards it.

We would argue that the analysis in this chapter provides academics and practitioners with an awareness of both the richness and the limitations of the evaluation process, its methods, and its outcomes.

References

Angell, I. O., and Smithson, S. (1991). *Information Systems Management: Opportunities and Risks*. London: Macmillan Press.

Avgerou, C. (1995). 'Evaluating Information Systems by Consultation and Negotiation'. *International Journal of Information Management*, 15/6: 427–36.

Banker, R. D., and Kemerer, C. F. (1992). 'Performance Evaluation Metrics for Information Systems Development: A Principal–Agent Model'. *Information Systems Research*, 3/4: 379–400.

Barthes, R. (1967). *Elements of Semiology*. New York: Noonday.

Baudrillard, J. (1983). 'The Ecstasy of Communication', in H. Foster (ed.), *The Anti-aesthetic: Essays on Postmodern Culture*. Port Townsend, WA: Bay Press, 126–34.

—— (1988). *Selected Writings*. Stanford, CA: Stanford University Press.

Beenhakker, A. (1980). *A System for Development Planning and Budgeting*. Aldershot, UK: Gower.

Boehm, B. W. (1988). 'A Spiral Model of Software Development and Enhancement', *IEEE Computer*. 21/5: 61–72.

Callon, M. (1986). 'Elements of a Sociology of Translation: Domestication of the Scallops and the Fishermen of St Brieuc Bay', in J. Law (ed.), *Power, Action and Belief: A New Sociology of Knowledge?*, London: Routledge, 196–233.

Ciborra, C. U. (2000) (ed.). *From Control to Drift: The Dynamics of Corporate Information Infrastructures*. Oxford: Oxford University Press.

—— and Lanzara, G. F. (1994). 'Formative Contexts and Information Technology: Understanding the Dynamics of Innovation in Organizations'. *Accounting, Management and Information Technologies*, 4/2: 61–86.

Cooper, R. (1991). 'Formal Organization as Representation: Remote Control, Displacement, Abbreviation', in M. Reed and M. Hughes (eds.), *Rethinking Organization*. London: Sage, 254–72.

—— (1993). 'Technologies of Representation', in P. Ahonen (ed.), *Tracing the Semiotic Boundaries of Politics*. Berlin: de Gruyter, 279–312.

Deleuze, G., and Guattari, F. (1988). *A Thousand Plateaus: Capitalism and Schizophrenia*. London: The Athlone Press.

Derrida, J. (1982). 'Sending: On Representation'. *Social Research*, 49: 295–326.

Dror, Y. (1971). *Ventures in Policy Sciences: Concepts and Applications*. New York: Elsevier.

Farbey, B., Land, F., and Targett, D. (1993). *How to Assess your IT Investment. A Study of Methods and Practice*. Oxford: Butterworth Heinemann.

—— —— —— (1995). 'A Taxonomy of Information Systems Applications: The Benefits' Evaluation Ladder'. *European Journal of Information Systems*, 4/1: 41–50.

—— —— —— (1999). 'Moving IS Evaluation Forward: Learning Themes and Research Issues'. *Journal of Strategic Information Systems*, 8/2, 189–207.

Gregory, A. J., and Jackson, M. C. (1992). 'Evaluation Methodologies: A System for Use'. *Journal of Operational Research Society*, 43/1: 19–28.

Hawgood, J., and Land, F. (1988). 'A Multivalent Approach to Information Systems Assessment', in N. Bjorn-Andersen and G. Davis (eds.), *Information Systems Assessment: Issues and Challenges*. Amsterdam: North Holland, 103–24.

Heidegger, M. (1977*a*). *Basic Writings*. New York: Harper and Row.

—— (1977*b*). *The Question Concerning Technology and other Essays*. New York: Harper and Row.

Hetherington, K. (1997). 'In Place of Geometry: The Materiality of Place', in K. Hetherington and R. Munro (eds.), *Ideas of Difference: Social Spaces and the Labour of Division*, Oxford: Blackwell, 183–99.

Hirschheim, R. A., and Smithson, S. (1988). 'A Critical Analysis of Information Systems Evaluation', in N. Bjorn-Andersen and G. Davis (eds.), *Information Systems Assessment: Issues and Challenges*, Amsterdam: North Holland, 17–37.

Irani, Z., and Love, P. E. D. (2001). 'Information Systems Evaluation: Past, Present and Future'. *European Journal of Information Systems*, 10/4: 183–88.

—— —— (2002). 'Developing a Frame of Reference for *ex-ante* IT/IS Investment Evaluation'. *European Journal of Information Systems*, 11/1: 74–82.

Jain, H. K., Tanniru, M. R., and Fazlollahi, B. (1991). 'MCDM Approach for Generating and Evaluating Alternatives in Requirements Analysis'. *Information Systems Research*, 2/3: 223–39.

Jones, S., and Hughes, J. (2001). 'Understanding IS Evaluation as a Complex Social Process: A Case Study of a UK Local Authority'. *European Journal of Information Systems*, 10/4: 189–203.

Kallinikos, J. (1993). 'Identity, Recursiveness and Change: Semiotics and Beyond', in P. Ahonen (ed.), *Tracing the Semiotic Boundaries of Politics*. Berlin: de Gruyter, 257–79.

—— (2002). 'Recalcitrant Technology: Cross-contextual Systems and Context-embedded Action', Working Paper 103, London: Department of Information Systems, London School of Economics.

Knorr-Cetina K. (1981). *The Manufacture of Knowledge. An Essay on the Constructivist and Contextual Nature of Science*. Oxford: Pergamon Press.

Korac-Kakabadse, N., Kouzmin, A., and Korac-Kakabadse, A. (2001). 'Emerging Impacts of Online Over-connectivity', in S. Smithson, J. Gricar, M. Podlogar, and S. Avgerinou (eds.), *Proceedings of 9th European Conference on Information Systems*, Bled, Slovenia, June 27–9. Kranj, Slovenia: Moderna Organizacija, 89–97.

Kumar, K. (1990). 'Post Implementation Evaluation of Computer-based Information Systems: Current Practices'. *Communications of the ACM*, 33/2: 203–12.

Land, F. (1976). 'Evaluation of Systems Goals in Determining a Design Strategy for a Computer-based Information System'. *Computer Journal*, 19/4: 290–94.

Larsen, M. A., and Myers, M. D. (1999). 'When Success Turns into Failure: A Package-driven Business Process Re-engineering Project in the Financial Services Industry'. *Journal of Strategic Information Systems*, 8/4: 395–417.

Latour, B. (1987). *Science in Action: How to Follow Scientists and Engineers Through Society*. Cambridge, MA: Harvard University Press.

—— (1988). *The Pasteurization of France*, Cambridge, MA: Harvard University Press.

—— (1992). 'Where are the Missing Masses? Sociology of a Few Mundane Artefacts', in W. Bijker and J. Law (eds.), *Shaping Technology, Building Society: Studies in Sociotechnical Change*. Cambridge, MA: MIT Press: 225–58.

—— (1999*a*). 'On Recalling ANT', in J. Law and J. Hassard (eds.), *Actor Network Theory and After*. Oxford: Blackwell, 15–25.

—— (1999*b*). *Pandora's Hope. Essays on the Reality of Science Studies*. Cambridge, MA: Harvard University Press.

—— and Woolgar, S. (1986). *Laboratory Life, the Construction of Scientific Facts*, Princeton, NJ: Princeton University Press.

Law, J. (1999). 'After ANT: Topology, Naming and Complexity', in J. Law and J. Hassard (eds.), *Actor Network Theory and After*. Oxford: Blackwell, 1–14.

Legge, K. (1984). *Evaluating Planned Organisational Change*. London: Academic Press.

Lindblom, C. E. (1959). 'The Science of Muddling Through'. *Public Administration Review*, 19: 79–88.

Linstone, H. A., and Turoff, M. (1975) (eds.). *The Delphi Method: Techniques and Applications.* Reading, MA: Addison Wesley.

McGrath, K. (2001). 'The Golden Circle: A Case Study of Organisational Change at the London Ambulance Service', in S. Smithson, J. Gricar, M. Podlogar, and S. Avgerinou (eds.), *Proceedings of 9th European Conference on Information Systems*, Bled, Slovenia, June 27–9. Kranj, Slovenia: Moderna Organizacija, 1137–48.

Mason, R. O., and Swanson, E. B. (1981). *Measurement for Management Decision.* Reading, MA: Addison Wesley.

Monteiro, E. (2000). 'Actor-network Theory and Information Infrastructure', in C. U. Ciborra (ed.), *From Control to Drift: The Dynamics of Corporate Information Infrastructures.* Oxford: Oxford University Press, 71–83.

Mumford, L. (1934). *Technics and Civilization.* London: Harvest/HBJ.

—— (1952). *Art and Technics.* New York: Columbia University Press.

—— (1970). *The Myth of the Machine: The Pentagon of Power, Vol. 2.* New York: Columbia University Press.

—— (1986). *The Future of Technics and Civilization.* London: Freedom Press.

Orlikowski, W. J., and Gash, D. C. (1994). 'Technological Frames: Making Sense of Information Technology in Organizations'. *ACM Transactions on Information Systems*, 12/2: 174–207.

—— and Iacono, C. S. (2001). 'Desperately Seeking the "IT" in IT Research: A Call to Theorizing the IT Artefact', *Information Systems Research*, 12/2: 121–34.

Parker, M. M., Benson, R. J., and Trainor, H. E. (1988). *Information Economics: Linking Business Performance to Information Technology.* Englewood Cliffs, NJ: Prentice Hall.

Radcliffe, R. (1982). *Investment: Concepts, Analysis, Strategy.* Glenview, IL: Scott Foreman.

Remenyi, D., and Money, A. (1993). *A Guide to Measuring and Managing IT Benefits.* Manchester: NCC Blackwell.

Robson, W. (1994). *Strategic Management and Information Systems: An Integrated Approach.* London: Pitman.

Sauer, C. (1993). *Why Information Systems Fail: A Case Study Approach.* Henley-on-Thames, UK: Alfred Waller.

Serafeimidis, V., and Smithson, S. (2003). 'Information Systems Evaluation as an Organizational Institution—Experience from a Case Study'. *Information Systems Journal*, 13/3: 251–74.

Smith David, J., Schuff, D., and St. Louis, R. (2002). 'Managing Your IT Total Cost of Ownership'. *Communications of the ACM*, 45/1: 101–6.

Symons, V. (1990). 'Evaluation of Information Systems: IS Development in the Processing Company'. *Journal of Information Technology*, 5: 194–204.

—— (1991). 'A Review of Information Systems Evaluation: Content, Context and Process'. *European Journal of Information Systems*, 1/3: 205–12.

Walsham, G. (1993). *Interpreting Information Systems in Organisations.* Chichester, UK: John Wiley.

Weick, K. E. (1985). 'Cosmos vs. Chaos: Sense and Nonsense in Electronic Contexts'. *Organizational Dynamics*, 15: 51–64.

Willcocks, L. (1992). 'IT Evaluation: Managing the Catch 22'. *European Management Journal*, 10/2: 220–29.

—— and Lester, S. (1994). 'Evaluating the Feasibility of Information Systems Investments: Recent UK Evidence and New Approaches', in L. Willcocks (ed.), *Information*

Management: The Evaluation of Information Systems Investment, London: Chapman and Hall, 49–75.

Willcocks, L. and Lester, S. (1996). 'Beyond the IT Productivity Paradox', *European Management Journal*, 14/3: 279–90.

Notes

1. This case has been previously analysed in terms of evaluation roles (Serafeimidis and Smithson 2003).
2. The authors wish to acknowledge the contribution of Vassilis Serafeimidis in collecting the original data for the case study.

12

Reflections on information systems strategizing

ROBERT D. GALLIERS

Introduction

This chapter has the aim of reflecting on developments in the area of information systems strategy and, more particularly, on the process of information systems strategizing. It does so against the background of something of a hiatus in the treatment of the topic in the Information Systems literature, especially since the heightened interest in this area of research up to the early 1990s. A further motivation arises from the relative paucity of serious reflection on Information Systems issues in much of the Strategic Management and Organizational Behaviour literatures on strategy and strategizing (Orlikowski 2000 being a notable exception). The chapter also aims to take account of key advances in the early twenty-first century in information and communication technologies, knowledge management, and the rapidly-changing nature of the business environment.

Surely, few would argue that the strategic management of data, information, and knowledge—and associated ICT—represents a major strategic challenge and opportunity for organizations in the twenty-first century. The market for ICT products and services can be measured in tens of billions of dollars/euros. It has been estimated that companies in the developed world spend something in the region of two per cent of turnover annually on hardware and software alone (Willcocks 1992, 1999). This figure would no doubt grow considerably if the costs associated with staff development, maintenance, and the management of change associated with the implementation and ongoing operation of ICT-based systems were taken into account. But we still talk glibly of the information age, of the networked society, of globalization, of knowledge management—each in its own way enabled and facilitated by ICT. It is therefore surprising how little we strategize about these issues.

Although attitudes differ, there is little doubt that ICT is here to stay (Land 1996). While some see the advent of this 'brave new world' as being nothing other than a boon, others mutter their discontent at the spiralling costs involved, at 'techies' who fail to understand the subtleties of organizational life, at the disruption created, at the invasion of privacy, and so on (Galliers 1992). Notwithstanding, the impact of ICT is likely to be felt increasingly as its power and reach continue to outstrip even the wildest predictions. This impact is felt by individuals, organizations, national

This chapter arises from a paper (Galliers 2001) presented at the EGOS Colloquium, Lyon, France, July 2001. A further version appears as Galliers and Newell (2003*a*).

governments, and society as a whole. What more need be said to argue that this is a topic worthy of our attention in any strategy discourse?

Given the above, it would seem strange that information systems strategy barely rates a mention in most business strategy courses. Strange that the topic most often appears as an optional course, at best, in MBA curricula or in Master's courses in Management or Organizational Behaviour. Strange that many firms rush, lemming-like, to avoid the pain of managing their information resource and the related technologies by outsourcing their ICT or information services departments (Lacity and Willcocks 2000). Strange that we reel from one bandwagon, one fad to the next with apparent abandon, often to rue the consequences later.[1] Strange that we simultaneously revel in, and yet revile, the industry that plies us with one solution after the next—an industry that, nonetheless, appears not to ask what questions its 'solutions' are meant to be answering.

The purpose of this chapter, then, is to counter these cavalier attitudes and provide a serious commentary on some of the key issues associated with strategizing in the context of managing organizational information and knowledge, and the related ICT. This will not be a technologically-oriented, nor indeed a technologically-deterministic, treatment of the topic although, inevitably, developments in ICT have had a profound effect on the scope and orientation of information systems strategy. Rather, it will deal with developments in our thinking and practice in Information Systems from a strategy—or, rather, strategizing—perspective. Even more important, it will provide a critical commentary on some of the more trite treatments of the topic that tend to appear in the popular media.

The chapter is organized as follows. First, an attempt is made to provide something of a tutorial on developments in the theory and practice of information systems strategy from the early days of commercial data processing (DP) up to the 1990s (e.g. Somogyi and Galliers 1987, 2003). Secondly, it examines some of the key concepts and frameworks that have underpinned much of information systems strategy theory during this period. We then proceed to consider some of the more recent developments and new thinking in the field that have emerged over the last decade or so, with a view to pointing out future directions and current concerns, culminating in a proposed inclusive framework for information systems strategizing.

Background history: from data processing to competitive advantage

There have, of course, been many developments in ICT since the earliest days of business computing. In parallel with these innovations, and with an increasingly sophisticated understanding of the role of these technologies in organizations, our understanding of information systems strategy has grown too during this period. Figure 12.1 provides a simplified framework within which to situate some of these developments. It suggests that we might usefully view such developments in four phases that have differed in terms of:

(*a*) the degree to which the information systems strategy might be viewed as a business-driven, 'top-down' process—as against more technology-driven, 'bottom-up' concerns; and

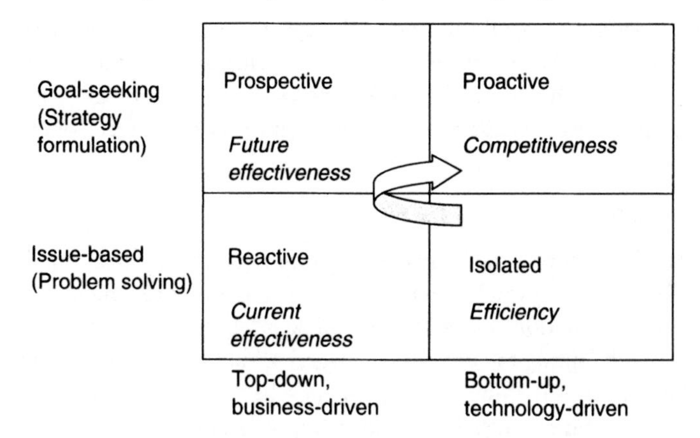

FIG. 12.1. Tracing the developments in information systems strategizing
Source: Adapted from Galliers (1987: 226).

(b) the extent to which such strategies have been based on short-term problem-solving as against more long-term strategic goal-setting.

The model in Figure 12.1 suggests that the focus of information systems strategizing may be seen to have gone through four phases, during which it has shifted away from—and back to—ICT, and from matters of efficiency to matters of effectiveness and competitiveness. This is clearly a highly-stylized and overly-simplistic view of developments, but the framework helps to provide something of an overview of the changes that have taken place since the 1960s. In some respects, we might suggest that current information systems strategizing incorporates aspects of each of these phases. For example, there is evidence of what has come to be been termed 'storage resource planning', characterized by a concern for the efficient storage of data across an enterprise to improve current and future efficiency, effectiveness, and competitiveness.

In the first phase, in the early days of commercial computing, information systems strategy was predominantly concerned with issues of the day and the efficient utilization of the technology for mainly operational purposes. From this perspective, information systems strategy may be viewed as having been being fairly *isolated* from the rest of the business. There followed a period where more formal, 'top-down', business-driven strategies were commonplace, with the emphasis being for the most part on *reactive* effectiveness. Such strategies took as read the existing business plans and objectives, and attempted to identify information systems applications to meet those business needs. Over time, information systems strategies became more forward-looking, bearing in mind the need to invest in technology that would stand the test of time despite changing information requirements. Such strategies may be seen as being essentially *prospective* in character. A move towards the *proactive* use of ICT for competitive advantage emerged during the 1980s and 1990s, applying concepts for the most part developed by Michael Porter and his Harvard Business School colleagues such as Warren McFarlan (e.g. McFarlan 1984;

Cash and Konsynski 1985; Porter and Millar 1985). This was superseded by Business Process Redesign or Re-engineering (BPR), which aimed to automate streamlined processes in line with customer requirements (e.g. Hammer 1990). The following subsections provide further detail of such developments, during each of these phases.

Operational efficiency: the isolated phase

In the first phase, during the early days of commercial data processing, hardly any strategic thought was given by senior managers to the uses to which ICT could be put in their organization, other than to think in terms of improving operational efficiency or attempting to cut costs. Managers would leave it to their information systems colleagues to develop and implement what was thought to be necessary in terms of computing systems. Targets for computerization (automation by another word) were simple production processes and record keeping, such as accounting systems. Little, if any, thought was given to the impact of the 'new' technology to ongoing operations, little concern was expressed over the kinds of skills that might be required to get the best out of the investment, and most developments or acquisitions were undertaken on a piecemeal basis. What little management of information systems there was tended to be considered the province of what we now call the Information Technology function, and its management. In short, there was little planning for information systems, let alone strategizing.

Current effectiveness: the reactive phase

Senior management increasingly became concerned that DP was not delivering the promised efficiency gains, nor focusing on key business concerns and imperatives. From the days when DP was seen as almost entirely the province of the technologist, we gradually saw the emergence of business-driven IS planning approaches. One such was IBM's Business Systems Planning (BSP) methodology (Zachman 1982), a service IBM provided to its customers that was meant to identify not only how the organization could harness ICT to meet business needs but also, of course, to demonstrate the need for more computing. In essence, BSP was developed to identify key business processes and their associated information requirements. A comparison with the data output from existing information systems would then lead to the identification of additional required information systems applications—and additional hardware and software too.

The idea that ICT and business needed to be aligned was first introduced during this 'reactive' era. Alignment is an issue that has remained with us ever since, as discussed later in the chapter. At the time of this era, organizations had to rely on mainframe technology, with so-called 'dumb' terminals on employees' desks usually providing periodic output for control purposes. This was commonly known as 'batch processing', as data were processed in batches rather than on a continuous, real-time basis. For example, weekly or monthly management reports—forming what were called management information systems (MIS)—would be produced on reams of paper. This usually required much additional human analysis to provide anything meaningful.

Future effectiveness: the prospective phase

The advent of database systems in the late 1970s and 1980s not only led to the development of executive information systems (EIS), where managers could ask the database for answers to specific questions, but also to a major rethink of information systems strategy. The thought here was that, rather than identifying particular information systems applications, organizations would simply have to identify the key data entities with which they were dealing (e.g. customer; product) and their attributes (e.g. name, address; product code, size). These could then be mapped to demonstrate their linkages, as a precursor to database design. A champion of this approach was James Martin (1982). Something of a 'garbage can' model (see Cohen, March, and Olsen 1972) for information systems strategy, with database technology in mind, thus appeared on the scene. It was thought that organizations would no longer have to concern themselves with issues of prioritizing information requirements associated with particular functions, managers, or processes. Rather, the database would enable the delivery of whatever information was required, wherever and whenever it was needed. In some cases, the error in this line of reasoning was not realized until after the invoice had been received for the massively increased computing power necessary to run the resultant database.

In some ways, this era may be seen to have spawned the so-called critical success factor (CSF) approach (Rockart 1979). Under the guise of executives defining for themselves their own critical data needs, the approach was rapidly appropriated by managers and consultants alike, since it enabled prioritization to take place. The approach was also welcomed because it brought an element of control back to harassed executives, who had seen their ICT budgets expanding at a time when they were being promised increased computing power for their limited financial resources—but nonetheless were becoming increasingly concerned with budget overspends. In outline, the approach centred on the identification of key objectives for the organization or strategic business unit (SBU) concerned, followed by the identification of key management processes necessary to enable the achievement of the stated objectives. CSFs associated with these processes were then pinpointed as a means of identifying the data that had to be made available for executives to manage and control the processes within their spheres of responsibility. The CSF concept was utilized by various approaches, such as Programme Quality Management (PQM)—another IBM methodology (Ward 1990)—and has continued to be incorporated into management thinking to this day.[2]

Competitiveness: the proactive phase

As we moved into the 1980s, the concepts of Porter and colleagues at Harvard had an enormous impact on thinking regarding the competitive advantages to be gained by firms from the astute application of ICT. Utilizing such concepts as the 'Five Forces' and 'Value Chain' models, they demonstrated how ICT, and the information it produces, could: provide added value to goods and services; retard competition from both traditional rivals and new entrants; and be used to leverage relationships with suppliers and customers alike (Porter 1980, 1985; McFarlan 1984; Porter and Millar 1985). A considerable amount of consultancy activity was spawned by this kind of thinking, and a great deal of literature was written on the topic throughout the 1980s and into the 1990s.

In line with this style of thinking, there emerged in the 1990s another approach to the strategic utilization of ICT, but this time focusing more on internal processes. The movement was spawned by the likes of Michael Hammer and Tom Davenport and became known as BPR (e.g. Hammer 1990; Davenport and Short 1990; Davenport 1993). A basis for their argument was that the mere computerization of a messy situation will lead to nothing more than a computerized mess. They argued for a clean-slate approach that identified and streamlined the key business processes. The trick was then to identify which of these processes could be automated, thereby improving efficiency and cutting costs. In addition, by focusing on customer requirements, the processes would lead to improved effectiveness.

While success rates were reported as being quite low (e.g. Davenport 1996), and advocates of the process were at pains to warn organizations of the risks involved, BPR was big business and was attempted by most major corporations in the English-speaking world. For example, the market for BPR services in 1995 was estimated to be in excess of $50 billion (ibid.). By 1996, however, the bubble had begun to burst when one of the founding fathers of the movement, Tom Davenport, finally recognized the loss of considerable organizational knowledge through the swathes of redundancies brought about by the downsizing strategies that accompanied many BPR efforts. BPR had become, in his words, 'the fad that forgot people' (ibid.: 70).

In some respects, then, we had come full circle. When we first began to think of information systems planning and strategy, the focus was primarily on the technology itself, since managerial concerns regarding the application of computing were mainly about matters of operational efficiency. We then moved into an era during which business-driven approaches were prevalent, with concern shifting to matters of effectiveness, and prioritization. As we entered the 1980s, and then into the 1990s, the focus moved to ICT for competitive advantage, and subsequently to BPR. In this era, attention shifted once more to a concern for how the technology could be harnessed proactively to increase competitiveness, at first through an analysis of the competitive environment and, later, by an analysis of internal processes. Throughout the whole 'competitive' phase, however, approaches to information systems strategy might reasonably be characterized as being based on a rational, deliberate paradigm, rather than the kind of emergence discussed by Mintzberg (e.g. Mintzberg and Waters 1985), among others. Additionally, little attention had been paid to more pluralistic and innovative strategizing.

This characterization of information systems strategy theory and practice as predominantly rational, objective, and unitary is illustrated by Figure 12.2, which is

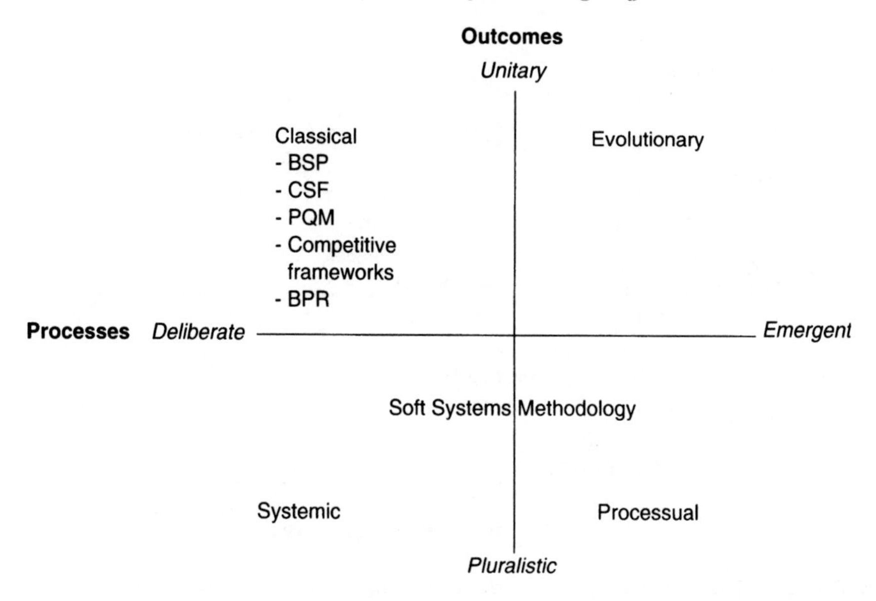

FIG. 12.2. Locating common information systems strategy approaches
Source: Adapted from Whittington (1993: 3).

based on Whittington's (1993) framework for mapping the developments in strategic thinking in the latter half of the twentieth century. It soon becomes clear that much information systems strategizing has been of the traditional school, with strategy formulation based on profit maximization as the primary, if not sole, objective. What is more, there has been a tendency, certainly in practice, to assume the equivalence of data, information, and knowledge. Latterly, however, both tendencies have been brought into question, as we shall see later in the chapter. A contrast can be found with the traditional school of information systems strategizing in the soft systems methodology (Checkland 1981; Galliers 1993*a*; Stowell 1995). Here, the outcome of the analysis is not predetermined and an ICT 'solution' is by no means a foregone conclusion. Additionally, alternative outcomes will be the subject of debate and further iteration. The process of strategizing, with a view to gaining a shared appreciation of the context in which this strategizing is taking place, is just as important, if not more so, than the decisions made as a result. Thus, soft systems methodology might be seen as spanning the two quadrants in the lower portion of Figure 12.2.

From localized exploitation to business scope redefinition

A somewhat different framework, but nonetheless one that also provides a perspective on the changes in information systems strategic thinking, arose from a major research programme conducted during the late 1980s, coordinated at the Massachusetts Institute of Technology (MIT) under the title 'Management in the

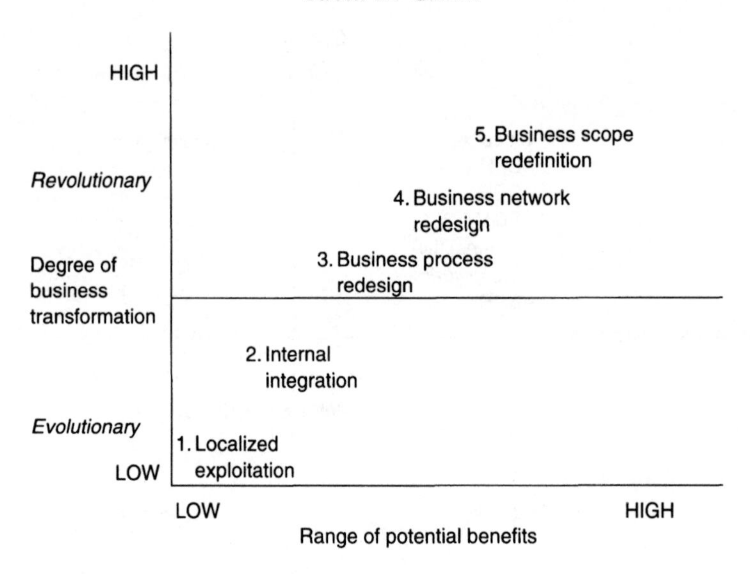

FIG. 12.3. The MIT Management in the 1990s Programme: 'IT-based revolutionary change leads to major benefits'

Source: Adapted from Venkatraman (1991: 127).

1990s' (Scott Morton 1991). Funded by major corporations from both sides of the Atlantic, it sought to uncover the means by which ICT could be harnessed to provide truly significant advances in terms of business performance. The framework is reproduced here as Figure 12.3.

One conclusion drawn by the MIT research team was that many companies were obtaining only relatively low business benefits from their investment in, and application of, ICT. They argued that this was due mainly to the fact that a relatively low level of business transformation had been attempted, with most companies operating at levels 1 and 2 of Figure 12.3. The researchers argued that such evolutionary approaches would not deliver the requisite order-of-magnitude improvements being sought after, which they deemed necessary in highly-competitive markets. This, they argued, could occur only via revolutionary change of the style proposed by the BPR advocates (level 3).

'Don't automate, obliterate' was the uncompromising title of a famous *Harvard Business Review* article by Hammer (1990). But, as we have seen, BPR focused for the most part on internal process redesign. The MIT team extended the focus of BPR, in much the same way as the Porterian school had done with the value-chain concept, to include what they termed 'business network redesign' (level 4). This extended the process analysis to ensure electronic links provided along the value chain included suppliers and customers, in order to form electronically-mediated strategic alliances (Rayport and Sviokla 1995). At one stage, this would have involved utilizing electronic data interchange (EDI) technology. Nowadays, the World Wide Web and the Internet would be used.

The MIT team concluded that truly significant business benefits would emerge only from redefining the very scope of the business through the utilization of the full power of ICT to create new products and services (level 5). Case examples that have entered the mythology of strategic information systems include: the Apollo and Sabre airline reservation systems of United and American Airlines; Thomson Holidays; Frito-Lay; Otis Elevators; American Hospital Supply; and Mrs Field's Cookies (Galliers 1993*a*). Senn (1992) and Ciborra (1994), among others, have argued that these systems were introduced initially with a view to increasing efficiency, but subsequently underwent various enhancements that—somewhat serendipitously—provided the companies concerned with a competitive advantage.

While Figure 12.3 stresses only the revolutionary potential of ICT when used proactively, it is clear that it is not always sensible to base one's business strategy on such an aggressive use of the technology. Indeed, Figure 12.3's 'range of potential benefits' axis might reasonably be re-labelled 'degree of business risk', given that revolutionary change can bring with it much greater risks than would be the case with a more incremental approach (Galliers 1997). A means of assisting in deciding whether there is a potential strategic advantage by providing added-value services based on information and ICT is provided by the Information Intensity Matrix (Porter and Millar 1985), which is depicted in Figure 12.4.

Figure 12.4 asks us to consider the extent to which information forms a critical part of the value-chain activities and of the product itself. In situations where this 'information intensity' is high, it can be concluded that ICT is integral to the delivery of goods and services. Where it is low, the potential use of ICT is more limited. Competing on the basis of providing additional information in terms of the product itself, or in relation to value-chain processes, can thus be considered by using this framework.

Distinguishing the components of information systems strategies

Much of the MIT research—and indeed a great deal of mainstream thinking on information systems strategy—suggests that the key issue is to align ICT with the business strategy, as might be supposed from the earlier approaches such as BSP, CSF

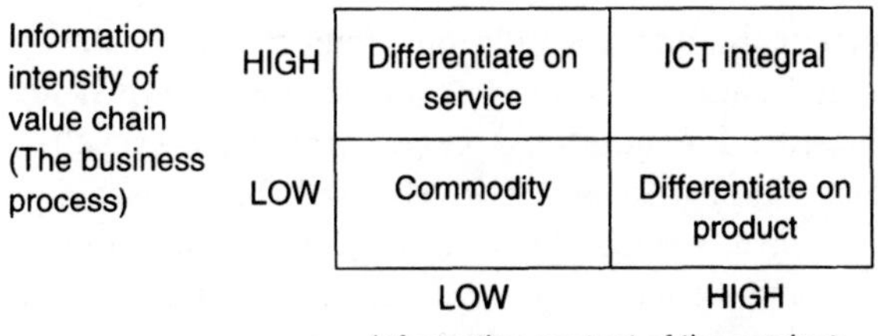

FIG. 12.4. Applying the Information Intensity Matrix

Source: Amended from Porter and Millar (1985).

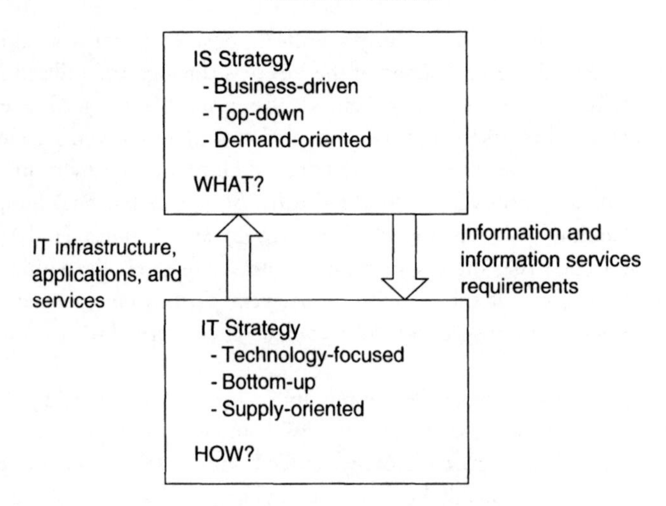

FIG. 12.5. Earl's distinction between information *systems* and information *technology* strategies

Source: Amended from Earl (1989: 63).

and PQM.[3] However, there is quite a conceptual gap between a business strategy and the necessary IT infrastructure to support it. As illustrated in Figure 12.5, Earl (1989) makes a distinction between information *systems* and information *technology* strategy, arguing that the former is essentially concerned with the 'What?' of the information required, while the latter is concerned with the 'How?' questions about the use of ICT to provide that information.

Note that, as indicated in Figure 12.5, Earl proposes that the information systems strategy is essentially business-led and demand-driven: it can be seen as a 'top-down' process, feeding off the business strategy. He further argues that information systems strategy should be the concern of the business executive—not the IT Director. Conversely, the information technology strategy is seen as being driven more by technology and supply, in that it depends to an extent at least on the existing technological infrastructure—on what is feasible from a technological standpoint within the current planning horizon. This is much more within the province of the IT Director.

Earl's distinction also brings with it some implications for the concept of alignment. For example, information systems strategy is viewed here as being about strategizing because it is ongoing and process-based ('processual'). Conversely, the information technology strategy is relatively fixed. This makes alignment difficult, as explored later in the chapter. Earl developed this line of thinking further by adding another component to the information systems and information technology strategy, namely the information management strategy. Having asked the 'What?' and the 'How?' questions, the information management strategy, in Earl's (1989: 64) formulation, asks the question 'Wherefore?'—to find answers to 'Why?' questions such as: 'Why this particular strategy as against any other?'.

The field of Information Systems is generally replete with terms that mean different things to different people. For instance, 'information technology' and 'information systems' are often used synonymously. The 'information management' term is another such example. This can sometimes connote a much broader concept than in Earl's amended model by encompassing the general field associated with the management issues concerned with information and ICT. Galliers (1991) noted this terminological confusion when building on the earlier work of Earl to produce a more comprehensive framework for information systems strategizing (see Figure 12.6). This framework included the questions related to 'What?' (in Earl's terms, the *information* strategy) and 'How?' (*information technology* strategy), but added the question 'Who?', relating to the *information services* strategy—the organizational arrangements for the provision of IS-related services. It also included considerations associated with the implementation of the strategy, with all its attendant management-of-change issues.

In terms of the 'Who?' question, the framework emphasizes the importance of developing an integrated information services strategy. This would need to include the kind of information systems staffing and skills needed to facilitate the strategy, including training requirements. In particular, a key question to consider, as an integral aspect of information systems strategizing, would be whether to outsource ICT provision—a topic that was particularly popular in the 1980s and 1990s. IT outsourcing refers to the 'significant contribution by external vendors in the physical and/or human resources associated with the entire or specific components of the ICT infrastructure in the user organization' (Loh and Venkatraman 1992).

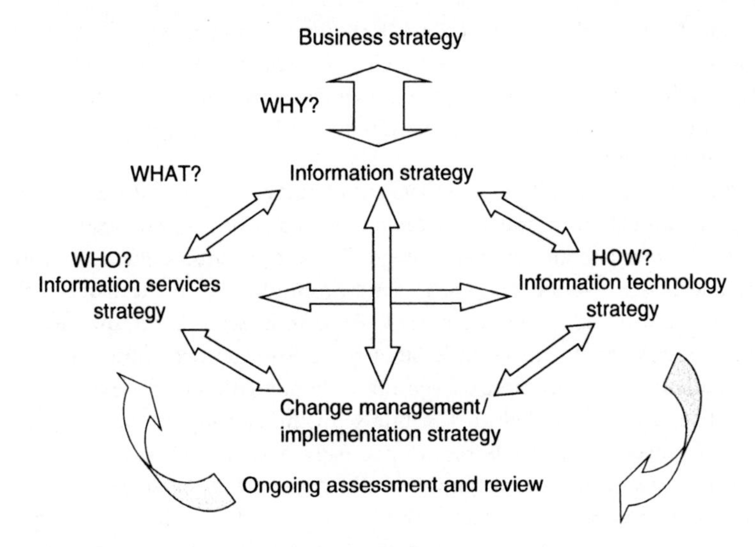

Fig. 12.6. Components of information systems strategy

Source: Amended from Galliers (1991: 60; 1999: 230).

As Lacity and Willcocks (2000) remind us, however, the appropriate question is not whether to outsource *per se*, but what would be the appropriate sourcing arrangements.

Another additional element in the Figure 12.6 framework is the explicit recognition of the importance of managing the change process associated with the implementation strategy. Galliers had become very much aware from empirical research and consultancy assignments that the outcome of many information systems strategy projects was what might be termed 'shelfware', as plans for such projects often collected dust on the office shelf because such little information systems implementation occurred as a result of such projects.

It also appeared that few lessons had been learned from the mainstream literature on strategizing. From the start, this was particularly the case in relation to the consideration of implementation and change management issues (Wilson 1992). Other issues that required attention included: the emergent quality of strategies and strategizing (Mintzberg and Waters 1985); the unanticipated consequences of any ICT implementation (e.g. Brown and Eisenhardt 1995; Robey and Boudreau 1999), and what Weick (2001) terms interpretative flexibility. As a result—and also drawing on Systems Theory (e.g. Checkland 1981)—the model depicted in Figure 12.6 incorporated features that demonstrated the need to monitor and learn from the emergent features of strategic decisions. It also takes account of the unintended consequences of these decisions, and the various interpretations of, and reactions to, events and innovations expressed by different stakeholders. 'Change management' and 'ongoing review and feedback' were therefore incorporated into the model.

The framework can be used in analysing information systems strategies in organizations by considering the extent to which each of the components is in place. This may provide an insight into the orientation of any particular organization towards information systems strategy. For example, does the organization emphasize ICT strategy to the detriment of identifying strategic information requirements? Or does the organization consider implementation and change management issues as part of their strategizing?

In addition, however, it suggests that each component of the information systems strategy is mutually dependent on each other component. For example, questions can be asked about whether strategic decisions regarding the organization of information systems services (e.g. whether they should be centralized or distributed; whether to outsource or not) are considered as an integral part of the information systems strategy, or whether—as is often the case—they are considered in isolation. Similarly, questions can be asked not only in relation to the extent to which required information is identified in line with the existing business strategy, but also if information is available that can actually question whether the strategy is appropriate or not, given changing business circumstances and as a consequence of the ongoing assessment and review of outcomes. This is the 'Why?' question that appears in Figure 12.6. The framework therefore envisions information systems strategy to be more all-encompassing than the distinction between IS and IT strategies provided by Earl in Figure 12.5.

Assessing information systems capability

Deciding on an appropriate strategy depends, in part at least, on an organization's ability to carry out that strategy successfully. However, Figure 12.7 shows there are occasions when it may well be advisable to follow a more conservative line, notwithstanding the arguments of those who follow the 'revolutionary change' school of thought. For example, a more evolutionary approach would appear to be called for if an analysis of the information intensity (Figure 12.4) of an organization's business processes and products and/or services found that the opportunities for adding value through information are limited. Nevertheless, if the opportunities are there but the capability is limited, then such an aggressive approach may well present too great a risk without outside assistance, or the development of internal human and technological resources.

The problem is that many organizations find themselves in the 'Catch-22' position of the lower right-hand quadrant of Figure 12.7, where—in a sense—they are damned if they do and they are damned if they don't. In such circumstances, organizations have to beware of the aggressive strategies of a competitor that might well have greater information systems capability than themselves. In response, an organization may well attempt a similar strategy itself, but fail in the attempt due to a lack of internal information systems resources—human as well as technical. Should the organization decide the risk is too great and do little in response, it is likewise open to attack.

How, then, might an organization evaluate its current information systems capability? One approach is to assess its current information systems strategy using the framework illustrated in Figure 12.6. But this provides an overview only. A more detailed positioning framework, which is explained below, is based on the so-called

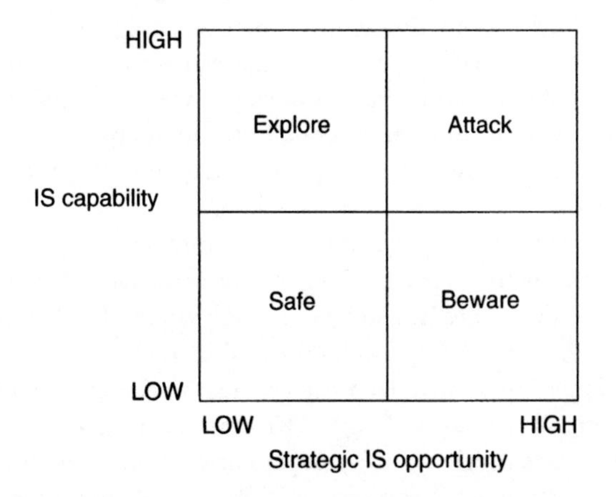

FIG. 12.7. When, and when not, to pursue an aggressive business strategy based on information technology

Source: Amended from McLaughlin, Howe, and Cash (1983).

'Stages of Growth' thesis first enunciated by Nolan (Gibson and Nolan 1974; Nolan 1979) and on the well-known '7-S' framework of McKinsey & Co. (Pascale and Athos 1981). Nolan's Stages model has its roots in Greiner's (1972) earlier work, essentially positing that firms will grow in maturity through recognizable 'stages' in terms of their management and use of ICT.

Nolan first formulated a four-stage model, but later extended this to six stages to take account of the database technology that was becoming available at the time—a technology that enabled firms to integrate their systems across functions and business units in a manner that had previously been impossible (see the second 'internal integration' stage of the MIT model in Figure 12.3). His six stages were (using Nolan's numbering system, that I also adopt below for a revised model):

I Initiation;
II Contagion;
III Control;
IV Integration;
V Data Administration; and
VI Maturity.

The story told through these stages unfolds as follows. At first, organizations are relatively unaware of the capabilities and potential uses of new and emerging ICT (Stage I). But once they have a few adherents, a kind of 'me too' mentality sweeps through the organization and demands increases almost exponentially (Stage II). As a result, management becomes increasingly concerned that things—especially budgets—are getting out of control, and they therefore impose tighter controls on ICT expenditure (Stage III). As management becomes increasingly aware that the looked-for business benefits from the ICT investment are escaping them because of lack of compatibility between different systems and a lack of information flow across processes and functions, further investment occurs in technologies that enable greater systems integration (Stage IV). This stage leads into one during which greater efforts are expended in ensuring the consistency of the data being shared across the organization, for example in terms of definition and interpretation (Stage V). The final stage of maturity is reached once integration is complete and compatibility is assured (Stage VI).

As is implied by the above, patterns of expenditure on ICT give a clue to which stage an organization has reached. Expenditure accelerates during Stages II and IV/V and tapers off in Stages III and VI—thus following a kind of double-S curve. While Nolan's (1979) model has been criticized in academic circles for its lack of conceptual underpinnings and its failure to provide an accurate prediction of growth empirically (Benbasat *et al.* 1984; King and Kraemer 1984), it was nonetheless highly popular and used extensively by many major corporations in the English-speaking world. Indeed, it spawned a consultancy company—Nolan Norton and Co.—which was eventually taken over by KPMG. Despite this popularity, it clearly had its limitations, particularly in relation to its technological focus. An extended Stages model was therefore developed by Galliers and Sutherland (1991), following case study research in Europe and Australia. This model, shown here as Table 12.1, focused on broader information management issues and borrowed the McKinsey 7-S framework that was in widespread circulation at the time (see first column in Table 12.1).

The framework depicted by Table 12.1 may be difficult to take in at first glance, but it essentially parallels the Nolan (1979) model in terms of the six stages of growth, which it renames (keeping the same numbering system):

I Ad hocracy;
II Starting the foundations;
III Centralized dictatorship;
IV Democratic dialectic and cooperation;
V Entrepreneurial opportunity; and
VI Integrated harmonious relationships.

Referring to our account earlier in this chapter of the developments in thinking and practice with respect to information systems strategy, we can trace this development through the six stages of strategy growth. We can see, for example, that information systems strategy develops from what is little more than the acquisition of IS products and services on more or less an *ad hoc* basis, though to top-down, business-led planning (see Earl's model, Figure 12.5)—and on to competitive advantage. The sixth stage is characterized by a strategy that integrates information systems considerations into the business strategy itself. Similarly, we can trace developments in the kind of staff and skills that are available to the organization (whether in-house or through a sourcing arrangement).

Managerial attitudes towards the strategic aspects of information systems can also be traced. From the bewilderment and confusion of the early stages of growth (Stages I and II), there has been a tendency for management to adopt the somewhat negative and adversarial stance associated with Stage III. This has tended to be as a result of past disappointments and concerns over spiralling ICT expenditure—with sometimes little in the way of perceived business benefits in return. The latter stages are characterized by a more positive, but informed, perspective. More specifically, with growing cooperation and a realization that greater integration across functions and SBUs is called for, a more concerted approach towards integration is evident in Stage IV. A more outward-facing perspective characterizes Stages V and VI, with an entrepreneurial and opportunistic stance being in evidence. A number of lessons emerged from the application of the Table 12.1 Stages of Growth framework, including the following.

First, it should be noted that the model is no more than a model—it is a positioning framework only. The foregoing discussion might unwittingly give the sense that all this development is preordained and is followed in every instance. This is far from being the case. The model has been found to be useful as a means of facilitating shared understanding as a result of posing a series of questions in relation to aspects of information systems management, based on the 7-S list. It certainly does not provide any answers. And shared understanding does not necessarily mean

Table 12.1. *An extended Stages of Growth model*

Stage:	I Ad hocracy	II Starting the foundations	III Centralized dictatorship	IV Democratic dialectic and cooperation	V Entrepreneurial opportunity	VI Integrated harmonious relationships
Element:						
Strategy	Acquisition of IT (services)	Audit of IT provision	Top-down analysis	Integration, coordination	Competitive advantage	Interactive planning, collaboration
Structure	Informal	Finance controlled	Centralized IS department	Information Centre	SBU coalition	Coordinated solutions
Systems	*Ad hoc* operational, accounting	Gaps/duplication, large backlog, heavy maintenance	Uncontrolled end-user computing versus centralized systems	Decentralized approach, some Executive Information Systems	Coordinated centralized and decentralized IS, some strategic IS	Inter-organizational systems, IS/IT-based products and services
Staff	Programmers, contractors	Systems analysts, data manager	IS planners, IS manager, database specialists	Business analysts, information resource manager	Business and IS planners integrated	IS/IT Director (Board level)

Style	Unaware	'Don't bother me, I'm too busy'	Abrogation, delegation	Partnership, benefits management	Individualistic (product champions)	Multi-disciplinary teams (key themes)
Skills	Individual, technical, low-level	Systems development methodology, cost-benefit analysis	IS awareness, project management	IS/business awareness	Entrepreneurial marketing	Lateral thinking (IT/IS potential)
Shared values	Obfuscation	Confusion	Senior management concern, IS defence	Cooperation	Opportunistic	Strategy making and implementation

Sources: Amended from Galliers and Sutherland (1991: 111), with elements from Pascale and Athos (1981).

consensus. It is a subjective measure, and opinions will sometimes diverge, but it at least provides a kind of benchmark against which to assess matters, and to begin to understand why certain views are held by some, but not others. The model is an aid to sense-making (Weick 1990); used judiciously, it can be of assistance in gaining a shared appreciation of key information systems management issues on the part of management teams.

Second, there is no intrinsic right for organizations to move inexorably through the stages towards Stage VI. Indeed, some companies have realized that they have occasionally moved 'backwards'. A series of discussions as to why movement has or has not occurred may provide further insight. Third, different parts of the organization may each present a different profile. As a result, assessments can be made as to whether these differences are harmful and need to be dealt with—or that the company can live with them, or indeed, that they are entirely appropriate. Fourth, organizations will not find themselves at a particular stage with respect to all the elements, but will find that some of these will lag 'behind' while others will be further 'ahead'. Again, assessments can be made as to what these differences mean in terms of strategic directions and imperatives. Further, it will seldom be the case that an organization's profile will fit neatly into the stages, as there will be elements that exhibit characteristics of more than one stage. This is an imprecise 'science'.

Fifth, it may prove useful to map the implied profile of a proposed strategy and contrast this with the existing situation. If there is considerable distance between the two, an assessment of the risks involved in attempting the proposed strategy can be made. Sixth, as a result of these kinds of deliberations, the shared understanding reached should lead to the identification of change projects designed to move the organization to a desired position. Finally, what constitutes 'maturity' (as referred to in the earlier Nolan models) will be changing and contextual, so Stage VI should not be viewed as an end in itself. Other elements to the model could also be incorporated; for example, an eighth 'S' might usefully be concerned with security issues.

Information systems strategic thinking in the 1990s[4]

As we have seen, the field of information systems strategy had come some distance in the latter part of the twentieth century. From a relatively isolated, narrow, and technologically-oriented activity, it had become much more business-oriented and competitively minded. There had been increasing realization, too, that the management of change and people issues are a significant—perhaps the key—aspect of what is required.

In some respects, though, IS strategy had not come very far at all. It had reached a point at which current thinking might reasonably be summarized by another framework from the MIT Management in the 1990s Programme (see Figure 12.8). For example, we had learnt our lessons from the many BPR failures: IS strategy and change was more, much more, than focusing on business processes and technology alone. People mattered, and their capabilities and knowledge had to be nurtured. Information systems needed to be seen as social systems, admittedly with

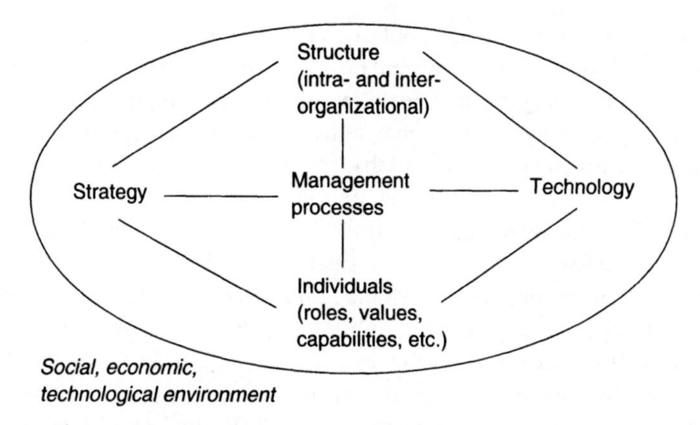

FIG. 12.8. The MIT model of strategic change and fit

Source: Adapted from Scott Morton (1991: 20) and Sauer, Yetton, and Associates (1997: 281).

an increasingly technological component—but not as technological systems *per se*. While this model moves us well beyond the technological focus of earlier information systems strategy approaches, it is also similar to Leavitt's (1965) 'diamond' of the mid-1960s. Leavitt argued that organizations could be viewed as complex systems, consisting of four interacting variables: objectives, structure, technology, and people. These variables clearly bear a remarkable resemblance to those identified in Figure 12.8.

Despite this, information systems strategy had indeed come a long way, but it also had a very long way to go to catch up with other strategy discourses. This emphasizes the point already made with regard to Figure 12.2, that builds on the framework developed by Whittington (1993) for identifying different schools of thought relating to strategy and strategizing. That point is also illustrated in the next section, which questions some prevailing myths about the strategic potential of ICT.

Uncovering the myths of strategic ICT

There have, of course, been many developments in ICT in recent years. In this section, a number of these recent developments will be considered in relation to the various strategy issues. Specifically, it will be argued that—despite the developments in thinking about information systems strategy discussed earlier—many myths about ICT continue to be promulgated: myths about how to develop ICT strategically, how to use ICT to support knowledge management, and about ICT and competitive advantage.

Myths about how to develop ICT strategically

There are essentially two related elements to the myth about ICT strategic developments. The first is that ICT systems should align with the business strategy and the second is that ICT systems should be rationally planned. As noted earlier, a central tenet of much of the theory and practice of information systems strategy has been the concept of alignment. The notion of alignment suggests that information systems strategy is a rational and deliberate activity. Intuitively appealing, alignment has been a taken-for-granted concept that remained largely unchallenged for many years. Earl's (1989) distinction between information systems and information technology (or ICT) strategies (Figure 12.5) is very helpful in terms of demarcating the two terms, as we have seen. However, it can too easily hide a key issue with respect to alignment, concerning the fact that the information needs for the great majority of organizations are in constant flux. Of course, there is a subset of information requirements that remains reasonably constant over time but, with fast-changing competitive environments, that subset is by no means representative of the totality.

Conversely, organizations are investing in ICT that will remain with them for quite some time, and will have to serve the test of time. Additionally, the view we have put forward here is that information systems strategy is ongoing and processual. ICT decisions, while they may be cumulative, are nevertheless one-off. The question of alignment is therefore a vexed one, as it is about changing requirements and (relatively) unchanging technology. There is often a dynamic involved, with strategies falling in and out of alignment over time (Sabherwal, Hirschheim, and Goles 2001). The introduction of Internet technology internally, in the form of 'intranets', has assisted considerably in providing the requisite flexibility. However, most firms are having to deal with significant problems in upgrading their so-called legacy systems—both in terms of meeting changed information needs and of integrating them with new systems and technology. Enterprise Resource Planning (ERP) systems are sold partly on the basis of the need to replace such legacy systems (CACM 2000). Presumably, however, in time, ERP systems will themselves become legacy systems. Increasingly, too, alignment is required along the 'virtual' value chain—with electronic links to suppliers and customers alike (Rayport and Sviokla 1995). The open nature of the Internet, with new customers and sources emerging constantly, can complicate matters enormously. 'Alignment with whom?' becomes a more significant and increasingly difficult question to answer.

The second ICT strategic development myth, as we have seen, is that most of the approaches to information systems strategy suggest a rational analysis of ICT needs. For example, the radical approach championed by the MIT Management in the 1990s team (Scott Morton 1991), or as articulated by the advocates of the BPR approach, both start from the premise that a rational analysis of business needs should be undertaken. Indeed, as we have seen, the very notion of alignment suggests that information systems strategy is a rational and deliberate activity. However, following Mintzberg (e.g. Mintzberg and Waters 1985), there is an increasingly strong school of thought that talks of the 'emergent' nature of information systems strategy and of strategic information systems (e.g. Ciborra 1994). Neither should we forget the essentially political nature of most technological appropriations (Swan and Clark 1992). Moreover, as mentioned previously, many of the successful ICT systems that have been developed, and lauded as being 'strategic'—for example, the Apollo and Sabre airline reservation systems—have emerged though a process of gradual enhancement (Senn 1992) and improvisation (e.g. Ciborra 1994, Chapter 1 this volume; Galliers 1991, 1993*b*). Ciborra uses terms like 'bricolage' and 'tinkering' to signify the bubbling up of innovative ideas within organizations. This is in stark contrast to the kind of radical approach championed by the MIT team, or as articulated by the advocates of BPR (e.g. Davenport and Short 1990; Hammer 1990).[5]

This analysis suggests that no amount of rational planning can ever hope to create an ICT system that aligns with the business strategy, even in the short term. ICT system development is thus best considered as an interactive process, constantly ongoing and emergent as new information needs arise and new opportunities are identified. This conclusion is somewhat in line with the analysis of alignment conducted by Sabherwal, Hirschheim, and Goles (2001), when they talk of 'punctuated equilibrium'.

Myths about how ICT can support and enable knowledge management

Knowledge management is one of the latest fads to be adopted by management (e.g. Abrahamson 1991). The emergence of this concept followed the recognition that knowledge is perhaps *the* key resource of organizations, allowing them to innovate and compete. Ironically, perhaps, this recognition occurred at about the same time as the BPR revolution, when much valuable knowledge was lost through companies' back doors, along with the legions of middle-ranking executives made redundant in the name of efficiency—often as a direct result of BPR initiatives. The ERP systems subsequently developed were claimed to be more in tune with the recognition of the importance of capturing organizational knowledge. Such systems have been diffused and adopted widely during the late 1990s and early 2000s, sold on the premise that they will assist to improve efficiency by integrating knowledge about business processes that cut across functions in SBUs and locations.

Efficiency and innovation. Importantly, ERP is promoted as a means of helping to transfer 'best practice' knowledge. Thus, a key feature of an ERP system is that it has built-in processes which force an organization using it to adapt itself and its processes to the exigencies of the ERP software. These inbuilt processes are, supposedly, based on 'best practice' industry models. In this instance, then, ICT can be seen as a force for standardization, and therefore for speeding competitive convergence, given that the models remain more or less constant irrespective of the organization implementing the system. The myth is thus created that the adoption of an ERP system will enable an organization to transfer to itself the 'best practice' industry knowledge of how best to organize various processes.

Further, it is interesting to consider such systems in relation to the earlier discussion on alignment which explained how ERP systems are implemented partly to replace legacy systems—but themselves eventually become a legacy. Moreover, by advocating the copying of 'best practices' to improve efficiency, organizations are, potentially at least, running the risk of reducing their capacity to create the new knowledge that is needed to innovate and creatively respond to their ever-changing environment—a key concern of business strategy, surely. Another way of putting this might be to think of the issue in terms of the long-standing dilemma between efficiency and innovation, or between exploitation and exploration (Clark and Staunton 1989; March 1991; McElroy 2000).

The above distinction between efficiency and innovation is important in attempting to understand the role ICT can play in an information systems strategy that seeks to harness the increasing power of the technology, while facilitating innovation and knowledge creation in organizations—especially those that operate on a global basis. Information systems strategy, as we have seen, attempts to square the circle between efficiency, effectiveness, and competitiveness. The latter increasingly relies on constant innovation for it to be sustained. If the Internet, ERP systems, and other ICT have been a force for competitive convergence, as Porter (2001) argues, then how can we claim that ICT such as this provides firms with new means of competing?

These problems and myths surrounding ERP systems can be related to more general myths that have emerged about knowledge management and, more particularly, knowledge management systems (KMS). Most importantly, the myth has been created that suggests KMS can store and transfer knowledge, thus supporting and facilitating knowledge exploitation (the reuse of knowledge across time and space, for example by the transfer of 'best practices') and knowledge creation. The software solutions that were peddled as EIS or mere database systems at the end of the last century have been metamorphosed by marketing executives into the KMS of the twenty-first century. Such systems are based on the view that knowledge is 'out there', ready and available to be harvested or mined. A contrary perspective is provided here.

Data, information, and knowledge. To comprehend the argument here more fully, it is perhaps useful to go back to basics and understand the distinction between data, information, and knowledge—terms that tend to be used synonymously in everyday parlance. Data become informative for a particular purpose to human beings by the way people interpret the world about them through their own individual lenses, and by applying their memory and personal knowledge to each new situation they confront. This is how we innovate and adapt. Data are context-free and can be interpreted in many different ways for different purposes. For example, the results of a government election in any country in the world will doubtless be interpreted in different ways by the victor and the vanquished. So-called *information* technology therefore processes data, not information. We should, as a result, revert to the original name used for information systems in the 1960s and 1970s: *data* processing systems (Galliers and Newell 2003*b*).

Individuals inform themselves in order to undertake some particular task or make a particular decision. Information is therefore context dependent, and information systems have to include human beings and the act of interpretation for the term to be

Table 12.2. *Key characteristics of data, information, and knowledge*

Data	Information	Knowledge
Explicit	Interpreted	Tacit/embedded
Exploit	Explore	Create
Use	Build/construct	Rebuild/reconstruct
Accept	Confirm	'Disconfirm'
Follow old recipes	Amend old recipes	Develop new recipes
No learning	Single-loop learning	Double-loop learning
Direction	Communication	Sense-making
Prescriptive	Adaptive	Seminal
Efficiency	Effectiveness	Innovation/redundancy
Predetermined	Constrained	Flexible
Technical systems/ networks	Socio-technical systems/networks	Social networks
Context-free	Outer context	Inner context

Source: Reproduced from Galliers and Newell (2003*a*: 189; 2003*b*: 11).

at all meaningful. Knowledge, on the other hand, is tacit and embedded. It resides within our brains, and enables us to make sense of the data we capture. Knowledge is individuals' 'justified belief'—a belief that allows them to interpret and take purposive action in the world around them.[6] The distinction between the terms is made clearer in Table 12.2, although the latter should be interpreted with some care—given that the characteristics are provided merely to assist in sense-making (see Weick 1990).

The above characterization of knowledge, or rather, 'knowing' (Blackler 1995) suggests that knowledge sharing is facilitated through discourse and dialogue (von Krogh, Ichijo, and Nonaka 2000). Thus, the emphasis is on developing communities of practice (Brown and Duguid 1991; Lave and Wenger 1991) and project teams where individuals interact over time to develop shared understandings that can lead to innovation and creativity. ICT systems can support this dialogue, at least partially, but they cannot store or communicate knowledge as such. ICT systems store and transfer *data* that can be interpreted in each context by individuals who make sense of these data, for a particular purpose, based on their personal knowledge, experiences, and predilections.[7]

Myths about ICT and competitive advantage

Undoubtedly, the growth and impact of the Internet has been the most noticeable ICT development around the turn of the century, spawning the so-called dotcom companies and a considerable degree of hyperbole concerning e-business. In a *Harvard Business Review* article, Michael Porter (2001) argues that firms should view the Internet as a complement to, rather than something that cannibalizes, more traditional forms of organization and organizational ICT. He claimed that while some have argued that 'the Internet renders strategy obsolete ... the opposite is true ... it is more important than ever for companies to distinguish themselves through strategy' (Porter 2001: 63). His argument echoes what he was saying twenty years before: it is not the technology itself (in this case, the Internet) that will create competitive advantage, but the uses to which it is put that may do so. As ever, he sees the two fundamental factors that will ultimately determine profitability as being industry structure and sustainable competitive advantage. The former determines the profitability of the average competitor. The latter allows a firm to outperform the average competitor.

Porter goes on to argue that, although the Internet has created new companies and even industries (e.g. online auctions and financial institutions), its impact will be felt most in enabling 'the reconfiguration of existing industries that had been constrained by high costs for communicating, gathering information, or accomplishing transactions' (ibid.: 66). He gives, as examples, distance-learning programmes, catalogue retailers, and automated fulfilment centres, and contends that the Internet 'only changes the front end of the process' (ibid.: 66).

Porter maintained his belief in his Five Forces analysis (Porter and Millar 1985), stating that these 'still determine profitability even if suppliers, channels, substitutes, or competitors change' (ibid.: 66). However, because the impact of each force varies from industry to industry, he argues that it would not be appropriate to attempt to draw any general conclusions regarding the Internet's impact on long-term profitability. He does point to some general trends, though. For instance, he notes that ICT tends to: bolster buyer bargaining power by providing easier access to information on products and services; reduce barriers to entry by circumventing existing channels; and create substitute products and services. Rivalry intensifies because of the open nature of the Internet and the resultant difficulties that firms confront in retaining proprietary offerings. Rivalry also intensifies because of the global reach of the new technology. Finally, he argues that the Internet's tendency to reduce variable costs leads to pressure to engage in price competition. He observes: 'The great paradox of the Internet is that its very benefits—making information widely available; reducing the difficulty of purchasing, marketing, and distribution; allowing buyers and sellers to find and attract business with one another more easily—also make it more difficult for companies to capture those benefits as profits' (ibid: 66).

This analysis leads Porter to foresee greater competition due to increased numbers of competitors and pressure on prices, exacerbated by growing customer power. With the average profitability of most industries falling, the need for individual firms 'to set themselves apart from the pack' grows considerably. This leads to the conclusion that advantages must be gained in terms of cost and price, through improved operational efficiency and effectiveness, strategic positioning, and by doing things differently from the competition. He notes (ibid.: 70): 'The Internet affects operational effectiveness and strategic positioning in very different ways.

It makes it harder for companies to sustain competitive advantages, but it opens new opportunities for achieving or strengthening a distinctive strategic positioning.'

It should be clear from the foregoing why Porter (ibid.: 78) argues that the Internet has not altered the basic principles of competitive advantage:

> In our quest to see how the Internet is different, we have failed to see how the Internet is the same. While a new means of conducting business has become available, the fundamentals of competition remain unchanged. The next stage of the Internet's evolution will involve a shift in thinking from e-business to business, from e-strategy to strategy. Only by integrating the Internet into overall strategy will this powerful new technology become an equally powerful force for competitive advantage.

Porter sees competitive advantage as being gained by those companies that can integrate uses of the Internet with traditional means of doing business. He contends that it is easier for 'traditional' companies to do this than for dotcoms to adopt and integrate traditional approaches. But the traditional strengths of any company remain the same, with or without the Internet, such as unique products, superior knowledge of products and customers, strong personal service, and effective relationships.

Thus, we can raise serious concerns about ICT's impact on firms' long-term competitive business strategy. In essence, perhaps, the problem is that in each instance companies are utilizing new developments in ICT to promote efficiency. But, as already noted, in doing this they are—potentially at least—running the risk of reducing the capacity to innovate and to respond creatively to their ever-changing environment. Again, then, we return to the dilemma between efficiency and innovation (e.g. Clark and Staunton 1989; March 1991; McElroy 2000).

Synthesis: towards an inclusive framework for information systems strategizing

Where is the argument followed in this chapter leading in terms of our reflections on the concept of information systems strategy? One would hesitate to propose an all-encompassing framework that captures the essence of the above—and, indeed, the very concept of such a framework might well seem antithetical to the arguments immediately preceding this. Having said that, and as Weick (1990) might argue, frameworks do help with respect to sense-making, and do provide something of a benchmark against which informed debate and communication might take place. It is in this spirit that the following is presented for consideration.

Figure 12.9 builds on Figure 12.6, but is an attempt to incorporate some of the more recent thinking that we have just introduced. For example, the concept of an information infrastructure strategy—or what might be termed an information 'architecture'—is adopted and incorporated in an attempt to connote an enabling socio-technical environment for both the exploitation of knowledge (efficiency) and the exploration of knowledge (innovation). The debate was previously often couched in terms of exploration *versus* exploitation. Increasingly, however, we see

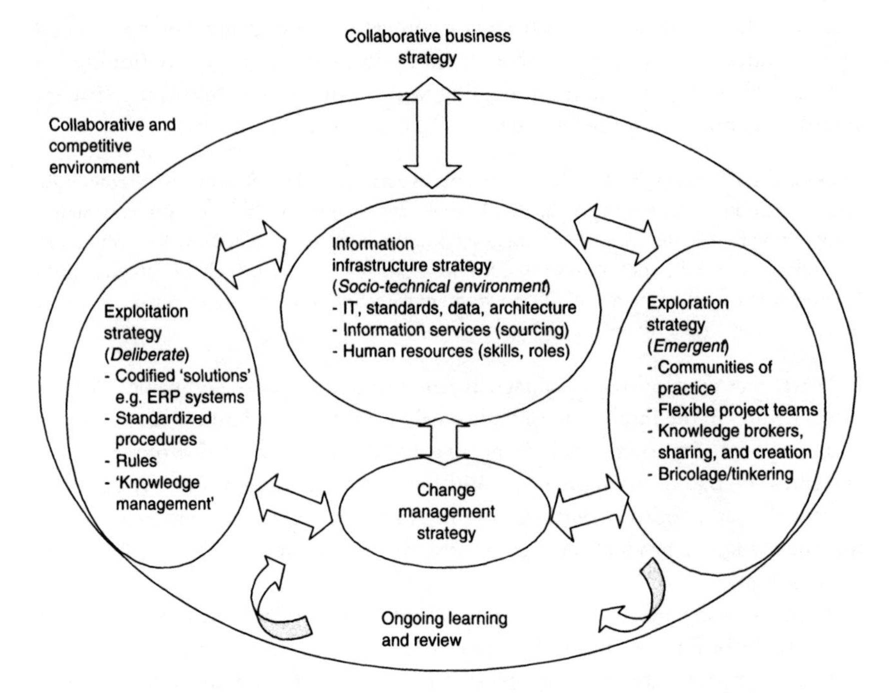

FIG. 12.9. Towards a more inclusive framework for information systems strategizing
Source: Reproduced in Galliers and Newell (2003*a*: 193).

different ICT initiatives, such as ERP and KMS, being implemented in tandem in an attempt to foster the simultaneous development of organizational efficiency and flexibility (Newell *et al.* 2003).

The concept of an information infrastructure (or architecture) has developed in response to the need for greater flexibility, given changing information requirements (Ciborra 2000). In the 1980s and 1990s, the term information infrastructure usually connoted the standardization of corporate ICT, systems, and data, with a view to reconciling centralized processing and distributed applications. Increasingly, however, Figure 12.9 depicts how the concept has come to relate not just to data and ICT systems, but also to the human infrastructure (roles, skills, capabilities, viewpoints, etc.)—and this is where knowledge creation, and sharing and innovation, play a crucial role. Star and Ruhleder (1996) unbundled the concept still further by talking of infrastructures in terms of, for example, their embeddedness, transparency, reach, links with conventions of practice, and installed base. Infrastructures are thus seen as being heterogeneous and socio-technical in nature.

As depicted here, then, information systems strategy, incorporating an information architecture strategy, is meant to be interpreted as being a part, albeit an increasingly important part, of collaborative business strategizing. It is collaborative because the focus will not be related just to internal matters but will also, crucially, involve partner organizations, such as customers, suppliers, and other organizations, for example those with whom sourcing arrangements are in place. The implication here is that the very boundary of an organization will become increasingly porous, debatable, and changing. Therefore, strategy needs to take this into account, especially information systems strategizing, given the virtual nature of many collaborative arrangements. This means that information systems strategizing has both a location and temporal dimension (Adam 1990)—the latter, in particular, being as yet under-researched.

Information systems strategy should also be seen as being ongoing and processual, crucially dependent on learning from 'below', from tinkering and improvisation, and from the emergent and unintended consequences of strategic decisions, as well as from the more deliberate, designed, and codified ICT 'solutions' that have been implemented. Figure 12.9 attempts to incorporate the embedded, socio-technical characteristics of information architectures—architectures that provide the kind of environment in which knowledge sharing and knowledge creation may be fostered, in tandem. Strategic information, therefore, not only supports existing strategic processes, but also questions the kind of taken-for-granted assumptions on which existing information systems strategies may be based. ICT is there too: not as *the* answer, not as a 'solution', but as a means of capturing data that may be interpreted in a purposeful manner, with which to make sense of phenomena in unique circumstances.

Conclusions

This chapter has attempted to reflect on developments in the thinking and practice associated with information systems strategy and the process of strategizing. In so doing, and in drawing on a fairly broad literature base, we have been able also to question some of the more taken-for-granted concepts found in mainstream accounts of IS, and reflect on the appropriate role of ICT in modern-day organizations. Concepts such as the alignment of business and ICT strategies, ICT and competitive advantage, 'best practice', knowledge management, and—more particularly—KMS have all been called into question in what Robey and Boudreau (1999) term a 'logic of opposition'. Indeed, the very nature of information and knowledge have been examined in a fresh light. In addition, Information Systems, as a field of study, may be seen to suffer from an element of faddishness, similar to the world of practice and ICT-based 'solutions'. By providing something of a historical account, an attempt has been made to draw together lessons from the past into the kind of cumulative account that has often continued to be missing from Information Systems discourse (Keen 1990). It is hoped that such reflection may prove useful to those interested in the social study of ICT, and not just those who share an interest in information systems strategy itself.

References

Abrahamson, E. (1991). 'Managerial Fads and Fashions: The Diffusion and Rejection of Innovations'. *Academy of Management Review*, 16/3: 586–612.

Adam, B. (1990). *Time and Social Theory*. Oxford: Polity Press.

Benbasat, I., Dexter, A., Drury, D., and Goldstein, R. (1984). 'A Critique of the Stage Hypothesis: Theory and Empirical Evidence'. *Communications of the ACM*, 27/5: 476–85.

Blackler, F. (1995). 'Knowledge, Knowledge Work and Organizations: An Overview and Interpretation'. *Organization Studies*, 16/6: 1021–46.

Brown, J., and Duguid, P. (1991). 'Organizational Learning and Communities of Practice: Toward a Unified View of Working, Learning, and Innovation'. *Organization Science*, 2/1: 40–56.

Brown, S., and Eisenhardt, K. (1995). 'Product Development: Past Research, Present Findings, and Future Directions'. *Academy of Management Review*, 20/2: 343–78.

CACM (2000), Communications of the ACM. Special issue on Enterprise Resource Planning (ERP) Systems. *Communications of the ACM*: 43/4.

Cash, J. I., and Konsynski, B. R. (1985). 'IS Redraws Competitive Boundaries'. *Harvard Business Review*, 63/2: 134–42.

Checkland, P. B. (1981). *Systems Thinking. Systems Practice*. Chichester, UK: John Wiley.

Ciborra, C. U. (1994). 'From Thinking to Tinkering. The Grassroots of IT and Strategy', in C. U. Ciborra and T. Jelassi (eds.), *Strategic Information Systems: A European Perspective*. Chichester, UK: John Wiley, 3–24.

—— (2000) (ed.). *From Control to Drift: The Dynamics of Corporate Information Infrastructures*. Oxford: Oxford University Press.

Clark, P. A., and Staunton, N. (1989). *Innovation in Technology and Organization*. London: Routledge.

Cohen, M. D., March, J. G., and Olsen, J. P. (1972). 'A Garbage Can Model of Organizational Choice'. *Administrative Science Quarterly*, 35/1: 1–25.

Davenport, T. H. (1993). *Process Innovation: Re-engineering Work through Information Technology*. Boston: Harvard Business School Press.

—— (1996). 'Why Re-engineering Failed. The Fad that Forgot People'. *Fast Company*. Premier Issue: 70–4.

—— and Short, J. E. (1990). 'The New Industrial Engineering: Information Technology and Business Process Redesign'. *Sloan Management Review*, 31/4: 11–27.

Earl, M. J. (1989). *Management Strategies for Information Technology*. London: Prentice Hall.

—— (1993). 'Experiences in Strategic Information Systems Planning'. *MIS Quarterly*, 17/1: 1–24.

Galliers, R. D. (1987). 'Information Systems Planning in the United Kingdom and Australia—A Comparison of Current Practice', in P. I. Zorkoczy (ed.), *Oxford Surveys in Information Technology*, 4. Oxford: Oxford University Press, 223–55.

—— (1991). 'Strategic Information Systems Planning: Myths, Reality and Guidelines for Successful Implementation'. *European Journal of Information Systems*, 1/1: 55–64.

—— (1992). 'Information Technology: Management's Boon or Bane?'. *Journal of Strategic Information Systems*, 1/2: 50–6.

—— (1993*a*). 'Towards a Flexible Information Architecture: Integrating Business Strategies, Information Systems Strategies and Business Process Redesign'. *Journal of Information Systems*, 3/3: 199–213.

—— (1993*b*). 'IT Strategies: Beyond Competitive Advantage'. *Journal of Strategic Information Systems*, 3/4: 283–91.

—— (1997). 'Against Obliteration: Reducing Risk in Business Process Change', in C. Sauer, P. W. Yetton, and Associates, *Steps to the Future: Fresh Thinking on the Management of IT-based Organizational Transformation*. San Francisco: Jossey-Bass: 169–86.

—— (1998). 'Reflections on BPR, IT and Organizational Change', in R. D. Galliers and W. R. J. Baets (eds.), *Information Technology and Organizational Transformation: Innovation for the 21st Century Organization*. Chichester, UK: John Wiley, 225–43.

—— (1999). 'Towards the Integration of e-Business, Knowledge Management and Policy Considerations within an Information Systems Strategy Framework'. *Journal of Strategic Information Systems*, 8/3: 229–34.

—— (2001). 'Rethinking Information Systems Strategy: Towards an Inclusive Framework for Business Information Systems Management?'. Paper presented at EGOS Conference, Lyon, France, July.

—— and Newell, S. (2003*a*). 'Strategy as Data + Sense Making', in S. Cummings and D. C. Wilson (eds.), *Images of Strategy*. Oxford: Blackwell: 164–96.

—— —— (2003*b*). 'Back to the Future: From Knowledge Management to Data Management'. *Information Systems and e-Business Management*, 1/1: 5–13.

—— and Sutherland, A. R. (1991). 'Information Systems Management and Strategy Formulation: The "Stages of Growth" Model Revisited'. *Journal of Information Systems*, 1/2: 89–114.

—— and Swan, J. A. (1999). 'Information Systems and Strategic Change: A Critical Review of Business Process Re-engineering', in W. L. Currie and R. D. Galliers (1999) (eds.), *Rethinking Management Information Systems: An Interdisciplinary Perspective*. Oxford: Oxford University Press, 361–87.

Gibson, C. F., and Nolan, R. L. (1974). 'Managing the Four Stages of EDP Growth'. *Harvard Business Review*, 52/1: 76–88.

Greiner, L. E. (1972). 'Evolution and Revolution as Organizations Grow'. *Harvard Business Review*, 50/4: 37–46.

Hammer, M. (1990). 'Don't Automate, Obliterate'. *Harvard Business Review*, 68/4: 104–12.

Hansen, M., Nohria, N., and Tierney, T. (1999). 'What's your Strategy for Managing Knowledge?'. *Harvard Business Review*, 77/2: 106–16.

Keen, P. G. W. (1990). 'MIS Research: Reference Disciplines and a Cumulative Tradition', in E. R. McLean (ed.), *Proceedings: 1st International Conference on Information Systems, Philadelphia, PA*. Atlanta, GA: Association for Information Systems, 9–18.

King, J., and Kraemer, K. (1984). 'Evolution and Organizational Information Systems: An Assessment of Nolan's Stage Model'. *Communications of the ACM*, 27/5.

Lacity, M. C., and Willcocks, L. (2000). *Global Information Technology Outsourcing: In Search of Business Advantage*, Chichester: Wiley.

Land, F. (1996). 'The New Alchemist: Or How to Transmute Base Organisations into Corporations of Gleaming Gold'. *Journal of Strategic Information Systems*, 5/1: 5–17.

Lave, J., and Wenger, E. (1991). *Situated Learning: Legitimate Peripheral Participation*, Cambridge: Cambridge University Press.

Leavitt, H. J. (1965). 'Applying Organizational Change in Industry: Structural, Technological and Humanistic Approaches', in J. G. March (ed.), *Handbook of Organizations*, Chicago: Rand McNally.

Loh, L., and Venkatraman, N. (1992). 'Information Technology Outsourcing: A Cross-sectional Analysis'. *Journal of Management Information Systems*, 9/1: 7–24.

McElroy, M. (2000). 'Integrating Complexity Theory, Knowledge Management and Organizational Learning'. *Journal of Knowledge Management*, 4/3: 195–203.

McFarlan, F. W. (1984). 'Information Technology Changes the Way You Compete'. *Harvard Business Review*, 62/3: 98–102.

McLaughlin, M., Howe, R., and Cash, J. I. (1983). 'Changing Competitive Ground Rules—The Impact of Computers and Communications in the 1980s'. Unpublished Working Paper. Boston: Graduate School of Business Administration, Harvard University.

March, J. (1991). 'Exploration and Exploitation in Organizational Learning'. *Organization Science*, 2/1: 71–86.

Martin, J. (1982). *Strategic Data Planning Methodologies*. Englewood Cliffs, NJ: Prentice Hall.

Mintzberg, H., and Waters, J. A. (1985). 'Of Strategies, Deliberate and Emergent'. *Strategic Management Journal*, 6/3: 257–72.

Newell, S., Huang, J. C., Galliers, R. D., and Pan, S. L. (2003). 'Implementing Enterprise Resource Planning and Knowledge Management Systems in Tandem: Fostering Efficiency and Innovation Complementarity'. *Information & Organization*, 13: 25–52.

Nolan, R. L. (1979). 'Managing the Crises in Data Processing'. *Harvard Business Review*, 57/2: 115–26.

Nonaka, I., and Takeuchi, H. (1995). *The Knowledge-creating Company: How Japanese Companies Create the Dynamics of Innovation*. Oxford: Oxford University Press.

Orlikowski, W. (2000). 'Using Technology and Constituting Structure: A Practice Lens for Studying Technology in Organizations'. *Organization Science*, 12/4: 404–28.

Pascale, R. T., and Athos, A. G. (1981). *The Art of Japanese Management*. London: Penguin.

Porter, M. E. (1980). *Competitive Strategy: Techniques for Analyzing Industries and Competitors*. New York: The Free Press.

—— (1985). *Competitive Advantage: Creating and Sustaining Superior Performance*. New York: The Free Press.

—— (2001). 'Strategy and the Internet'. *Harvard Business Review*, 79/3: 63–78.

—— and Millar, V. E. (1985). 'How Information Gives You Competitive Advantage'. *Harvard Business Review*, 63/4: 149–60.

Rayport, J. F., and Sviokla, J. J. (1995). 'Exploiting the Virtual Value Chain'. *Harvard Business Review*, 73/6: 75–85.

Robey, D., and Boudreau, M. C. (1999). 'Accounting for the Contradictory Organizational Consequences of Information Technology: Theoretical Directions and Methodological Implications'. *Information Systems Research*, 10/2: 167–85.

Rockart, J. F. (1979). 'Chief Executives Define Their Own Data Needs'. *Harvard Business Review*, 57/2: 81–93.

Sabherwal, R., Hirschheim, R., and Goles, T. (2001). 'The Dynamics of Alignment: Insights from a Punctuated Equilibrium Model'. *Organization Science*, 12/2: 179–97.

Sauer, C., Yetton, P. W., and Associates (1997). *Steps to the Future: Fresh Thinking on the Management of IT-based Organizational Transformation*. San Francisco: Jossey-Bass.

Scarbrough, H., Swan, J., and Preston, J. (1999). *Knowledge Management and the Learning Organization*. London: IPD.

Scott Morton, M. S. (1991) (ed.). *The Corporation of the 1990s: IT and Organizational Transformation*. Oxford: Oxford University Press.

Senn, J. A. (1992). 'The Myths of Strategic Systems: What Defines True Competitive Advantage?'. *Journal of Information Systems Management*, 9/3: 7–12.

Somogyi, E. K., and Galliers, R. D. (1987). 'Applied Information Technology: From Data Processing to Strategic Information Systems'. *Journal of Information Technology*, 2/1: 30–41.

—— —— (2003). 'Information Technology in Business: From Data Processing to Strategic Information Systems', in R. D. Galliers and D. E. Leidner (eds.), *Strategic Information Management: Challenges and Strategies in Managing Information Systems*. Oxford: Butterworth-Heinemann, 3–26.

Star, S. L., and Ruhleder, K. (1996). 'Steps Towards an Ecology of Infrastructure: Design and Access to Large Information Spaces'. *Information Systems Research*, 7/1: 111–34.

Stowell, F. (1995). *Information Systems Provision: The Contribution of Soft Systems Methodology*. London: McGraw-Hill.

Swan, J. A., and Clark, P. A. (1992). 'Organisation Decision-making in the Appropriation of Technological Innovation: Cognitive and Political Dimensions'. *European Work and Organisational Psychologist*, 2/2: 102–27.

Venkatraman, N. (1991). 'IT-induced Business Reconfiguration', in M. Scott Morton (ed.), *The Corporation of the 1990s: IT and Organizational Transformation*. Oxford: Oxford University Press, 122–58.

von Krogh, G., Ichijo, K., and Nonaka, I. (2000). *Enabling Knowledge Creation. How to Unlock the Mystery of Tacit Knowledge and Release the Power of Innovation*. Oxford: Oxford University Press.

Ward, J. (1990). 'Planning for Profit', in T. Lincoln (ed.), *Managing Information Systems for Profit*. Chichester, UK: John Wiley, 103–46.

Weick, K. E. (1990). 'Technology as an Equivoque: Sensemaking in New Technologies', in P. Goodman and L. Sproull (eds.), *Technology and Organizations*. San Francisco: Jossey-Bass.

—— (2001). *Making Sense of the Organization*. Oxford: Blackwell.

Whittington, R. (1993). *What is Strategy?—and Does it Matter?* London: Routledge.

Willcocks, L. (1992). 'IT Evaluation: Managing the Catch 22'. *European Management Journal*, 10/2: 220–9.

—— (1999). 'Managing Information Technology Evaluation: Techniques and Processes', in R. D. Galliers, D. E. Leidner, and B. S. H. Baker (eds.), *Strategic Information Management: Challenges and Strategies in Managing Information Systems*. Oxford: Butterworth-Heinemann, 271–90.

Wilson, D. C. (1992). *A Strategy of Change: Concepts and Controversies in the Management of Change*. London: Routledge.

Zachman, J. A. (1982). 'Business Systems Planning and Business Information Control Study: A Comparison'. *IBM Systems Journal*, 21/1: 31–54.

Notes

1. Examples of recent fads include business process redesign and re-engineering, enterprise resource planning, and knowledge management systems—concepts that will be introduced and discussed later in the chapter.
2. PQM was a further refinement of the BSP and CSF approaches. Again, essential business processes were identified in line with business objectives. These processes were assessed in terms of the number of CSFs impacting on them, and the quality and cost of IT-based systems in place to support them. Further developments deemed to be necessary were identified on the basis of criticality (in business terms) and performance (both business and technological, current and future).

3. A more considered approach to the question of alignment is provided in Sabherwal, Hirschheim, and Goles (2001), where they consider how ICT and business strategies move into, and out of, alignment over time.
4. A taxonomy of approaches to strategic information systems planning prevalent in the early 1990s is provided by Earl (1993).
5. For critiques of the BPR approach, see for example: Davenport (1996); Sauer, Yetton, and Associates (1997); Galliers (1998); and Galliers and Swan (1999).
6. Nonaka and Takeuchi (1995), after Plato, talk of knowledge as 'justified true belief'. Given the emphasis here on the process of applying knowledge to data in order to make informed judgements about the world in which we live, the word 'true' has been dropped from the definition.
7. The contrast provided here is similar to the personalization–codification distinction of Hansen, Nohria, and Tierney (1999), and the community–codification distinction made by Scarbrough, Swan, and Preston (1999).

13

Socially self-destructing systems

Richard L. Baskerville and Frank Land

Introduction

This chapter is about systems and organizational destruction, which is rather atypical for an information systems study. The study of IS has usually been shaped by its 'vocational' mission: the construction of practical and beneficial applications of information technology (see Banville and Landry 1989; Keen 1991). Not surprisingly, therefore, an important role in our research is played by the constructive aspect of IS: how people can, and do, construct practical systems. Knowledge of the construction of IS is broad, for example based on an understanding of the interrelationship between the information system and the social-organizational system. However, there are many open debates about this relationship. For instance, discussions regarding determinism are concerned with issues such as whether IT determines organization structure or vice-versa (Markus and Robey 1988)—or whether there is some process of co-evolution, as the complexity theorists maintain (Lewin and Volberda 1999; Lewin, Long, and Carroll 1999; McKelvey 2002).

The often tacit assumption underlying the vocational mission of IS is that the construction of IT-based systems supports and benefits the organization in its evolution. Indeed, many IS researchers go further and explicitly state that the implementation of information systems is a precondition for competitive success (Scott Morton 1991). The possibility that the application of IT can harm, and even destroy, the organization has received far less attention. However, information systems can be destructive, in the sense that they can prevent necessary social activities which hold the organizational elements together. Any introduction of IS that does not take into account the social-cultural dynamics in the organization setting in which it is used has the potential to destroy the organization it was intended to support. Thus, there is a need to understand better how short-term technological successes can emerge as long-term failures, a phenomenon that has been detected, for example, in strategic systems (Kettinger *et al.* 1994).

Socially-destructive systems may operate in subtle, easily overlooked ways. A stronger understanding of how these destructive systems achieve their counter-intuitive and dysfunctional outcomes can help to shape the requirements definition process, and may lead to fewer long-term system failures (Abdel-Hamid and

The authors gratefully acknowledge the observations of Cliff Freund on the case used here, and the helpful comments of Urooj Amjad and Chrisanthi Avgerou on the chapter.

Madnick 1990). Improving understanding of such systems, illustrated and informed by a case study, is the main aim of this chapter. But, first, it is necessary to make a clear distinction between the structure in a human organization and the artefacts that are intended to support that structure.

Organizational structures and their supportive artefacts

An artefact is an object that is generally made by people, usually with skill, for subsequent use. Most artefacts have their origins in uses determined in an earlier time or cultural stage. This origin suggests that an artefact has a physical persistence. These connotations are important. The way they distinguish the making of the object from its use imply that such objects are cultural icons, and that their existence may persist through to later periods of time and later cultural stages.

Artefacts can be physical objects, for instance a hammer, or structural artefacts in the form of abstract representations of the world, such as an organization chart. Actual human organizations may conflict with their structural artefacts. Indeed, as time passes, there can be a widening divergence between the human organization and the artefact that appears to define that organization (Leifer 1989). This divergence occurs because human organizations are emergent, undergoing continual change driven by social and environmental dynamics, while structural artefacts remain constant without intervention (Truex, Baskerville, and Klein 1999).

There is an inevitable tension between the new emergent elements and older structures and processes. Similar problems can arise when new information systems are inconsistent with organizational emergence. As a consequence of emergence, new IS features may cause people to alter their behaviour in ways that avoid the new system. New practices emerge which are not represented in the structural artefact. An example in many organizations is the reality that the CEO's secretary wields real power, for example when taking responsibility for assigning tasks 'further down the line.' Every person in the organization will be aware of this line authority, yet it almost never appears in the organizational chart or position descriptions. This illustrates how human organization differs in reality from the artefacts that supposedly define it. Hence, it is sometimes cast as the 'informal' organization. Similarly, tools designated for a specific use in the 'rule book' may actually be used in quite a different way for a very different purpose.

This distinction between the structure of the human organization and the artefacts intended to define that structure are important, because change is often benchmarked against the artefacts. If the artefacts do not accurately reflect organizational reality, change management may be destructive. Given its importance, this distinction requires a more precise terminology.

The terms 'organization', 'structure', and 'system' will appear in this chapter as icons for fairly strict dictionary concepts. Organization is used to encompass a collective of people who are working together for some common purpose. 'Organized' is used to represent an entity consisting of interdependent or coordinated parts that are formed into a whole especially for taking united action with predetermined goals. Organization is defined recursively: a group of persons or smaller organizations that are organized for some end or work. 'Organizational structure' is used here for the predefined and persistent relationships between the people (or smaller groupings) in organizations (Baskerville 1996). 'Persistent' implies repeated instances of relationships occurring with regularity.

These careful definitions blunt the common distinction between formal and informal organizations. The formal organization may be imaginary, in the sense that it may fail to correspond to actual, persistent human relationships. In such cases, these aspects of the formal organization are of little consequence. The informal organization shares prominence with the non-imaginary elements of the formal organizations. However, whilst a structural artefact, such as an organization chart, articulates the formal organization, the informal organization has no such articulation and hence can be understood and described very differently by different people in the organization. For example, the structure of the informal organization can be explicated by reference to the organizational culture rather than any structural artefact.

Information systems and organizational destruction

Concepts regarding the structure, construction, and destruction of organizational and information systems have all evolved from the Latin term *struere*, meaning 'to pile up, or arrange'. 'Structure' refers to the way things are put together; 'construction' regards the process of putting things together; and 'destruction' (or sometimes 'deconstruction') concerns the process of taking things apart. In the field of IS, structure and construction have been studied. Rarely have the generally negatively perceived aspects of systems and organizational destruction been considered, although this is often an equally important element of organizational change.

In this chapter, we will distinguish between 'inwardly-destructive' systems that come apart internally and 'outwardly destructive' ones that take apart their environment. Inwardly-destructive systems are unstable and unable to persist. Outwardly-destructive systems are a hazard to the persistence of the host organization itself. For example, consider a new IS that is introduced into an organization. If this system is inwardly destructive, it will begin to break down or fail as it deconstructs itself structurally. However, the organizational structures may survive relatively untouched by the failure. An inwardly-destructive IS brings about its own degraded performance or system failure. For example, a system that incurs data capture errors because of poor human interface design will eventually become error-ridden and irrelevant. On the other hand, if that system were outwardly destructive, the organizational structures will begin to break down, or fail as the IS forces the organizational components apart. Nevertheless, the IS structures would remain relatively intact. An outwardly-destructive IS brings about the degradation or failure of the organization. For example, a system that unwaveringly enforces a hopelessly inefficient organizational workflow in the face of tight competition will eventually force the organization into bankruptcy.

We can also distinguish a totally-destructive system as one that forces apart both its own structures and those of its environment. A totally-destructive IS brings about its own degradation or failure, precipitating an equally important organizational degradation or failure. A combination of inwardly- and outwardly-destructive systems would illustrate such a system. We can also distinguish externally-destroyed systems that are destroyed by some outside agency. For example, systems can be destroyed by a virus or by importing information which has been designed by its originator to deceive its recipient. These systems appear to have some of the characteristics of all the destructive systems discussed above, and to an outside observer may be indistinguishable from them. However, this chapter does not take further an analysis of externally-destroyed systems further.

The distinctions outlined above will operate differently in the case of embedded subsystems, i.e. when viewed at different levels of theoretical systems abstraction. For example, an outwardly-destructive IS can force the destruction of the organization into which it has been introduced. From the perspective of the organization, however, the introduction of the destructive IS simply converted the organizational system into an inwardly-destructive system.

Understanding IS foundations: systems structure, organizational structure, and failure

A considerable amount of research has been concerned with the relationship between systems structure and organizational structure. From the earliest foundations of management as a discipline, organizational structure has been richly studied, with a focus on the question of what constitutes a proper organizational structure. These studies reach back to Fayol (1984 [1949]), Taylor (1911), and others (e.g. Pugh, Hickson, and Hinings 1985). Organizational structure is considered to be one of the four key concepts underlying the study of organizations, the other three being organizational processes, organizational behaviour, and organizational design (Gibson, Ivancevich, and Donnelly 1976).

Our perspective for approaching the destructive nature of IS is shaped by the large body of literature on organizational and systems structure. To a certain degree, any study of this destructive quality represents a 'reversal of field' of the existing structural concepts, a notion arising in the post-modern deconstructionist literature (Cooper 1989). This reversal means that, in order for one opposing concept to be considered, a range of opposing concepts must be collectively erected; one system of ideas must be opposed by a mirror system of opposing ideas. The field of concepts must be reversed in order to develop an adequate antithesis (dialectic).[1] Consequently, we will very briefly survey the field of organizational and systems structural thinking, considering the kinds of insights into destructive systems that inhabit each of the predominant theories.

Organizational structure

Classic management theory holds a deterministic view of organizational structure: management determines the structure, with organizational processes effectively and efficiently bound together through accompanying procedure manuals and rulebooks. If management declares the structure to be a hierarchy, then the organization will behave like a hierarchy. If management determines that the role of an employee is to act like an automaton on an assembly line, then work will be organized on that basis. Early work in organizational behaviour also adopted these assumptions that structure and process are interdependent. In light of these classical theories, destructive information systems are most likely to be seen as outwardly destructive. If the system design is incompatible with organizational structure, the organizational actors will do the wrong sorts of things, and market forces will compel the organization to redesign or fail.

The sociology of science has reshaped our thinking about organizational structures by the concept of social constructivism (Berger and Luckmann 1966), which recognizes that organizational structure is negotiable in various ways. Two theories illustrate this reshaping.

First, Argyris and Schön's (1978) theory of organizational learning holds that organizations learn in a nested series of two learning 'loops'. In the inner loop, organizational actors learn to enact activities that lead to the achievement of the organizational goals. In this inner loop, the assumptions underlying the goals and activity are not questioned. To a certain extent, the actors adjust to the 'espoused-theory' often inculcated by the organization. This adjustment is frequently achieved by adopting different theories which reflect the actual actions of the actors (the 'theories-in-use'), until the outcomes match the theories that are reflected in what the actors say they do (their inculcated 'espoused theory'). While the outcomes may match, the two theories rarely will. In the outer learning loop, the organization learns through challenges to the assumptions underlying organizational goals and activities. Actors question their espoused-theory and adjust their theories-in-use until new goals and activities become embodied as espoused theories. The outer loop is related to official organizational structures and organizational memories, while the inner loop represents informal organizational structures and does not penetrate the organizational memory. From this theoretic base, destructive IS will also be seen as outwardly destructive. An incompatible system design represents a departure of espoused-theory from theory-in-use, i.e. a shift of organizational goals and activities away from the espoused assumption base on which previous organizational goals and activities rested. Conflicts then arise between desired actor behaviour (theory-in-use) and the IS constraints (espoused-theory), because the in-use and espoused assumption spaces become incompatible. Unless the organization recognizes that the system design is incompatible and must be adapted, outer-loop organizational learning becomes blocked. An inability to adapt the system design, and thus enable outer-loop learning, will cause the organization to degrade, and possibly fail.

Secondly, Giddens' (1979; 1984) structuration theory holds that structures emerge through the actions of social organizations, which are then confined and bound by the structures. The behaviour of organizations is determined by their previous behaviour in shaping their own structures. These structures then confine future behaviour and determine future structures. In a manner similar to organizational learning, structuration theory favours the identification of outwardly-destructive information systems. A mismatch between a new systems design and the existing organizational structure can result in a dysfunctional restructuring, with the outcome being organizational degradation. These structures will persist during restructuring, leading to further—although less severe—degradation. For instance, Enterprise Resource Planning (ERP) software installations often shift information flows, producing different patterns of reporting than may have been historically present in the organization. In some cases, an old IS might have produced a production report intended for a business analyst, while the new ERP system is designed to produce a different format of production report intended for a sales executive. The existing organizational structure will have to adjust to the change in information flow, or else the old organizational structures may persist dysfunctionally. Either way, the transition period, at least, will produce extra work—and therefore degrade organizational performance.

Systems structure

Organizations can be viewed variously as social systems, organizational systems, or as the setting for IS. Consequently, systems theories have also become closely linked to research into organizational structures. Work in general systems theory (e.g. Boulding 1956) and autopoietic systems theory (e.g. Luhmann 1986) has focused on the systemic structures of social organizations. Autopoietic systems are self-remaking, containing all of the necessary elements for recreating themselves as a revised system. This is a higher-level view that could encompass organizational learning and structuration theories, since all of these theories are founded on a certain degree of self-referential behaviour in systems. An autopoietic view allows for a destructive IS that could become self-justifying, despite its unsuitability. As a consequence, the organization (as a host) would degrade or fail. Such a view sees these systems as outwardly destructive. For example, a mainframe-based legacy system that is expensive to maintain, absorbing the organizational IS budget, is sometimes justified on the basis that it is too expensive to be replaced. If the system retards organizational adaptation and its competitive posture, its preservation may indeed cripple its host organization.

Certain linguistic theories have provided useful analogies for understanding organizational IS, for example using notions of grammars based on the work of Noam Chomsky (Wand and Weber 1995; Truex, Baskerville, and Klein 1999) or John Searle's Speech Acts (Goldkuhl and Lyytinen 1982). In contrast to previous theories, these approaches tend to see destructive IS as inwardly destructive. Wrong system designs would violate speech-act protocols, or be 'senseless' in terms of Chomsky-like deep structures. The systems cannot function within the organization because the system processes cannot be understood in terms of the organizational processes. Incompatible system designs would cause IS degradation and failure, and the organization would reject the system. For example, an exchange of information (a speech act) often involves something equivalent to grammar. This type of grammar could require that an invoice always includes an order number. A new business practice, such as just-in-time supply chains that deliver products without orders, could present such a speech act that is not grammatically correct, for example an invoice without a known order number. If the system is incapable of processing the information, either the system or the practice must go. Since competition and organizational survival rely on the new practice, the system would have to be rejected by the organization.

IS theory has drawn particularly on several additional theoretical streams. These include: structured systems design (e.g. Yourdon and Constantine 1979); socio-technical design (e.g. Mumford 2003); and soft systems design (e.g. Checkland 1981). Critical social theory has also been used as a basis for understanding IS in the social-organizational context (e.g. Lyytinen and Klein 1985). In general, these theories promote the view that problematic information systems are inwardly destructive.

In structured systems design, heavy analysis is advocated to prevent incompatible systems design. Poor analysis could result in failure of the new system, not the organization. Similarly, socio-technical systems design places its emphasis on involving users to prevent socially-unacceptable systems from reaching implementation. At the same time, socio-technical theory avers that it is possible to design a system that engages in a trade-off between its social and technical elements. One objective when management adopts the socio-technical approach is to eliminate user resistance that would lead to system rejection and system failure. Although organizational destruction is not the major issue, socio-technical theorists and practitioners see it as a threat their methodologies can prevent. The soft systems methodology recognizes the threat of designers creating destructive systems and attempts to overcome it by an iterative process of re-evaluating the impact of the design on the organizational setting. Designers and users interact in order to make incompatible systems evolve gradually. The soft system process is designed to prevent the production of either inwardly- or outwardly-destructive systems being implemented.

Critical social theory, as its name implies, places systems design in its context as a social act whose meaning is instrumental, and often politicized and communicative. This theory exposes the social consequences of traditional systems design; in the long run, this can lead to the destruction of the system and, indeed, the organization. Any destructive characteristics of the design are the consequence of the asymmetry in the power of the organizational actors. Critical theory has been powerful in showing the shortcomings of current practices, but has been less successful in offering constructive proposals for overcoming the problems inherent where power is not equably distributed.

Actor network theory (ANT) places emphasis on the relationships that influence designers (Callon 1991). While structure could seem to arise from a rational act by a designer, the designer and the design act are enmeshed in the web of an intricate network of influence. The design and its subsequent organizational structures result from the actors' negotiation with, and through, their surrounding influences.

Indeed, the organizational structure is a social order comprised of a diverse network of artefacts, contracts, arrangements, skills, practices, etc. From this viewpoint, systems designs are seen as inscribing new material and visions on this order, more than defining the order (Monteiro and Hanseth 1995). The response of the actor network to such new inscriptions can be destructive when the new material and visions are incompatible with the goals of the actor network. The incompatible visions will decouple the influence of the system from the network, or alternatively the network itself will begin to disengage and dissipate.

The management literature also forms the basis for certain design approaches where both the information and organizational systems are co-dependent on redesign. This includes the strategic IS approaches which depend on concepts like strategic fit (e.g. Venkatraman 1993), and embrace the re-engineering perspective that advocates radical restructuring of organizational processes (e.g. Hammer and Champy 1993). In these approaches, the major focus is on organizational design, and IT is merely an enabling factor permitting certain types of design, for instance 'flattened', 'pipeline', and 'customer-oriented' organizations. A destructive IS is characteristically seen as outwardly destructive, since its shortcomings would eliminate a key enabling feature of the redesigned organization. The organization would come apart, degrade, and fail even though the IS may continue to function, albeit somehow delivering the 'wrong' information.

Information systems failure

Some discussion about the destructive nature of information systems arises in literature about IS failures. For the most part, this literature considers the failure of IS in a rather 'self-contained', almost circular sense. That is, a system fails because it is defective. Systems can be defective because requirements are inadequately articulated by the user or misinterpreted by the designer (Land 1986, 1989; Land and Somogyi 1986). Often, in a setting where the system is technically adequate, its failure is attributed to its inability to match the idealized organizational structure. For example, systems failure is characterized as dependent on technical or social factors (Bostrom and Heinen 1977*a*, *b*). When technological success is achieved but the system still fails, the problem is attributed to a failure of the IT to achieve necessary acceptance among its user community (Lyytinen and Hirschheim 1987). An IS can also be both a technical and a social success, and yet still fail because it depends on an organizational form that is unsuitable for its mission (Miles and Snow 1984). All of the destructive concepts are seen as inwardly directed.

Other research into IT failure has explored the tendency for system champions to ignore early warnings of a forthcoming IS failure (Ginzberg 1981). When such early warnings are noted, the solutions may involve the application of simplistic technological 'fixes' in attempts to overcome complex settings that cause IS failure. Such technical fixes typically do not address the essential causes of the failures (Brooks 1987). In each case, the destructive aspect of an IS design is framed by an assumption of inwardly-destructive characteristics (Symons 1992).

Much of our thinking about the success and failure of systems is shaped by underlying assumptions regarding the potential compatibility of the system and organizational structures. However, both these structures can be fluid and interdependent. When incompatible system designs attempt to position system structures that do not match the organization structures, something must unravel: either the system structures (inward destruction) or the organization structures (outward destruction).

Structural artefacts and inwardly-destructive organizations

Organizational systems, IS, and systems failure literatures do not adequately explore the close relationship between organizational structure and potentially destructive systems. In particular, we know very little about how an IS becomes outwardly destructive, or how that correlates with inwardly-destructive organizational systems. This is because research into organizational structures and systems are not generally concerned with IS *per se*, but with organizational behaviour. Research into IS design and systems failure also tends to focus on IS performance and use. Consequently, the usual IS perspective is that systems may become inwardly destructive. The fact that an IS might have been outwardly destructive does not assume a major theoretical role, since it is assumed that healthy organizational structures should be sustained through such technological problems.

As a result of these past biases in IS research, we know very little about settings in which the IS is constructed to match an organizational structure that proves to be less-than-ideal, inadequate—or even defective. This is especially possible in cases where the real organizational structure is not properly reflected in the human artefacts that are created to reflect organizational structure.

IT and other structural artefacts

There are varied viewpoints about the relationship between IT and organizational structure. For instance, Markus and Robey (1988) distinguish three views:

(*a*) restructuring IT will cause a subsequent restructuring of the organization;
(*b*) restructuring IT is a consequence of the restructuring of the organization; and
(*c*) an emergent view in which IT and organizational restructuring are interdependent processes.

All these views assume that the IT is somehow an elemental component of the organizational structure. However, organizational structure is concerned with persistent relationships between people. While the IS may enable or reflect this relationship, it does not embody the relationship itself. Therefore, it is important to take care in distinguishing organizational structure (human relationships) from the IT artefacts (mechanical products) that are intended to sustain or articulate those relationships.

Box 13.1. Examples of formal and informal artefacts that reflect their organizational structures

- *Organization charts*: graphical depiction of organizational roles, their incumbents, and their different kinds of relationships.
- *Personnel policies*: reporting lines (who is whose boss), job descriptions, rewards structures. and payroll policies.
- *Union agreements*: the relationships between union members and others in the organization.
- *Standard Operating Procedures* (SOPs): detailed functional policies that define important coordinated actions in the organization.
- *Resource access policies*: travel justification, modes of travel (private jet or tourist-class), company cars, cellular phones, etc.
- *Workspace division*: the size of an individual or group work-area, collocation of organizational members (whose office is next to whose), etc.
- *Workspace attributes and resources*: decoration, furniture (quality, quantity, and size), privacy (corner-office or open-plan cubical), dining facilities, etc.
- *Information systems*: the systems that determine access to computer accounts, the Internet, automatic channelling of inputs and outputs, screen and paper form designs, and other information resources.
- *Methods*: the development or purchasing of information resources, which determine who sets the goals, who participates in the design or evaluation, what issues are considered, and how the system elements are represented.
- *Health and welfare regulations:* the conditions for work, for example the maximum hours a truck driver can drive without a rest period, which often reflect laws and agreements with trade unions.
- *Cost and Financial Accounting schemas*: definitions of organizational structures for both the internal and external world.

Together with IT, there are many different artefacts that people create to reflect their organizational structures. Box 13.1 lists some examples of the artefacts that reflect, or encode, the organizational structure; these should not be construed to *be* the organizational structure. Such artefacts represent the different rules and protocols by which the members of the organization choose to behave. The accuracy of these representations is, of course, variable. Their rules and protocols have been likened to grammars in languages (Wand and Weber 1995). The grammar metaphor is useful as an analogy because linguistic grammars vary among communities, undergo change, naturally conflict among versions, and may be accurately or inaccurately represented by grammatical texts. The 'rulebook', which is similar to the Standard Operating Procedures in Box 13.1, may be slavishly conformed-to in one organizational unit, but regarded as a joke in another. It is instructive that the expression 'working to rule' is often regarded as the first phase in industrial action against employers.

Conflicting versions of organizational structure

Different structural artefacts overlap, and will sometimes encode the organization's structure in conflicting ways. For example, the payroll policies, personnel policies, and organization chart (see Box 13.1) could encode an individual as being without influence in organizational strategy; yet the workspace division, workspace attributes, and information systems may encode a structure in which that same individual bears essential responsibility for shaping organizational strategy. What, then, is the 'real' organizational structure? For the purposes of this chapter, it is not defined by any particular set of organizational artefacts, but by the roles individuals adopt, how these roles are perceived by the individual's peers, and the relationships between the individuals. When the artefacts suggest conflicting structures, the reflection of the real structure is blurred and the determination of its shape is made more difficult. This definition underlines how important it is to understand the tensions between formal and informal organization, and the potentially dysfunctional impact of this tension on the relationship between the organization and its IT-based systems.

The important distinction between the structural artefacts and the organization is not well pronounced in social constructionist thought. The organizational artefacts can tell different stories about the organizational structures. Conflicts between these stories seem to arise most often when the actual organizational structure differs from an 'official' version of the organization. In a sense, the official version defines the normative. But these official organizational artefacts reflect the version of the story of the organizational structure as told by a particularly privileged class of organizational members: relatively senior management, usually. In a similar manner to the priest-class in a theocracy, this class of individuals is widely accepted as owning the authority to determine organizational structure. Accordingly, this class controls a large set of overt organizational artefacts. It also often believes, and wants others to believe, that the official version does describe the organization—as in the case of the global oil firm Enron's corporate deception at the turn of the century. Hence, the systems designer who relies on information received from this class may well be on the way to designing both inwardly- and outwardly-destructive systems. The public face of the organization is frequently and intentionally reflected by its organizational artefacts.

Again, the structural artefacts were constructed in the context of a past organizational reality. The real organization has to respond to a wide range of pressures in order to survive. As a result, a host of subtle changes in roles, responsibilities, and processes are constantly taking place. There is no similar pressure to maintain the structural artefacts in congruence with the changing reality. However, when the structural story suggested by these official artefacts conflicts with the structural story suggested by many other organizational artefacts, then the reality of organizational structure is indistinct. Informal artefacts that may be beyond the influence of management also include: those that are too menial for official control, such as workspace collocation; those that can be replaced by alternative artefacts, for instance a lower-authority policy taking effect even though it contradicts and countermands a higher-authority policy; or those in which important functional artefacts ignore the specification artefacts, such as when the real information flows violate operating policies.

Conflicting realities of organizational structure

An important aspect of the differences between structural artefacts that purport to represent the same organizational structure is the broad acceptance of 'official' artefacts even when these conflict blatantly with the majority of other representations. An example of this acceptance is an approval form that specifies an authorized signature block, for example specifying a 'Director's Signature', although the actual affixed signature in the block is nearly always that of a subordinate of the director. This aspect illustrates the important, almost priestly, power of the privileged organizational classes to interpret and pronounce organizational reality. As a result, management itself, many organizational scientists, and those chosen to do the systems design will not look beyond these selected structural artefacts. They accept as reality those organizational structures that are described by an officially-sanctioned subset of organizational artefacts.

The process by which an imaginary belief becomes accepted as being real is described by Berger and Luckmann's (1966) well-known concept of the 'social construction of reality', which we apply here. This suggests that the structural reality of an organization arises from routine relationships that become habitualized and explained in a symbolic universe. This symbolic universe in organizations can be strongly influenced by a powerful set of managers (the priestly class), or by reference to long-established precedent. It can justify roles and relationships symbolically, such as by reference to an ancient organization chart, even when these roles and relationships become unnecessary or harmful, for example when government clerks follow patently absurd bureaucratic rules because they justify some other part of the bureaucracy. Sometimes, the structural artefact is used as a cover to hide reality from perceived threats, including competitors, regulators, trade unions—or even systems analysts. However, the cover-up may equally deceive benign inquisitors, such as academic researchers.

Inwardly-destructive organizations

From the viewpoint of structural artefacts, we see that it can be difficult to distinguish organizational structure from the conflicting renditions offered by various artefacts. If the selected artefacts become the benchmarks against which change is measured, it can be equally difficult to discover that a process of inward destruction is taking place, since the reality of organizational structure (and therefore its destruction) may not actually be reflected by the structural artefacts. Hence, a new outwardly-destructive IS can cause the organizational structures to crumble, and this process may be undetectable when registered against the official artefacts. It would simply seem that things have stopped working in reality, while at the same time things appear to be working normally when measured by the usual organizational artefacts. This can be illustrated by recalling the 'signature block' example, in which a director in a large organizational division delegates signatory responsibility to subordinates. If, based on the official organizational artefacts, the systems designers actually require the director to review and electronically sign the form, the director could become overloaded and the organization begin to suffer because its directors are becoming distracted by this kind of trivial duty.

Those facing such situations have little theory available. The existing power structure tends to draw the analysts into the problem, rather than exposing the problem to them. Since the artefacts frame the analysts' perceptions of anomalies, the reality of organizational structure may be increasingly hidden because the artefacts are assumed to be the 'normal' context, not the cause of the anomaly. If the analysts were able to recognize potentially inaccurate structural artefacts, they would be able to question their validity, and hence the standards by which performance has customarily been judged. While the new IS is probably not the only destructive trigger, it could have greatly facilitated an inwardly-destructive organization, using automation to speed up the process by which the organization self-destructs. An example would be an overly centralized organization that collapses when a new IS mandates 'official' voluminous decision-making by overloading central managers.

Perhaps a worse problem is an IS that facilitates the slow self-destruction of an organization. Such an IS might well have been hailed as an enormous success at its outset. Yet, the organization may slowly degrade and fail through a process undetectably worsened by the changes in its IS. For example, if the overloading of the central managers was slight, the decline in performance could be transparent at the outset, but grow slowly and insidiously with time. In terms of Leifer's (1989) dissipative structures, the decline represents growing entropy, as the central managers increasingly struggle to match the information system to the needs of organizational reality.

There is nothing particularly special about the information system as an artefact that can induce an inwardly-destructive organization. Such an inducement could be made by changing any of the structural artefacts suggested by Box 13.1 above. For example, an overall organizational decline could be triggered by renegotiating a union agreement that centralizes minor decisions, with the effect of overloading middle management while making union members feel powerless. However, changes involving a computer-based IS may provide one of the most severely constraining changes in an organization's structural artefacts.

Destructive information systems

Destructive information systems are those that lead organizations or systems to come apart, with the structures disassembled. A recent example of a destructive system is the one devised by economists to auction radio-spectrum frequencies for future communication systems (Wigfield 2001). Many countries have adopted such auction systems and congratulated themselves on the lucrative revenue collections. However, the unintended consequence was that the bidders paid much more than they could afford in their efforts to stave off competition. This led to a number of bankruptcies and near-bankruptcies, destroying a large degree of the competitiveness the system had intended to promote.

Earlier, we distinguished between systems that are inwardly destructive (an unstable system unable to persist), outwardly destructive (hazardous to the persistence of its host organization), and totally destructive (forcing apart both its own structures and those of its environment). We also allowed that an outwardly-destructive IS and its inwardly-destructive host organization system are corresponding; to an outside observer, they are perhaps even indistinguishable events. We can further differentiate two ways in which these destructive processes are enabled: as a 'constructive inhibitor' or 'destructive facilitator'.

A constructive inhibitor is a characteristic of the information system that prevents some necessary constructive function. Constructive inhibitors in an IS may be inward or outward. For example, an inwardly-constructive inhibitor may interfere with system maintenance. A typical example is poor documentation that could cause the IS to degrade and finally fall apart. Another example would be intentionally designing limits into a system, like a restricted number of customers, in order to limit its lifespan and promote sales of future versions. An outwardly-constructive inhibitor would demonstrate how the IS might degrade, or even prevent some necessary organizational process to be completed. A good example is an organizational system that has no mechanism permitting a lower-level employee to report a structural system dysfunction to those in authority. Disabling, or simply not providing, a mechanism for the exchange of information between two or more individuals could isolate knowledge and allow critical organizational learning to be inhibited, causing the organization to begin falling apart.

A destructive facilitator is a characteristic of the IS that enables some destructive function. While constructive inhibitors prevent systems and organizations from adapting to change, destructive facilitators actively promote degradation, even during periods of stability. Destructive facilitators may also be inward or outward. An inwardly-destructive facilitator may permit uncontrolled software modifications. For instance, a deficient software version-control function could enable conflicting software maintenance changes that cause the system to begin disintegrating. An example of an outwardly-destructive facilitator could demonstrate how the IS permits some dangerous organizational process to take place. At the time these facilitators are designed and implemented, they are assumed to add value to the process. Many performance measurement systems have the effect of distorting overall performance, in an attempt to reach performance targets that are regarded as adding value and are defined by the measurement system. Allowing workers to have control over confidential information about their own performance could enable poor performance to be disguised, thus destroying training and performance incentives across the workforce, further causing the organization to begin to crumble.

No doubt, every newly developed or redeveloped IS has certain characteristics that could be considered inward or outward, or as constructive inhibitors or destructive facilitators. The challenge for the systems designer is to identify such features, and to either eliminate them from the design or, at least, to prevent them from being implemented. In addition, we can distinguish between active and passive destructive characteristics.

Passive destructive characteristics are most likely to arise because the system requirements are derived from structural artefacts that do not accurately reflect organizational reality. Active destructive characteristics are introduced into the IS either for the purpose of forcing system or organizational destruction. Characteristics that are actively introduced usually involve a motive. For example, a disgruntled or dispirited worker may work-to-rule and place destructive elements into a system, even though these are clearly harmful. This action represents a form of sabotage of the system design. More often, it is put in place as a way of drawing attention to a grievance, and the full destructive force of the feature is not recognized. Likewise, information warfare might involve an attack on an organization by promoting destructive characteristics in its newly developed IS.

The distinction between active and passive is helpful to analysts and designers in identifying and creating inventories of possible constructive and destructive elements in planned systems. For example: 'constructive active' could designate a designed capability that inhibits certain required actions and/or highlights problems with a reporting system that overloads the decision maker; and 'constructive passive' would highlight problems such as the poor maintenance example earlier. 'Destructive active' implies a specific design or implementation feature that has intentional destructive effects, such as a customs process tracking system that is intended to destroy a historically bribery-based organization. 'Destructive passive' implies that something necessary for the organization to regenerate is lacking, such as in the software modification example earlier. Table 13.1 summarizes such categorizations of destructive elements in information systems.

Table 13.1. *Destructive elements in information systems*

	Inwardly-constructive inhibitor	Outwardly-constructive inhibitor	Inwardly-destructive facilitator	Outwardly-destructive facilitator
Passive	Poor documentation	Not providing a mechanism for critical communication	Deficient version-control function	Self-controlled performance data
Active	Designing unnecessary limits into the system	Disabling a mechanism for critical communication	Work-to-rule demonstration of system limits	Setting targets that encourage game playing to achieve targets

Case study of a socially-destructive system

Much of the foregoing theory regarding destructive information systems emerged from the analysis of a holistic case study, based on Yin's (1984) formulation of case study research methods. This involved an approach that is typical for hypothesis-building and the theoretical explanations that emerge *ex post facto*: developing a case description, collecting data using participant observation, and applying an interpretive explanation-building mode of analysis.

The case we refer to involved an innovative decision-support IS that was initially a technical success, but gradually degraded over its three-year operational lifespan. A surface objective analysis, such as those that might be conducted periodically to evaluate system performance, would hold that the system slowly fell into disuse because its information was not kept current, and hence was perceived by decision makers as unreliable. This was the public reason for the removal of the system. A deeper, more ethnographic, and interpretive analysis shows that the system was outwardly destructive. The unspoken (and poorly understood) reason for the removal of the system was the shattering effect that it wreaked on the informal structure of the organization.

Organizational setting

The organization studied was a large naval command headquarters. The headquarters staff supported a senior admiral, who exercised command over national and international naval forces in a wide geographic area. An admiral who acted as both the deputy commander and the chief-of-staff headed the headquarters organization. The staff were organized as a tight military hierarchy, consisting in the main of six functional departments, such as logistics, intelligence, and operations. Each department was headed by a senior navy captain who acted as the deputy chief-of-staff for their departmental areas. These departments were also tight military hierarchies of uniformed and civilian naval personnel.

The mission of the peacetime headquarters was shaped by the need for the 'readiness' of several large naval shore installations and several naval fleets, which required maintaining prepared plans and capabilities for coping with crisis situations. A decision room was created that permitted the commander, the chief-of-staff, and the deputy chiefs to meet in an environment that included television and rear-screen projections of a variety of information. These senior officers met in this room daily for a briefing, and discussed the decisions to be taken. But the chief-of-staff had identified an important problem: the once-a-day meetings would be inadequate during crisis situations, but more frequent meetings of the entire senior staff would result in an impractical interference with other intense staff activities in such situations.

The new information system

The chief-of-staff championed a new television and computer-based IS that replicated the decision-room information. Information about readiness was used as input to software programs that generated graphics similar to those on the decision-room screens. An audio-video briefing, rather like a television news program, was prepared daily. This information was available on menu-driven screens on the desks of the commander, the chief-of-staff, and most of the deputies.

The system enjoyed immediate success. The commander and the chief-of-staff decided that it was no longer necessary for the senior staff to gather in the decision room daily during routine operations. Instead, the staff relied for their information on the computer screens and recorded briefings, and the commander consulted directly with the appropriate deputy when decisions in that deputy's area were needed. For example, if the commander was faced with a tough operational decision, the deputy in charge of the operations department might be asked to step into the commander's office for a quick conference on the matter.

Over time, the deputies began having problems with this system. Occasionally, they found themselves in the commander's office, hovering over the information screens—and being surprised by problems that were just emerging. Because the screens were being updated several times daily, the commander had access to the information as quickly as the deputies. If the deputy had been tied up in meetings recently, it was entirely possible that the commander would know more about an emerging readiness problem than the staff. This situation was not merely embarrassing, but also threatening, since the deputies were called upon to propose alternatives that had not been properly thought-out.

Gradually, key deputies ordered their departmental staff to withhold input data from the decision system until it had their personal approval. This mechanism ensured that the commander could not get the information before them, and that there was adequate time for them to consider their analysis of readiness problems and formulate recommendations. This mechanism also ensured that the system information was not as current as it needed to be for dealing with real emergencies, and hence was of less use to the commander. It also meant that the system information was of little use to the deputies, since they had reviewed it before it was used to generate the graphics. Although the system was a technical success that achieved its design goals, it therefore fell into disuse and was eventually removed.

Analysis

The structural artefacts in the organization studied depicted a tight hierarchy, but the human organization in senior military commands is much more social. In order to be effective, the deputies had to build a sense of trust and confidence with their commander. This involved professional and social activities, availability, and reliability. For example, the deputies ostensibly worked for the chief-of-staff, and might never deal directly with the commander. However, if the commander and the chief-of-staff both trusted the deputy, all formalities were dropped, and the deputy would work directly with the commander. This direct work also happened on occasions with members of the deputy's staff. As a consequence, components of the organizational structure were created through personal trust.

These components were not reflected in any of the organization's structural artefacts. It should be noted that the new IS reflected the structural artefacts, but did not reflect all of the components of working in the organizational structure. Although this inaccurate reflection probably made the system less effective, it was not a key destructive element in the system design. There were at least two outwardly-destructive elements in the new IS: the elimination of trust-building activities of the deputies and the creation of new opportunities for trust-destroying activities by these deputies.

The reduction in decision-room meetings that were a critical element in trust building exemplified the first outwardly-destructive element of the system. As the deputies had known what information had been provided by their departments to the decision-room staff, they could be prepared to answer questions authoritatively on likely topics of interest to the commander. Indeed, the staff knew it was easy to 'engineer' the commander's questions in the briefings by leaving out certain minor data in which the commander frequently took interest. The staff would have this minor data ready-to-hand, so they could answer instantly, authoritatively, and in complete detail.

The way these trust-building activities were curtailed by the reduction of daily decision-room meetings indicates that the system was a passive, outwardly-constructive inhibitor (see top-left entry in Table 13.2). The human organization could not remake itself in the sense of Luhmann's (1986) autopoietic social systems. When the IS was introduced, the organizational structure was held together by trust-building activities. However, the system also inhibited future trust-building activities. For instance, as senior officers shifted posts to other duties in other commands, the trust-building activities of new officers were inhibited. Likewise, if a deputy's trust status was diminished in some way, the curtailed decision-room meetings inhibited the re-establishment of such trust. This is how the new system gradually promoted the destruction of the trust-based components of the organizational structure. Its effect was undetectable when measured against the structural artefacts that did not reflect the presence of the trust-based component.

The creation of new opportunities for trust-destroying activities by the deputies is shown in relation to the commander's equal temporal access to each deputy's information. This was promoted by the formal organizational structure because the commander was responsible for the decisions, and ostensibly needed the most current information. However, this situation placed the deputies in a position of potential ignorance: the commander may have more time to think over a readiness problems and alternatives put forward by unprepared deputies might have already been considered and rejected by the commander. In this setting, the deputies appeared to exhibit poor reasoning and lack of attention to their responsibilities. This not only inhibited the building of trust by the deputies, but also eroded existing trust. In this way, the new system was a passive outwardly-destructive facilitator (see top right entry in Table 13.2).

The trust-based component of organizational structure began a process of gradual destruction. Again, since this structure was not reflected in the structural artefacts,

Table 13.2. *Destructive analysis of the case studied*

	Inwardly-constructive inhibitor	Outwardly-constructive inhibitor	Inwardly-destructive facilitator	Outwardly-destructive facilitator
Passive		Curtailed daily decision-room meetings		Timely access to readiness information
Active			Deputies retain control over input information	

it was impossible to detect when measured in the context of the 'official' structural artefacts. The deputies attacked and eliminated this facilitator by delaying the input data. The fact that the systems design process allowed, or failed to detect, these delays represents an active inwardly-destructive facilitator (bottom entry in Table 13.2). The existing trust-based structures were preserved in this way, although—as mentioned above—the autopoiesis of the organization was still inhibited until the system became entirely disused. Having been abandoned, it was later quietly removed and stored in a distant warehouse.

Conclusions

The new IS in the case studied was a destructive element in the organization. It enabled a rather subtle process by which elements of the organizational structure gradually and unintentionally began to come apart. Had the system survived, it might have destroyed the entire organizational structure, even those reflected in the structural artefacts. Such a total destruction may then have been justified as a 'restructuring' because the old systems did not work, and the root cause for the destruction might have never been detected. By undermining the new IS, the deputies preserved the organization. This act of preservation (or the halted act of destruction) may have been good or bad in terms of human society, but the ethics are beyond the scope of this chapter.

Importantly, the structural artefacts of the organization carried no information about the structural elements that underwent destruction. As a consequence, the underlying reason for the organization's eventual rejection of the system was virtually undetectable by an objective analysis of the 'official' organizational structure. This case also illustrates the problematic nature of designing systems that adhere to an organizational structure as reflected by the structural artefacts. New information systems often introduce strong new structural artefacts that could reflect other structural artefacts, yet conflict with elements of organizational structure in very essential ways. An implication of this is that organizational structure, independent of structural artefacts, must inform IS design.

Analysts and designers therefore need to be sensitive to the informal organization and informal artefacts that can expose destructive and constructive system features. Complete reliance on structural artefacts may lead to systems that work perfectly—but only in theory. Destruction of organizational elements is a result of the interaction between IS and the social context of the organization. If structural artefacts are used at face-value for modelling and implementing new information systems, highly destructive systems may result.

The subtle nature of this destructiveness can extend even to pre-existing systems. A change to one system may extend collateral change to another one. The two systems may be independent according to the structural artefacts, but an unrecognized element of interdependence could lead to inwardly-destructive changes to the pre-existing system. For example, consider a production control system developed for a subsidiary business unit (SBU) founded on an existing centralized distribution system. When central management implements a distributed warehouse scheme without the knowledge of SBU production management, an expensive SBU production control system could come apart.

These implications further draw out the practical need for an ethnographic analysis of organizational structure as one preliminary component of new IS design (Baskerville and Stage 2001). Such an ethnographic analysis would illuminate the social-cultural processes, both formal and informal, in the system design. The analysis may focus on conflicts between different structural artefacts, and conflicts between these artefacts and the underlying organizational structure.

Our analysis of the case focuses on the destructive elements of the system and how these affected both the smooth workings of the system and the organization. There are certainly other layers of meaning in the case. For example, actor network theory helps to surface the effects of the interdependent behaviour of the various stakeholders in the case. To some extent, the systems developers created a technological 'actant'—effectively an agent in their community—mainly to please the system's champion (the chief-of-staff). The system was also launched into the organizational network without a deep understanding of the complexity and delicate interdependence of the network. Although it functioned beautifully and effectively immediately after its launch, the system resulted in key changes in the network of actors, including information exchanges between the deputies and the commander. Even though the system was the actant of the chief, the deputies were able to disengage it by starving the system out of effective participation as an actant. Once the disruptive system had been effectively removed, the network was quickly restored to its previous, healthy state.

References

Abdel-Hamid, T., and Madnick, S. (1990). 'The Elusive Silver Lining: How We Fail to Learn from Software Development Failures'. *Sloan Management Review*. 32/1: 39–48.

Argyris, C., and Schön, D. (1978). *Organizational Learning: A Theory of Action Perspective.* Reading, MA: Addison-Wesley.

Banville, C., and Landry, M. (1989). 'Can the Field of MIS be Disciplined?'. *Communications of the ACM*, 32/1: 48–61.

Baskerville, R. (1996). 'Structural Artifacts in Method Engineering: The Security Imperative', in S. Brinkkemper, K. Lyttinen, and R. Welke (eds.), *Method Engineering.* London: Chapman and Hall, 8–28.

—— and Stage, J. (2001). 'Accommodating Emergent Work Practices: Ethnographic Choice of Method Fragments', in B. FitzGerald, N. Russo, and J. DeGross (eds.), *Realigning Research and Practice in IS Development: The Social and Organisational Perspective.* New York: Kluwer, 12–28.

Berger, P., and Luckmann, T. (1966). *The Social Construction of Reality: A Treatise in the Sociology of Knowledge.* New York: Anchor Press.

Bostrom, R., and Heinen, S. (1977*a*). 'MIS Problems and Failures: A Socio-technical Perspective, Part I: The Causes'. *MIS Quarterly*, 1/1: 17–32.

—— —— (1977*b*). 'MIS Problems and Failures: A Socio-technical Perspective, Part II: The Application of Socio-technical Theory'. *MIS Quarterly*, 1/2: 11–28.

Boulding, K. (1956). 'General Systems Theory—The Skeleton of Science'. *Management Science*, 2/3: 197–208.

Brooks, F. P. (1987). 'No Silver Bullet: Essence and Accidents of Software Engineering'. *Computer*, 20/4: 10–19.

Callon, M. (1991). 'Techno-economic Networks and Irreversibility', in J. Law (ed.), *A Sociology Of Monsters: Essays On Power, Technology And Domination.* London: Routledge, 132–61.

Checkland, P. (1981). *Systems Thinking, Systems Practice.* Chichester, UK: John Wiley.

Cooper, R. G. (1989). 'Modernism, Postmodernism and Organizational Analysis 3: The Contribution of Jacques Derrida'. *Organizational Studies*, 10/4: 479–502.

Fayol, H. (1984 [1949]). *General and Industrial Management.* Belmont, CA: Lake Books.

Gibson, J., Ivancevich, J., and Donnelly, J. (1976). *Organizations: Behavior, Structure, Processes.* Dallas, TX: Business Publishers.

Giddens, A. (1979). *Central Problems in Social Theory: Action, Structure and Contradiction in Social Analysis.* Berkeley, CA: University of California Press.

—— (1984). *The Constitution of Society: Outline of the Theory of Structure.* Berkeley, CA: University of California Press.

Ginzberg, M. J. (1981). 'Early Diagnosis of MIS Implementation Failure: Promising Results and Unanswered Questions'. *Management Science*, 27/4: 459–78.

Goldkuhl, G., and Lyytinen, K. (1982). 'A Disposition for an Information Analysis Methodology Based on Speech-act Theory'. Paper presented at the 5th Scandinavian Research Seminar on Systemeering, Stockholm.

Hammer, M., and Champy, J. (1993). *Reengineering the Corporation: A Manifesto for Business Revolution.* New York: HarperCollins.

Keen, P. (1991). 'Relevance and Rigor in Information Systems Research: Improving Quality, Confidence, Cohesion and Impact', in H.-E. Nissen, H. Klein, and R. Hirschheim (eds.), *Information Systems Research: Contemporary Approaches and Emergent Traditions.* Amsterdam: North-Holland, 27–49.

Kettinger, W., Grover, V., Guha, S., and Segars, A. (1994). 'Strategic Information Systems Revisited: A Study in Sustainability and Performance'. *MIS Quarterly*, 18/1: 31–58.

Land, F. F. (1986). 'A Critical Assessment of Software Engineering—its Claims, Achievements and Failures', in D. Ince (ed.), *Software Engineering: The Decade of Change*. London: Peter Peregrinus, 222–29.

—— (1989). 'From Software Engineering to Information Systems Engineering', in K. Knight (ed.), *Participation in Systems Design*. London: Kogan Page.

—— and Somogyi, E. K. (1986). 'Software Engineering: The Relationship Between a Formal System and its Environment'. *Journal of Information Technology*, 1/1.

Leifer, R. (1989). 'Understanding Organizational Transformation Using a Dissipative Structure Model'. *Human Relations*, 42/10: 899–916.

Lewin, A., and Volberda, H. (1999). 'Prolegomena on Coevolution: A Framework for Research on Strategy and New Organizational Forms'. *Organization Science*, 10/5: 519–34.

—— Long, C., and Carroll, T. (1999). 'The Coevolution of New Organizational Forms'. *Organization Science*, 10/5: 535–50.

Luhmann, N. (1986). 'The Autopoiesis of Social Systems', in F. Geyer and J. van de Zouwen (eds.), *Sociocybernetic Paradoxes: Observation, Control and Evolution of Self-steering Systems*. London: Sage Publications, 172–93.

Lyytinen, K., and Hirschheim, R. (1987). 'Information Systems Failures: A Survey and Classification of the Empirical Literature'. *Oxford Surveys in Information Technology*, 4: 257–309.

—— and Klein, H. (1985). 'The Critical Theory of Jürgen Habermas as a Basis for a Theory of Information Systems', in E. Mumford, R. Hirschheim, G. Fitzgerald, and T. Wood-Harper (eds.), *Research Methods in Information Systems*. Amsterdam: North Holland, 219–31.

McKelvey, W. (2002). 'Managing Coevolutionary Dynamics: Some Leverage Points'. Paper presented at the 18th EGOS Conference, Barcelona, Spain, July 4–6.

Markus, L., and Robey, D. (1988). 'Information Technology and Organizational Change: Causal Structure in Theory and Research'. *Management Science*, 34/5: 583–98.

Miles, R. E., and Snow, C. C. (1984). 'Fit, Failure and the Hall of Fame'. *California Management Review*, 26/3: 10–28.

Monteiro, E., and Hanseth, O. (1995). 'Social Shaping of Information Infrastructure: On Being Specific about the Technology', in W. Orlikowski, G. Walsham, M. R. Jones, and J. I. DeGross (eds.), *Information Technology and Changes in Organizational Work*. London: Chapman and Hall, 325–43.

Mumford, E. (2003). *Redesigning Human Systems*. Hershey, PA: IRM Press.

Pugh, D., Hickson, D., and Hinings, C. (1985). *Writers on Organizations*. New York: Sage.

Scott Morton, M. (ed.) (1991). *The Corporation of the 1990s: Information Technology and Organizational Transformation*. Oxford: Oxford University Press.

Symons, V. (1992). 'The Management of Change: Guidelines for the Successful Implementation of Information System', in A. Brown (ed.), *Creating a Business-based IT Strategy*. London: Chapman and Hall.

Taylor, F. (1911). *The Principles of Scientific Management*. New York: Norton.

Truex, D. P., Baskerville, R., and Klein, H. K. (1999). Growing Systems in an Emergent Organization. *Communications of the ACM*, 42/8: 117–23.

Venkatraman, N. (1993). 'Continuous Strategic Alignment: Exploiting Information Technology Capabilities for Competitive Success'. *European Management Journal*, 11/2: 139–49.

Wand, Y., and Weber, R. (1995). 'On the Deep Structure of Information Systems'. *Information Systems Journal*, 5/3: 203–23.

Wigfield, M. (2001). 'Big Carriers Dominate Cellular Auction, Despite Plan to Help Start-ups'. *Wall Street Journal*, B1, 5 January.

Yin, R. (1984). *Case Study Research: Design and Methods*. Beverly Hills, CA: Sage Publications.

Yourdon, E., and Constantine, L. (1979). *Structured Design*. Englewood Cliffs: Prentice-Hall.

Note

1. This dichotomization is also recognized as a continuum, since reversing the reversal of field does not always wrap around to the original concept. For example, construction may reverse to destruction, but destruction may reverse to renewal.

Index